AF621457

Unifying the Field of Digital Twins for Urban Management

Unifying the Field of Digital Twins for Urban Management

Theoretical foundations and practical applications

Ramy Elsehrawy and Bimal Kumar

Published by Emerald Publishing Limited, Floor 5,
Northspring, 21-23 Wellington Street, Leeds LS1 4DL.

ICE Publishing is an imprint of Emerald Publishing Limited

Other ICE Publishing titles:
Digital Twins for Smart Cities.
Li Wan, Timea Nochta, Junqing Tang, and Jennifer Schooling.
ISBN 978-0-72776-600-7
Automation and Manufacturing for the Built Environment
Steve Thompson. ISBN 978-1-83608-599-7
Digital Twins in the Built Environment.
Qiuchen Lu, Xiang Xie, Ajith Kumar Parlikad, Jennifer Schooling,
and Michael Pitt. ISBN 978-0-72776-580-2

A catalogue record for this book is available from the British Library

ISBN 978-1-83662-235-2

© Ramy Elsehrawy and Bimal Kumar 2025

Permission to use the ICE Publishing logo and ICE name is granted under licence to Emerald from the Institution of Civil Engineers. The Institution of Civil Engineers has not approved or endorsed any of the content herein.

All rights, including translation, reserved. Except as permitted by the Copyright, Designs and Patents Act 1988, no part of this publication may be reproduced, stored in a retrieval system or transmitted in any form or by any means, electronic, mechanical, photocopying or otherwise, without the prior written permission of the publisher, Emerald Publishing Limited, Floor 5, Northspring, 21-23 Wellington Street, Leeds LS1 4DL.

This book is published on the understanding that the author is solely responsible for the statements made and opinions expressed in it and that its publication does not necessarily imply that such statements and/or opinions are or reflect the views or opinions of the publisher. While every effort has been made to ensure that the statements made and the opinions expressed in this publication provide a safe and accurate guide, no liability or responsibility can be accepted in this respect by the author or publisher.

While every reasonable effort has been undertaken by the author and the publisher to acknowledge copyright on material reproduced, if there has been an oversight please contact the publisher and we will endeavour to correct this upon a reprint.

Cover photo: VideoFlow/Shutterstock.com

Commissioning Editor: Michael Fenton
Content Development Editor: Cathy Sellars

Production Editor: Benn Linfield
Typeset by: KnowledgeWorks Global Limited
Index created by David Gaskell

Printed and bound by CPI Group (UK) Ltd, Croydon, CR0 4YY

Dedication

From Ramy:
To my parents, whose love and guidance shaped me,
To Mayssa, my heart and steadfast companion,
And to Jude, my son, whose arrival has filled our lives with joy.

Contents

Preface

Global trends such as population growth, rapid urbanisation, and increasing urban complexity are creating substantial challenges for urban sustainability and quality of life. Recently, the world population reached 8 billion, with the United Nations projecting that 68% of people will live in urban areas by 2050, up from 55% in 2022. This surge, alongside the rising complexity of urban systems, makes urban management and sustainability efforts more critical than ever. Urban systems are evolving into complex networks of interconnected dynamics, often resulting in unexpected challenges and undesirable outcomes. Consequently, sustainable urban development (SUD) and a transformative approach to urban management (UM) have become central focal points.

UM focuses on planning and interventions that improve urban conditions for both people and nature. While achieving SUD is challenging, the recent emergence of digital twin (DT) technology has presented a promising tool for UM. DT technology links a physical system to its virtual counterpart, allowing real-time data exchange between the two, often without the need for human intervention, thus enabling informed decision-making and value creation. Implementing DT to enhance UM practices has led to the formation of a new paradigm: digital twins for urban management (*DT for UM*).

However, for any paradigm to thrive, its underlying cultural system – comprising theories, approaches and methodologies – must be coherent, organised and unified. The *DT for UM* field currently lacks this coherence, with philosophical perspectives varying widely and often resulting in contradictory approaches when applied to complex urban challenges. Practical DT methods and tools are largely unstandardised, complicating effective communication and collaboration across the discipline. The absence of a structured methodology for DT implementation further underscores the need for a unified framework.

This work aims to develop a structured foundation for the paradigm *DT for UM* to support its growth. Using a design science research methodology, the book introduces a theoretical framework called DATUM, intended as the cultural system that the *DT for UM* needs. DATUM consists of three key elements:

- **Philosophical element**: Grounded in critical realism (CR), a philosophy that embraces pluralism, this element enables effective interventions without theoretical contradictions.
- **Methodological element**: The data-driven multi-method methodology (DM2) is introduced to bridge the gap between the abstract philosophical framework and practical applications, offering a systematic approach to DT interventions.

- **Methodical element**: The digital twin uses and classification system (DTUCS) offers a structured, three-part approach for categorising DTs and use cases, comprising a standardised classification framework, a taxonomy of DT functions and a model for documenting DT scenarios using the unified modelling language (UML).

DATUM provides a much-needed, unified approach to propel *DT for UM* forward, equipping the discipline with a structured foundation to foster its development and maturity.

The contributions of DATUM are manifold. In practice, DATUM provides a standardised language for defining DT use cases, specifying required DT features, and facilitating clear, consistent communication across the DT industry. It also offers practitioners a structured methodology to guide each phase of a DT-based intervention. On the theoretical side, DATUM introduces a foundational philosophical discourse to the *DT for UM* field, exploring the underlying assumptions and worldviews shaping this emerging paradigm. Built upon the inherently pluralistic philosophy of CR and by applying its abstract principles practically, DATUM helps researchers and practitioners avoid atheoretical or inconsistent frameworks, enabling them to integrate and combine diverse DT approaches effectively.

DATUM's capacity to bridge theory and practice offers numerous benefits. It encourages reflexivity among practitioners by providing a robust foundation for explaining, justifying, or critiquing implementation decisions. Furthermore, drawing on the theoretical roots of various DT methods – from quantitative and tech-driven to qualitative and humanistic – supports genuinely pluralistic approaches to tackling complex, multifaceted real-world challenges. DATUM was rigorously evaluated through three types of research methodologies.

- **Abstract research**: Assessed the philosophical consistency of DATUM's foundational CR element, utilising focus group discussions to examine how well it unites different worldviews within the *DT for UM* paradigm.
- **Extensive research**: Evaluated the methodical component of DATUM for its generalisability and practical relevance through multiple case studies.
- **Intensive research**: Employed action research to assess DATUM as a comprehensive framework, considering all its elements to gauge overall suitability and usability.

These evaluation methods demonstrate DATUM's robustness, underscoring its value as a unified, practical and theoretically grounded framework for digital twinning in UM.

Acknowledgements

We would like to extend our deepest gratitude to Dr. Kay Rogage, Professor Rebecca Strachan and Dr. Martin Wonders. Without their invaluable guidance and support, the successful completion of both case study projects detailed in Chapter 8 would not have been possible. We are also immensely grateful to the focus group participants, interviewees and the 'TwinView' development team, whose valuable contributions have enriched this study.

Special thanks to Richard Watson for his constructive feedback, insightful comments and invaluable guidance. We also wish to express our sincere appreciation to Professor David Greenwood and Dr. James Charlton for their generous support and expert opinions, which were instrumental in refining the work.

About the authors

Ramy Elsehrawy is a Managing Consultant Digital at Mott MacDonald, a polymath engineer, and a thought leader at the intersection of smart sustainable cities, information management, digital twins for the built environment, systems thinking, and philosophy. He holds an MSc in Construction Management from Glasgow Caledonian University and a PhD from Northumbria University, where he focused on developing and utilising digital twins to support urban management. With a robust publication record spanning theoretical frameworks and practical applications of digital twins in urban and built environments, Ramy brings over a decade of industry experience. He advises governments, public sector organizations and city leaders worldwide on digital twin initiatives and smart sustainable city strategies.

Bimal Kumar is Professor of Innovative Digital Construction Technologies in the Department of Architecture, University of Strathclyde, Glasgow, UK. He has had a long and distinguished career in computer-based design and construction research for over 35 years. He completed a PhD in artificial intelligence applications in conceptual structural design in the 1980s at Edinburgh University and has continued working in this thematic area ever since. He has had long and active collaborations (mostly through funded visiting professorships) at leading research centres throughout the world in this area, including Stanford and Carnegie-Mellon Universities in the USA, NUS (Singapore) and Hong Kong University of Science and Technology. He has led several cross-disciplinary projects funded by the UK's research councils totalling over £2 million. He has published three books, around a dozen edited volumes and over 150 articles in major international journals and conferences, and has supervised around 30 postgraduate research students. He is a recipient of the Parkman Medal of the Institution of Civil Engineers, London.

Abbreviations

AECO	architecture, engineering, construction and operation
AI	artificial intelligence
ANPR	automated number plate recognition
AR	augmented reality
BDA	big data analytics
BIM	building information management
BLD	refers to the 'Building' studied in one of the case studies
BMF	Burrell and Morgan framework
BMS	building management system
CDBB	Centre for Digital Built Britain
CDE	common data environment
CM	cognitive maps
CO_2	carbon dioxide
CO_2e	carbon dioxide equivalent
CON	refers to the project consultant involved in one of the case studies
CReDo	Climate Resilience Demonstrator
CR	critical realism
CS	cultural system
CUT	connected urban twins
DATUM	designed amalgam of twinning for urban management
DM2	data-driven multi-method methodology
DREI(C)	Bhaskar's explanatory model suited to natural sciences when applying critical realism
DSR	design science research
DT	digital twin
DT for UM	digital twin for urban management new paradigm proposed and described in this book
DTUCS	DT uses and classification system
FDM	foundation data model
GAR	refers to the 'Gardens' studied in one of the case studies
GH	refers to the 'Glass House' studied in one of the case studies
GIS	geographic information system
GUC	general use case
IA	integration architecture
IAQ	indoor air quality
ICT	information and communication technology

IMF	information management framework
IS	information systems
IPU	intelligent planning unit
IoP	internet of people
IoT	internet of things
MEP	mechanical, electrical and plumbing
ML	machine learning
M/M	morphogenetic/morphostatic cycle
MR	mixed reality
NDT	National Digital Twin
NDTp	National Digital Twin programme
NIC	National Infrastructure Commission
OF	ontological flexibility
OR	operational research
PLM	product lifecycle management
PM	project manager
RDL	reference data library
RRREI (C)	Bhaskar's explanatory model suited to social sciences when applying critical realism
RUL	remaining useful life
S-C	socio-cultural interactions
SD	systems dynamics
SoS	system of systems
SUD	sustainable urban development
UCS	use case scenario
UM	urban management
UML	unified modelling language
UoT	unit of time
VR	virtual reality

Ramy Elsehrawy and Bimal Kumar
ISBN 978-1-83662-235-2
https://doi.org/10.1108/978-1-83662-234-520251001
Copyright © 2025 by Ramy Elsehrawy and Bimal Kumar. Published under exclusive licence by Emerald Publishing Limited

Chapter 1
Introduction

1.1. Background, rationale and motivations

In today's world, the pursuit of sustainable urban development (SUD) is increasingly recognised as both a crucial goal and a complex challenge. SUD, as a goal, is focused on ensuring 'a non-decreasing level of well-being in the long term' while also working towards 'reducing harmful effects of development on the biosphere' (Camagni, 1998: p. 17). However, achieving this goal presents numerous challenges. Currently, over half the world's population resides in urban areas, and this figure is expected to grow to two-thirds by 2050 (United Nations Department of Economic and Social Affairs, Population Division, 2015). This rapid urbanisation is worsening the sustainability crisis faced by cities (Bibri, 2018). As urban environments become the focal points for emissions, waste production and the depletion of natural resources, they also evolve into highly complex systems. These systems intertwine social, cultural, economic, environmental and built structures in unprecedented ways. If this complexity is not carefully managed, the resulting consequences could be irreversible, with profound impacts lasting for generations. Climate change is a stark example of such systemic undesirable outcomes, posing significant risks globally.

Urban management (UM) as a concept has been widely debated, with its definition varying across contexts (Engin *et al.*, 2020; Kearns and Paddison, 2000; Stren, 1993; Werna, 1995). However, there is broad agreement that UM involves planning and interventions aimed at addressing the challenges of SUD, with the ultimate goal of achieving sustainable urban environments. This makes the role of urban managers both demanding and essential. Despite these challenges, the modern world also offers opportunities that can greatly enhance efforts towards SUD. One such opportunity is the rise of information and communication technology (ICT) and, in particular, the emergence of digital twins (DTs), which hold significant promise for managing urban environments and addressing their challenges (National Infrastructure Commission, 2017).

A DT is defined as a 'digital representation of assets, processes or systems' (Bolton *et al.*, 2018: p. 10). The concept of DTs has gained considerable attention over the past decade (Grieves, 2005; Hochhalter *et al.*, 2014), and, more recently, it has shown potential to unlock value within UM (National Infrastructure Commission, 2017; Yang and Kim, 2021). By connecting data from various sources across sectoral and organisational silos, DTs enable more advanced analysis, providing deeper insights, creating new knowledge and enhancing our understanding of complex urban dynamics. These insights, in turn, facilitate better, more informed decision-making (Council and Lamb, 2022).

An increasing number of studies demonstrate the use of DTs for addressing UM challenges, marking the emergence of a new paradigm: *DT for UM*. In 2017, the report *Data for Public Good*

(National Infrastructure Commission, 2017) marked a significant initiative towards the development of a national DT for the UK's infrastructure system of systems. Subsequent studies from industry (e.g. Arup, 2019; IET: The Institution of Engineering and Technology, 2019; ITRC, 2020; Witteborg, 2021) and academia (e.g. Al-Sehrawy *et al.*, 2021; Deren *et al.*, 2021; Engin *et al.*, 2020; Ketzler *et al.*, 2020; Pregnolato *et al.*, 2022) further demonstrated the potential of DTs to deliver value and unlock a range of benefits for managing urban environments and their associated systems and assets.

The literature reveals considerable diversity at three levels of abstraction when analysing relevant studies. At the most concrete level – the methodical – DTs can differ in numerous ways, such as their area of application, spatial or temporal scale, and more. Depending on the use case, different DTs may employ various techniques and functionalities, such as simulation, prediction or visualisation to achieve specific objectives. At a more abstract level – the methodological – DT practice can take different forms, with distinct approaches that amalgamate various DT methods to fulfil the DT's purpose. At the most abstract level – the philosophical – researchers and practitioners, though often implicitly, subscribe to different philosophical worldviews. These worldviews encompass assumptions regarding the urban environment, including what is considered real, how knowledge is acquired, and how humans interact and make decisions. These philosophical perspectives vary and directly influence the methodological approaches and methods adopted in DT practice.

1.2. Core challenge

The evolving landscape of *DT for UM* is characterised by a diverse range of methods, approaches, and philosophical worldviews. As thoroughly discussed later in Chapter 3, this diversity, however, presents several significant challenges. At the most concrete level – the methodical – the features and uses of DTs lack standardisation and uniformity. At the most abstract level – the philosophical – the four major paradigms underpinning various DT approaches are incommensurable, inconsistent and often contradictory. This situation is further complicated by the absence of a comprehensive, systematic methodology that can offer clear guidance to DT practitioners, effectively bridging the gap between abstract philosophical worldviews and concrete DT methods.

As a result, the field lacks cohesion, with methods and approaches frequently chosen without robust justification or alignment with a sound philosophical foundation. Despite the inherent heterogeneity within the *DT for UM* paradigm, the absence of coherence and consistency hampers its development and maturity. The gap between philosophical thought and practical application remains a critical issue, impeding the growth and potential of this emerging paradigm.

1.3. The motivation behind this book

The motivation behind this book evolved from an interest in exploring the potential of DT to support UM practices to a critical evaluation of the lack of coherence and consistency within the emerging *DT for UM* paradigm.

When a paradigm exhibits high ideational diversity and deep inconsistencies, it can progress through either isolationism or pluralism (Jackson, 2019). In isolationism, a single DT approach is considered sufficient to address all problems within the paradigm. While isolationism may ensure theoretical consistency by adhering to a single philosophical stance, it is often inadequate when tackling the multifaceted, complex challenges urban managers face. The shortcomings of isolationism are explored further in Section 3.8.5.

Pluralism, however, presents a more viable path forward, but it can manifest in two distinct forms: discordant pluralism or complementarism. Discordant pluralism allows different, and sometimes contradictory, theoretical approaches to challenge and enrich each other. In contrast, complementarism seeks to integrate and rationalise the differences between various approaches, creating a new, pluralistic paradigm (Jackson, 2019).

This book advocates for complementarism, including the systematisation of the paradigm *DT for UM* and the unification of DT approaches, for three primary reasons:

(a) The existing DT approaches, discussed in Section 3.8.4, are separately conceptualised and critiqued. A unified approach that harnesses their strengths while avoiding their weaknesses offers a more promising solution. As Mingers (2006: p. 209) argues, it is unclear why 'we have to accept the validity and in some sense equality of currently existing paradigms rather than try to go beyond them.' He questions why, if these paradigms have been shown to have serious flaws or limitations, research or intervention should be carried out entirely within one such paradigm. Instead, it is far better to develop new paradigms and research methods that build on their strengths and mitigate their weaknesses.
(b) Unlike complementarism, discordant pluralism lacks a theoretical paradigm or framework that justifies its practices. Further details on this are provided in Section 5.3.2.2.
(c) Drawing on Archer's work (1995), the current state of *DT for UM* places its practitioners in a situational logic that promotes unification and systematisation. Kuhn's (1970) view of scientific paradigm development suggests that the pre-paradigm phase, marked by inconsistency and incoherence, is typically followed by unification and the establishment of a new paradigm. This new paradigm becomes the foundation for scientific progress. More on Archer's and Kuhn's perspectives can be found in Chapter 4.

Ultimately, this book aims to systematise this emerging paradigm (i.e. *DT for UM*), supporting and accelerating its growth and maturity.

1.4. The scope of this book

The aim of this book is to systematise and unify the emerging paradigm of *DT for UM*. This is accomplished by developing a theoretical framework known as the designed amalgam of twinning for urban management (DATUM). DATUM serves as a comprehensive and inherently pluralistic system, designed to support the development and implementation of DTs to address real-world challenges in UM.

The objectives of this work are as follows.

O1: Review the existing literature to identify gaps in the current cultural system, including theories, methodologies and methods within the developing *DT for UM* paradigm.

O2: Design the conceptual structure of a theoretical framework, DATUM, which forms a new cultural system for DT in UM. This involves identifying the key components of DATUM, the requirements it must meet and the evaluation criteria by which it will be assessed.

O3: Develop DATUM by building its three core elements: methodical, philosophical and methodological.

O4: Evaluate DATUM against the established criteria and recommend improvements to its design based on the findings.

1.5. Book structure

This book is structured as follows.

Chapter 1: Provides an introduction to the book, outlining the background, inspiration, primary aim, objectives and its overall structure.

Chapter 2: Discusses the authors' philosophical perspective, grounded in critical realism (CR), and introduces the adopted research methodology, Design Science Research, which informs the approach taken in the following chapters.

Chapter 3: Presents a literature review, identifying the gaps in the existing cultural system of the emerging *DT for UM* paradigm. These gaps are analysed at three distinct levels: methodical, methodological and philosophical.

Chapter 4: Recommends how the *DT for UM* paradigm should evolve and mature by systematising its cultural structure. This is built on an exploration of the historical development of UM. The chapter then proposes the conceptual framework of DATUM and defines the criteria for its evaluation.

Chapters 5, 6 and 7: These three chapters detail the primary contribution of this work, with the development of DATUM's three core elements: methodical, philosophical and methodological, respectively.

Chapter 8: Focuses on the evaluation of DATUM and its core elements against the criteria established in Chapter 4. This evaluation is carried out using a methodology aligned with the author's CR perspective, as explained in Chapter 2.

Chapter 9: Concludes by summarising the research, highlighting the key findings, contributions, limitations and proposing directions for future research.

REFERENCES

Al-Sehrawy R, Kumar B and Watson R (2021) A digital twin uses classification system for urban planning & city infrastructure management. *Journal of Information Technology in Construction (ITCON)* **26**: 832–862.

Archer MS (1995) *Realist Social Theory: The Morphogenetic Approach.* Cambridge University Press, Cambridge, UK.

Arup (2019) *Digital Twins: Towards a Meaningful Framework.* Arup. https://www.arup.com/perspectives/publications/research/section/digital-twin-towards-a-meaningful-framework

Bibri SE (2018) *Smart Sustainable Cities of the Future: The Untapped Potential of Big Data Analytics and Context–Aware Computing for Advancing Sustainability.* Springer, Cham, Switzerland.

Bolton A, Butler L, Dabson I *et al.* (2018) *Gemini Principles* [Report]. CDBB, 10.17863/CAM.32260

Camagni R (1998) Sustainable urban development: definition and reasons for a research programme. *International Journal of Environment and Pollution* **10(1)**: 6–27. 10.1504/IJEP.1998.002228

Council, G and Lamb, K (2022) *Gemini Papers: How to Enable an Ecosystem of Connected Digital Twins* [Report]. Centre for Digital Built Britain. 10.17863/CAM.82193

Deren L, Wenbo Y and Zhenfeng S (2021) Smart city based on digital twins. *Computational Urban Science* **1(1)**: 1–11.

Engin Z, van Dijk J, Lan T *et al.* (2020) Data-driven urban management: mapping the landscape. *Journal of Urban Management* **9(2)**: 140–150.

Grieves M (2005) Product lifecycle management: the new paradigm for enterprises. *International Journal of Product Development* **2(1–2)**: 71–84.

Hochhalter J, Leser WP, Newman JA *et al.* (2014). *Coupling damage-sensing particles to the digital twin concept* (pp. 1–9). NASA Technical Reports Server, USA.

IET: The Institution of Engineering and Technology (2019) *Digital Twins for the Built Environment: An Introduction to the Opportunities, Benefits, Challenges and Risks*. https://www.theiet.org/media/8762/digital-twins-for-the-built-environment.pdf

ITRC (2020) *A Sustainable Oxford-Cambridge Corridor? Spatial Analysis of Options and Futures for the Arc*. Infrastructure Transitions Research Consortium. https://www.itrc.org.uk/wp-content/uploads/2020/01/arc-main-report.pdf

Jackson MC (2019) *Critical Systems Thinking and the Management of Complexity*. Wiley. https://books.google.co.uk/books?id=1jmNDwAAQBAJ

Kearns A and Paddison R (2000) New challenges for urban governance. *Urban Studies* **37(5–6)**: 845–850.

Ketzler B, Naserentin V, Latino F *et al.* (2020) Digital twins for cities: a state of the art review. *Built Environment* **46(4)**: 547–573.

Kuhn TS (1970) *The Structure of Scientific Revolutions*, vol. 2. University of Chicago Press, Chicago, IL, USA.

Mingers J (2006) *Realising Systems Thinking: Knowledge and Action in Management Science*. Springer US, New York. https://books.google.co.uk/books?id=C-UcDQEACAAJ

National Infrastructure Commission (2017) *Data for the Public Good*. National Infrastructure Commission Report, UK.

Pregnolato M, Gunner S, Voyagaki E *et al.* (2022) Towards Civil Engineering 4.0: concept, workflow and application of Digital Twins for existing infrastructure. *Automation in Construction* **141**: 104421. 10.1016/j.autcon.2022.104421

Stren R (1993) 'Urban management' in development assistance: an elusive concept. *Cities* **10(2)**: 125–138.

United Nations Department of Economic and Social Affairs, Population Division (2015) *World Urbanization Prospects: The 2014 Revision* (ST/ESA/SER.A/366). https://population.un.org/wup/Publications/Files/WUP2014-Report.pdf

Werna E (1995) The management of urban development, or the development of urban management? Problems and premises of an elusive concept. *Cities* **12(5)**: 353–359.

Witteborg A (2021) Digital twins for wastewater infrastructure. *Royal HaskoningDHV Digital*. https://global.royalhaskoningdhv.com/digital/blogs-and-news/publications/digital-twins-for-wastewater-infrastructure

Yang S and Kim H (2021) Urban digital twin applications as a virtual platform of smart city. *International Journal of Sustainable Building Technology and Urban Development* **12(4)**: 363–379. 10.22712/susb.20210030

Ramy Elsehrawy and Bimal Kumar
ISBN 978-1-83662-235-2
https://doi.org/10.1108/978-1-83662-234-520251004
Copyright © 2025 by Ramy Elsehrawy and Bimal Kumar. Published under exclusive licence by Emerald Publishing Limited

Chapter 2
Our lens and approach

2.1. Introduction

Research, at its core, is the systematic investigation and analysis of evidence to reach new conclusions and create knowledge. This book reflects the commitment to do so. Research involves a series of decisions about selecting and using methods for collecting and analysing data. These choices are based on pre-existing beliefs and philosophical assumptions about how research can generate knowledge (Chua, 1986; Guba, 1990). The research methodology serves as a framework that combines these choices and assumptions into a cohesive design to achieve the research objectives (Bryman, 2016). Clearly stating the research methodology, including its philosophical foundation and methodological approach, is essential. This transparency shapes what the research can achieve and allows for post-research critique and reflection on its merits, limitations and contributions. This chapter first introduces the philosophical paradigm underpinning this book – critical realism. It then details the design science research methodology adopted, explaining its five key steps sequentially.

2.2. Our philosophy – critical realism

Creswell (2009: p. 6) describes a researcher's worldview as a 'general orientation about the world and the nature of the research that the researcher holds.' This worldview is shaped by philosophical beliefs, including ontological assumptions about the nature of reality and epistemological assumptions about the nature and validation of knowledge.

Initially, two prominent research philosophies were considered: positivism and interpretivism. Positivism emphasises predictability, repeatability of results, and the testing of hypotheses based on observable phenomena (Hempel, 1965; Popper, 1957). For positivists, scientific knowledge is derived solely from perceptible reality, focusing on observable and factual phenomena. Science aims to develop generalised, predictive theories that describe regularities in the world (Keat and Urry, 1978). In essence, positivism relies on empirical observations and the formulation of universal causal laws in the 'Humean'[1] sense, which are tested and potentially refuted if predictions are not met.

However, while positivism is centred on empirical phenomena, it lacks the depth needed to explore the underlying mindsets and worldviews that drive these observable phenomena. Additionally, positivism's strength in reductionism, typically applied in controlled environments, does not align with the complexity of the urban environment under study, characterised by open systems and intricate social interactions.

[1]David Hume, the philosopher, believed that 'causes and effects are discoverable not by reason, but by experience' (Hume, 2000).

Conversely, interpretivism considers the observer's perspective, focusing on the meanings individuals assign to their observations. Interpretivists recognise that humans interpret their experiences subjectively, influenced by their unique consciousness and context (Walsham, 1993). Despite this, interpretivism does not adequately address the objective reality out there central to digital twins – the systems being twinned. Furthermore, data generated through interpretivism is heavily influenced by personal experiences and values, which can limit the generalisability of findings. This limitation conflicts with the book's aim to establish a unified paradigm: digital twins for urban management (*DT for UM*).

As a result, this book adopts the philosophy of critical realism (CR). Developed by Roy Bhaskar (1975) and later expanded by critical realists like Danermark *et al.* (2005), Keat and Urry (1978) and Sayer (1992), CR addresses the limitations of both positivism and interpretivism. Derived from transcendental realism, CR focuses on descriptive causal laws as the core objects of scientific inquiry (Bhaskar, 1975). For realists, scientific theories describe the structures and mechanisms that causally generate observable phenomena, providing explanations for them (Keat and Urry, 1978). These generative mechanisms represent the inherent ways things act (Bhaskar, 1975). An object's causal powers or mechanisms arise from its structure, the 'set of internally related objects' that form a whole (Sayer, 1992).

In CR, the concept of causal laws shifts from regular conjunctions to detailed explanations of the tendencies of things, focusing on how and why observed phenomena occur. This perspective is captured by the 'epistemic imperative' (Mouton, 1996), which seeks to understand what the world must be like for a set of observed events to happen.

It is important to recognise that CR is a complex philosophical paradigm. The brief introduction provided here does not capture its full depth, which has been thoroughly explored in works by Bhaskar (1975), Collier (1994), Danermark *et al.* (2005), Keat and Urry (1978), Lawson (1997), Mingers (2006, 2014), Porpora (2015), and Sayer (1992). This overview highlights only a few key principles relevant to this book, followed by an explanation of their significance to it.

2.2.1 Ontological realism (independent reality)

At the core of CR is the clear distinction between an ontological reality and our knowledge of it. CR acknowledges the existence of an independent reality, irrespective of our perception or awareness of it. As Trigg (1980) notes, the nature of reality and our understanding of it are separate issues, given that many aspects of reality surpass our conceptual and linguistic capacities. Consequently, CR gives primacy to ontology and avoids the 'epistemic fallacy' that is collapsing ontology into epistemology, or as Bhaskar describes it – the mistake of letting the question 'what can we know? determine our notions of what exists' (Bhaskar, 1975: p. 36).

At the same time, CR acknowledges the role of social actors in generating fallible knowledge. Knowledge creation is inherently a human activity, shaped by context, values, social constructs and the inherent limitations of our perception. Our understanding is not formed in isolation but through socio-cultural interactions that involve beliefs, biases and 'matters of interpersonal influence' (Archer, 2005: p. 25). Knowledge is thus dependent on the observer, whose capacity for sensing and understanding is limited. Moreover, our knowledge of independent reality is mediated by social structures, culture and other conditions within which the observer is situated.

CR differentiates between two dimensions of knowledge: the intransitive and the transitive. The intransitive dimension consists of the objects of reality we aim to study, while the transitive dimension encompasses our fallible observations, theories and knowledge of this reality, developed through scientific inquiry (Collier, 1994).

2.2.2 Epistemic relativism

Human knowledge is inherently transitive, meaning it is finite, contextual, value-laden, socially constructed and always fallible. Knowledge depends on and is created by the observer, thus varying from one person to another. Our limited access to reality makes it impossible to possess true knowledge of all its aspects, rendering our accounts and theories of what exists 'out there' perpetually fallible.

2.2.3 Judgmental rationalism

Due to the unobservable layers of reality and its open nature, multiple plausible explanations can exist for empirical observations and experienced events, attributed to the generative mechanisms behind them. While all explanations are socially produccd, value-laden and fallible, not all are equally robust or valuable. Judgmental rationalism involves making rational judgments between competing explanations or accounts of reality. Therefore, even though our knowledge is contingent and fallible, we can still rationally evaluate and compare theories based on their explanatory power (Danermark *et al.*, 2005).

This principle prevents CR from descending into judgmental relativism, where science becomes merely a 'self-referential exercise' (Johnson and Duberley, 2000: p. 153) and morals or ethics are reduced to nothing more than an 'expression of taste' (Porpora, 2019: p. 277). More about CR's perspective on ethics will be discussed in Chapter 6.

2.2.4 Structure, hierarchy and emergence

Interestingly, CR aligns seamlessly with systems thinking (more on systems thinking in Section 3.2). Mingers (2014: p. 17) asserts that CR has fulfilled the aspiration of developing a 'systemic philosophy.' He explains:

> *'What is of special interest ... is that critical realism is deeply and fundamentally systemic in character. Although the main texts of critical realism make little reference to the traditional systems literature, the discourse itself ... is couched almost exclusively in systems terms.'*

This systemic nature of CR is most evident in its concepts of structure, hierarchy and emergence, which are fundamental to systems thinking (Section 3.2.3). A real entity consists of a structure composed of 'internally related objects or practices' (Sayer, 1992: p. 92). The way these parts are interrelated and organised endows the structure with unique properties and causal powers that are irreducible to, and independent from, its constituent parts. These causal powers emerge not from the aggregation of the properties or powers of the parts but from the synergistic effect of their interconnection and organisational pattern. Therefore, the ontological value of any structure lies in its emergent properties, causal powers or 'tendencies' that cannot be reduced to its components.

In the context of social or socio-technical systems, the 'explanation of why things social are so and not otherwise depends upon an account of how the properties and powers of the 'people' causally intertwine with those of the 'parts'' (Archer, 1995: p. 15), rather than focusing solely on individual actions in isolation (Easton, 2010).

2.2.5 Stratified reality

Bhaskar critiqued the limited value of knowledge generated by the traditional scientific method, which often provides detailed descriptions of what happens in highly controlled experiments but rarely offers scientific explanations of why these events occur. CR recognises that knowledge production is inherently concerned with describing different hierarchical levels of reality and explaining how properties at one level emerge through generative mechanisms at a deeper level. This process involves continually moving scientific research from one hierarchical level to a deeper one, opposite the direction of emergence. For critical realists, this stratification in the production and discovery of knowledge reflects a real ontological stratification in the world. Otherwise, 'the stratification of science must appear as a kind of historical accident, lacking any internal rationale in the practice of science' (Bhaskar, 1975: p. 161).

In CR, the stratified real world is divided into three domains: *real*, *actual* and *empirical* (Figure 2.1). The domain of the 'real' includes generative mechanisms, which are essentially the ways in which things act (Bhaskar, 1975). These mechanisms or causal powers are the potentials that enable objects to cause certain events in the world (Danermark *et al.*, 2005). Objects inherit these generative mechanisms and causal powers through their structure (Collier, 1994; Smith, 2006). This necessary relationship between structure and generative mechanisms helps scientists avoid tautological arguments that an object can do something simply because it has the power to do so (Sayer, 1992). These mechanisms exist regardless of whether they are activated or whether their effects are realised, which depends on the external environment and contextual conditions.

The domain of the actual encompasses the events produced by these generative mechanisms. An event is the result of one or more generative mechanisms being activated. However, events are ontologically distinct from the mechanisms that generate them (Bhaskar, 1975) – for example, a mechanism might not produce any observable events if its effects are counteracted by other mechanisms (Gambetta, 1998). Conversely, the effects of a mechanism can be amplified by the reinforcing effects of other mechanisms. Different combinations of interacting mechanisms can lead to the same set of events in various plausible ways.

Finally, the domain of the 'empirical' consists of the experiences and the subset of events that have been perceived or observed. In summary, real structures acquire properties that endow them with

Figure 2.1 The stratification of reality in critical realism (adapted from Mingers, 2014)

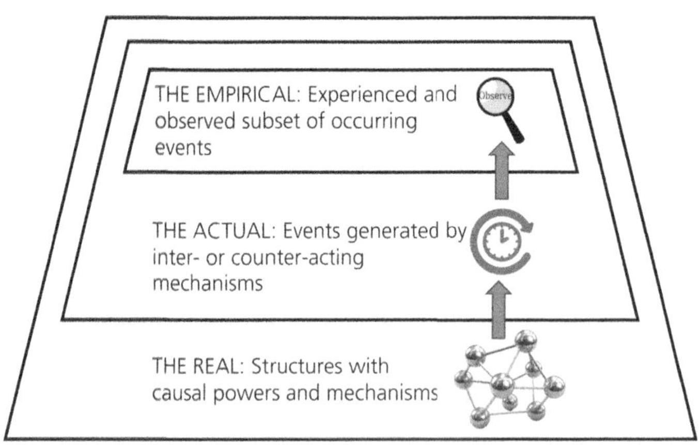

Figure 2.2 Stratified view of reality and the world of DT practice (adapted from Sayer, 1992)

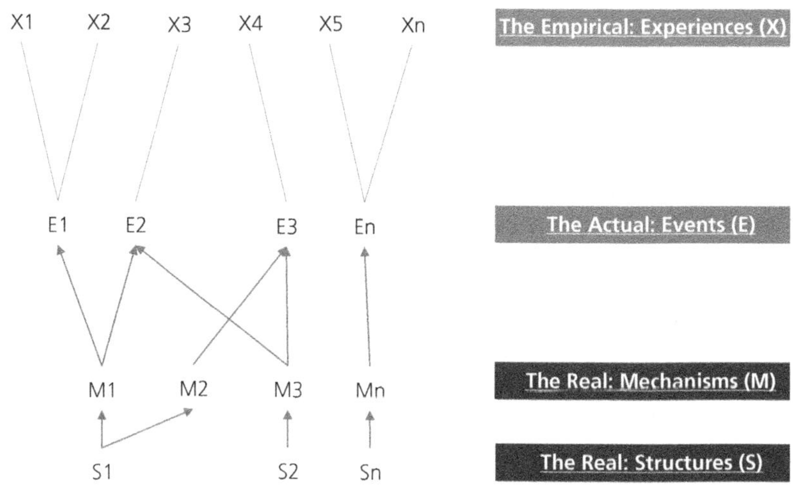

generative mechanisms, which, when interacting, lead to the emergence of actual events. However, we may only empirically experience or sense a subset of these occurring events. This stratification is illustrated in Figure 2.2, although it does not clearly indicate that our empirical experiences are limited to a subset of all events.

One profound implication of this stratified view of reality is that scientific investigations are now primarily focused on uncovering generative mechanisms. This approach aims to better understand and provide explanations – albeit fallible – of recorded observations (Volkoff *et al.*, 2007). The question of what the world must be like for a set of observed events to occur embodies the epistemic imperative of a critical realist (Mouton, 1996). Bhaskar (1975) argues that transitioning from knowledge of one stratum to a deeper one involves three levels of knowledge.

'Humean' level: Empirical regularities are observed, indicating the emergence of relatively regular patterns of events, which a critical realist believes result from active generative mechanisms.

'Lockean' level: Explanations of observed behaviour are posited in terms of necessarily operating causal powers and mechanisms that might exist and have been activated by contextual conditions. This level involves detailed explanations of how specific endogenous and exogenous factors collectively generate the observed events.

'Leibnizian' level: The system of interest is formally defined and recognised as the type of system (X) it is, based on its possession of the structure (S).

2.2.6 Analytical dualism

CR avoids both deterministic and voluntaristic positions. Determinism emphasises structural conditions' influence on people, while voluntarism stresses the role of free will. Instead, CR advocates for 'analytical dualism' (Archer, 1995), which involves a dynamic interplay between agency and social structure.

By applying CR's concepts of structure, hierarchy and emergence to social reality, both social structure and agency are seen as ontologically distinct strata, each with unique properties and powers. Archer (1995) critiques three ways social theories often conflate these strata.

Downwards conflation: From a deterministic perspective, agential action is trivialised to a mere epiphenomenon of structural conditions.

Upwards conflation: This reduces structure to individual activities, akin to individualism.

Central conflation: Exemplified by Gidden's (1979) structuration theory, it views humans and structure as mutually constitutive, presenting both as inseparable aspects of a duality.

The term 'dualism' suggests that structure and human agency exist at different strata, while 'analytical' indicates that this dualism is not observable in real social experiences but can be elaborated through scientific analysis of their relationship (Danermark *et al.*, 2005). Despite this duality, structure and agency are only analytically, not philosophically, separable. Social structure is activity-dependent; once produced, it retains emergent properties and causal powers, influencing[2] agents by shaping the conditions of human activities. In essence, 'Structure and agency are separate strata, that is, they possess completely different properties and powers, but the one is essential for how the other will be moulded' (Danermark *et al.*, 2005: p. 181).

Archer (2005) extends the concept of analytical dualism to 'culture,' encompassing ideas, theories, and beliefs. She defines culture as 'all intelligibilia, that is to any item that has the dispositional ability to be understood by someone – whether or not anyone does so at a given time' (Archer, 2005: p. 24). She conceptualises the interplay between the 'cultural system' (CS) and 'socio-cultural' interactions (S-C) within any culture, discipline or domain. The CS refers to the relations between cultural components, such as theories and ideational projects within a discipline, while S-C involves the relationships between cultural agents, including practices, actions and interactions among discipline members (Archer, 1995: p. 180).

2.2.7 Significance of critical realism to this book

Having outlined several key principles of CR, it is essential to highlight their significance for this book. The following three arguments justify adopting a critical realist stance and guide the research methodology detailed in the subsequent section.

(a) Multi-level analysis: The principles of hierarchy, emergence and CR's stratified view of reality provide the analytical depth necessary to examine relevant literature and offer profound answers to research questions and problems across multiple levels of analysis with varying degrees of abstraction.

(b) Separability of theory and practice: The principle of analytical dualism acknowledges the interplay between theory (CS) and practice (S-C), and their mutual influence. This book, which reflects on current *DT for UM* practices, gains significance only if we believe that these practices (S-C) can influence and transform the discipline's theory (CS). Conversely, contributions to the discipline's theory must influence the practices of its members, echoing Lewin's (1951: p. 169) assertion that 'there is nothing more practical than a good theory.'

[2]Influencing here implies mere conditioning and motivation and not a deterministic type of force. As Porpora (1989: p. 208) argues, the idea of agents having certain interests because of an existing structure 'doesn't mean that actors always with necessity act in their interests, but if they don't they are likely to suffer. A capitalist who shows no concern to maximise profit is liable to cease being a capitalist.'

(c) Plausibility of validation: The principles of ontological realism, epistemic relativism, and judgmental rationality support the idea of validating the theoretical propositions presented in this book by testing them in practice.
(d) Methodological pluralism: CR avoids the limitations of classical philosophies (positivism and interpretivism) while combining their strengths. This paves the way for adopting a pluralistic methodological approach, as explained in the following section.

2.3. Research methodology

Traditional research methodologies in management studies, including case studies and experiments, are also prevalent in recent studies related to digital twins. These methods are useful for building exploratory, explanatory and descriptive knowledge. However, they are less effective in solving problems and guiding the design and creation of artifacts, such as the cultural system of a scientific field.

This book, as outlined in Chapter 1, focuses on solving a problem and creating an improved paradigm for DT in UM with a systematised and consistent CS. Consequently, design science research (DSR) is identified as the most suitable methodology. DSR is 'a methodological approach concerned with devising artifacts that serve human purposes' (Dresch *et al.*, 2015: p. v). It evolved to fill the gaps in traditional scientific approaches, which are primarily exploratory, explanatory and descriptive, but not typically known for problem-solving. DSR facilitates producing solutions relevant to real-world problems and complex, multidisciplinary situations while simultaneously contributing to knowledge. In essence, traditional scientific approaches strive 'to know,' whereas DSR is implemented 'to do' (Checkland, 1999; Simon, 1996). 'An important outcome of this type of research is an artifact that solves a domain problem, also known as a solution concept, which must be assessed against criteria of value or utility' (Dresch *et al.*, 2015: p. v).

Given this book's aims and objectives, DSR is a suitable methodology to guide our endeavour. Figure 2.3 illustrates the DSR process, involving five key steps (Kuechler and Vaishnavi, 2011), the sub-methodologies used within each step, the output of each step as inspired by Manson (2006) and the corresponding relevant chapter in the book.

2.4. DSR [1]: Awareness of problem

The first step in DSR is to identify and deepen our understanding of the problem under investigation. Initially, the two notions most central to this book, UM and DT, are explored and defined. This is followed by a thorough systematic literature review to rigorously examine the current state of the new rising paradigm of *DT for UM*. The systematic literature review focuses on how DTs and the concept of digital twinning are currently implemented to support UM research and practices.

A critical realist views reality as stratified (Section 2.2.5). Accordingly, the same lens is used to conduct the systematic literature review. Using this theoretical perspective, the retrieved studies are analysed at three distinct levels (i.e. philosophy, methodology and methods), corresponding to the different strata of reality (i.e. real, actual, empirical), respectively, as illustrated in Table 2.1 and Figure 2.4 below. These three levels of analysis range from the most abstract to the most concrete, where problems relevant to each level are identified. Finally, the outputs of this step include detailed evidence and a comprehensive demonstration of the problems that need to be addressed.

As detailed in Chapter 3, DT practices face challenges at both the methodical and philosophical levels. Methodically, DT features and uses lack standardisation, while philosophically, the main paradigms behind common DT approaches are inconsistent and often conflicting. The issue is

Figure 2.3 DSR steps against outputs against relevant chapters (author's own)

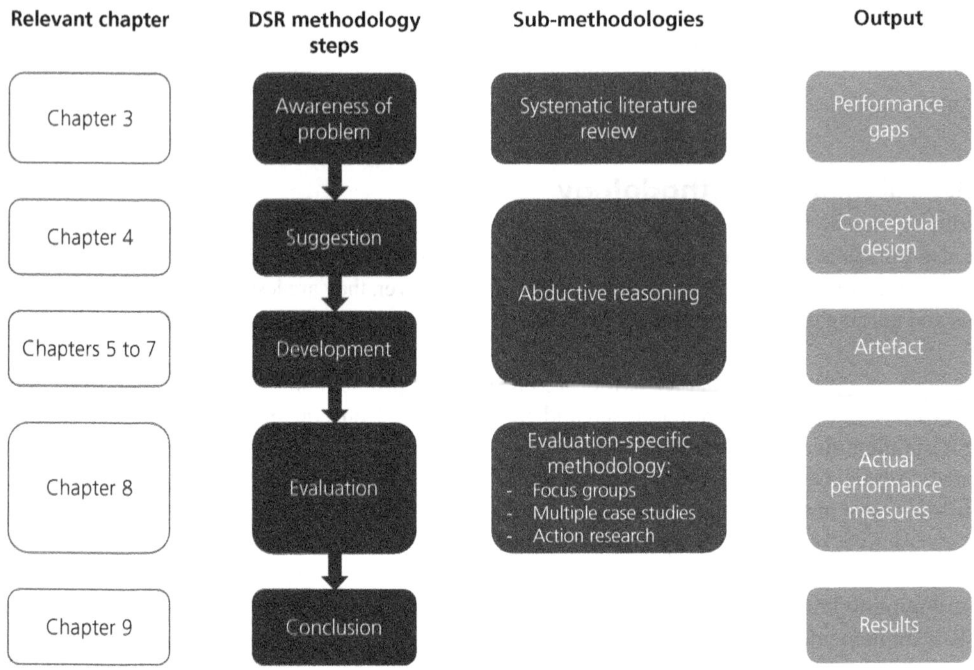

Table 2.1 Stratified reality of *DT for UM* research and practice

	Real	Actual	Empirical	*DT for UM* practice
Experiences (X)	X	X	X	*Methods*: Observable DT actions and practices
Events (E)	X	X		*Methodology*: Implemented DT approaches
Structures and Mechanisms (S)	X			*Philosophy*: Beliefs and philosophical worldview

compounded by the lack of a comprehensive methodology to guide practitioners in connecting abstract theories with practical DT applications.

2.5. DSR [2]: Suggestion

The second step in DSR requires using creativity and abductive reasoning to develop a conceptual proposal for an artifact that is potentially capable of addressing the problem highlighted in DSR's first step. While undertaking the suggestion step, it is crucial to have a solid understanding of the first step's outputs, including the performance gaps creating the problems with the current situation, as well as a vision of an envisaged desirable state that should ideally be achieved.

An artifact developed using the DSR methodology does not need to be a physical one. Consequently, DSR can be implemented to build theories and develop theoretical constructs. In this book, step 2 of DSR yielded two key outputs as shown in Figure 2.5 below. The first output is

Figure 2.4 Stratified reality of *DT for UM* research and practice (author's own)

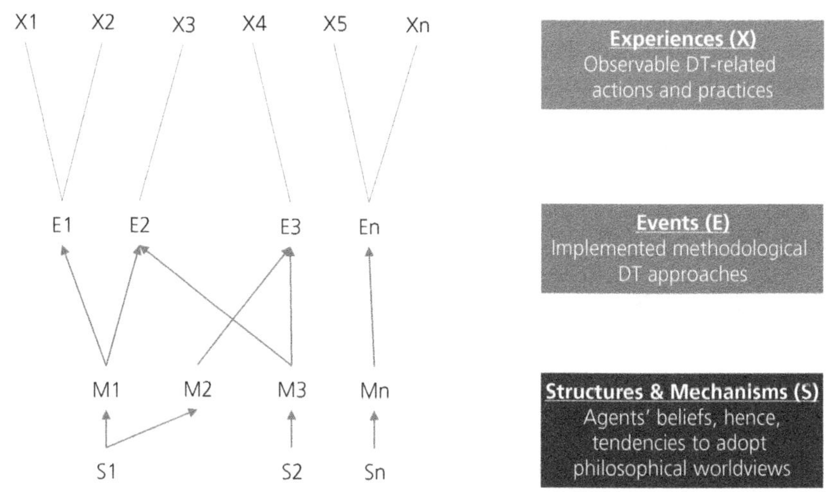

Figure 2.5 The criteria for evaluating DATUM (author's own)

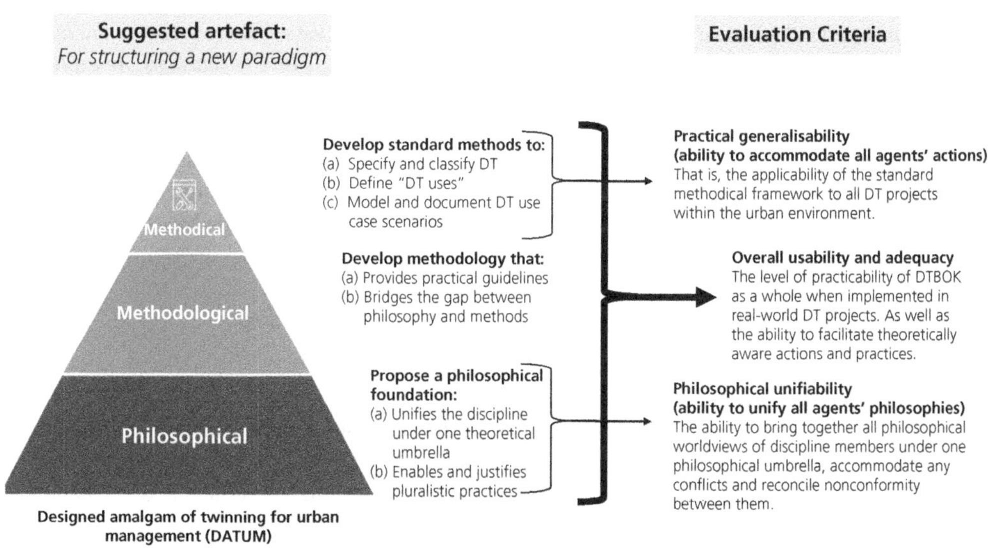

a conceptual design of the suggested artifact, which is the designed amalgam of twinning for urban management (DATUM) constituted by three key elements: philosophical, methodological and methodical. The second output of the suggestion step is a proposed set of evaluation criteria derived from and inspired by the overarching aim of this book and the understanding of the gaps identified in step 1. This set of evaluation criteria defines ways to measure the performance of the artifact and offers grounds for assessing its ability to realise the envisaged desirable state. More on the conceptual design of DATUM and the set of evaluation criteria is provided in Chapter 4.

2.6. DSR [3]: Development

Once the conceptual design of the artifact is suggested, the third step of DSR begins. During this step, the artifact and all its defined components are fully developed. The key output of this step is a functional, ready-to-use artifact. The development process must consider both the artifact's internal environment, including its constituent components and the interconnections between them, and the relationship between the artifact and its external environment, including the potential users and the context in which the artifact is designed to operate. In Chapters 5 to 7, the three main components constituting DATUM, as suggested in step 2 to address the problems identified in step 1, are developed. These components are the philosophical element, the methodological element and the methodical element.

2.7. DSR [4]: Evaluation

'Action without reflection and understanding is blind, just as theory without action is meaningless' (Reason and Bradbury, 2007: p. 4). As Checkland and Scholes (1990: p. xiv) asserted, 'theory which is not tested out in practice is sterile.' Therefore, after the artifact is designed and built in the world of theory, it is necessary, through the fourth step of DSR, to evaluate it in the world of practice. The main objective of this stage is to test the artifact and observe how well it can perform in providing a satisfactory solution to the research problem highlighted in the first stage. This is achieved by comparing the artifact's evaluation results to the requirements it was intended to meet, as proposed in step one, based on the evaluation criteria presented in step two. In this book, the methodology used for the evaluation step of DSR is the evaluation-specific methodology. This methodology comprises

> *'a set of principles (logic) and procedures (methodology) that guides the evaluation team in the task of blending descriptive data with relevant values to draw explicitly evaluative conclusions. An explicitly evaluative conclusion is one that says how good, valuable, or important something is rather than just describing what it is like or what happened as a result of its implementation. Evaluation-specific methodology is absolutely essential for answering truly evaluative questions such as whether a certain program, policy, or product is (a) just good enough to buy, fund, or support; (b) significantly better than that; (c) clearly better than the other two options we are considering (or might have considered); and/or (d) an excellent example of 'best practice''* (Davidson, 2005: p. xii).

The evaluation-specific methodology comprises the following detailed steps.

2.7.1 Define the purpose of evaluation

The first task in an evaluation process is determining the purpose of the evaluation. In this initial step, one needs to raise the 'big picture question(s)' (Davidson, 2005: p. 14). The broadest purpose of evaluating the artifact DATUM is necessarily linked to the overarching aim articulated in Chapter 1.

2.7.2 Identify evaluation criteria

At this stage, the evaluation criteria, also known as 'dimensions of merit' (Davidson, 2005: p. 23), are identified. These criteria represent the features or attributes of the artifact upon which its performance is assessed, determining whether it is performing adequately or inadequately. The evaluation criteria for DATUM were already identified during the second step of DSR.

2.7.3 Collection and analysis of evidence

The third step of the evaluation-specific methodology is the most crucial. It involves collecting sufficient evidence of how the developed artifact and its components perform when applied to address the problem of interest. This evidence is then analysed to assess the artifact's performance against each of the evaluation criteria.

This book adopts a mixed-method approach for the collection and analysis of evidence, advocating methodological pluralism based on two general arguments and a third specific to the nature of DATUM.

The first argument rooted in the critical realist principle of 'epistemic relativism' (Section 2.2.2) suggests that our perspectives on the world are fallible due to their limitations, as are our theories, including those concerning research methodologies. No method is considered epistemically superior to others; all are fallible and provide only a limited view of reality. As Smith (1975: p. 273) explains:

> *'we are really like blind men [sic] led into an arena and asked to identify an entity (say an elephant) by touching one part of that entity (say a leg). Certainly, we might make better guesses if we could pool the information of all the blind men, each of whom has touched a different part of the elephant.'*

Moreover, every perspective is unique. In other words, using Smith's metaphor above, different people can use the same method (i.e. touching), touch the same part of the elephant, and still come up with different guesses about what part of the elephant it is. To this end, triangulation, which implies using a mixed-method approach, is also advocated. Triangulation involves adopting distinct methods to collect data and evidence from various sources and analysing them using different analytical approaches in a complementary manner (Morse, 1991). This helps control bias by neutralising the various biases inherent in the researcher or research methods.

The second argument is based on the principle of judgmental rationalism, which commits us to making a rational judgment between competing answers to questions about reality. Due to the complexity and multi-dimensional nature of this inaccessible independent reality, some methods are more suitable than others for investigating particular phenomena. As Sayer (1992) argues, different methods have varying levels of 'practical adequacy' depending on the reality one is dealing with. For instance, some questions are better answered using quantitative methods, while others may be more appropriately investigated using qualitative methods. A mixed-method approach is crucial to account for how different aspects of reality require different research methods.

The third argument for adopting a mixed-method approach in this book arises from its nature. Since the book aims to systematise and unify the *DT for UM* paradigm, it is inevitable that the scope of the developed artifact (i.e. DATUM) will span all levels of reality (i.e. real, actual, empirical), as indicated in Figure 2.4. Therefore, different methods must be employed to adequately examine the various strata of reality. Consequently, based on the three arguments above, different types of research must be used to thoroughly evaluate the various elements of DATUM.

2.7.3.1 Types of research in critical realism

Using a critical realist's stratified perspective of the world, Sayer (1992) conceptualises three key types of research.

(a) Abstract Research.

This type of theoretical research focuses on investigating the underlying structures of reality and how they interact through generative mechanisms, which consequently give rise to different possible or potential events. It is concerned primarily with the deeper layers of reality and their causal mechanisms.

(b) Extensive Research.

In contrast, extensive research involves no abstraction and is interested solely in analysing events to identify generalisable patterns, common properties and regularities. This approach is more empirical, concentrating on observable phenomena and statistical regularities.

(c) Intensive Research.

Intensive research isolates a particular concrete event and deeply analyses, through abstract research, all the possible ways it could have been generated. It combines the depth of abstract research with the focus on specific instances or events, seeking to understand the mechanisms at play in particular cases.

As illustrated in Figure 2.6, these different types of research (i.e. abstract, extensive and intensive) were employed to evaluate DATUM based on the evaluation criteria. On the left-hand side of Figure 2.6, the theoretical artifact (i.e. DATUM), which resides in the world of theory (i.e. CS), is presented. It incorporates three key elements that vary in their level of abstractness: the philosophical element is the most abstract, the methodical element is the most concrete, and the methodological element falls in between.

The right side of the figure, however, represents the world of practice (i.e. S-C), involving the stratified reality of DT practices, as explained in the first step of the DSR methodology. The three research types allow for evaluating the philosophical and methodical elements of DATUM separately before evaluating DATUM as a complete whole by investigating the corresponding strata of reality. Below is a brief explanation of how these three research types are implemented in this book.

2.7.3.2 Abstract research

The goal of this part of the evaluation process (Figure 2.6 – research type A) is to assess the philosophical foundation of DATUM in terms of its ability to unify different perspectives. The main objective is to explore how effectively the philosophical element of DATUM can consistently integrate the distinct worldviews within the paradigm *DT for UM*. Achieving this requires a deep understanding of the various philosophical worldviews held by members of the discipline. The perspectives of DT practitioners and researchers shape how they perceive the concept of a DT, the urban environment and urban problems, which in turn influences their practical approaches. To uncover these worldviews and the philosophical assumptions implicitly endorsed by the discipline members – assumptions that are not directly observable – focus group discussions were conducted. These discussions aimed to explore and reveal these underlying beliefs. More details about the focus groups can be found in Section 8.2.

2.7.3.3 Extensive research

The worldviews and beliefs held by DT practitioners, examined in the previous evaluation step (i.e. Abstract research), lead to the emergence of various forms of DT practices. Influenced by their own beliefs and worldviews, as well as by the context and nature of the problems they address, DT practitioners tend to favour specific DT tools or methods during DT-based interventions. This results in the creation of different DTs with varying features and uses. Since DATUM aims to unify DT practices for UM, it should ideally provide a standard methodical framework that accommodates all kinds of DT features and uses, regardless of the underlying practitioners' worldviews. Thus, the 'extensive research' (Figure 2.6 – research type B) aims to evaluate DATUM's methodical element in terms of its practical generalisability, or its applicability across various DT-based projects within the UM context. To achieve this, a multiple-case study research method was employed. This approach allows for testing DATUM's methodical element across multiple projects

Figure 2.6 Evaluation of DATUM (author's own)

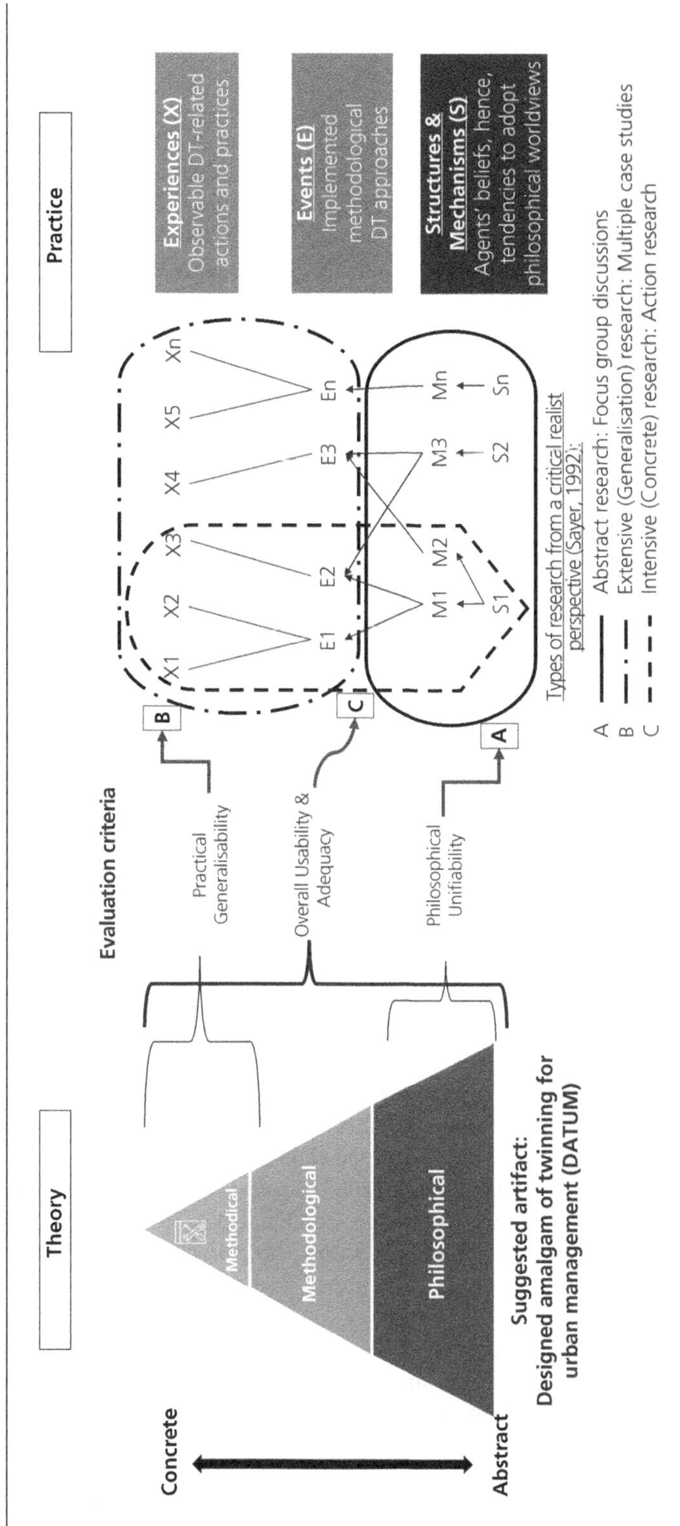

and different contexts, aiding in the formulation of more generalisable findings and assessing practical generalisability. Further details about the multiple-case studies are given in Section 8.3.

2.7.3.4 Intensive research

While abstract and extensive research were used to evaluate the philosophical and methodical elements of DATUM, respectively, this section focuses on evaluating DATUM as a whole artifact in terms of its overall adequacy and usability. This is accomplished through 'intensive' research, conducted using an action research method (Figure 2.6 – research type C). This involves implementing DATUM, including its three fundamental elements, to address real-world problem situations. Since the goal of this evaluation step is to assess usability and adequacy, action research was deemed the most appropriate method, as it is primarily focused on generating knowledge and 'learning by doing' (O'Brien, 2001). Additional details about the action research method and its implementation in the evaluation of DATUM are provided in Section 8.4.

2.8. DSR [5]: Conclusion

In the final step of DSR methodology, the research is summarised, conclusions are drawn and any limitations of the study are acknowledged. This step also emphasises the significance of constructing the designed artifact, explaining its effectiveness in addressing the original problems and highlighting the reasons behind its success. Furthermore, the new knowledge generated throughout this endeavour is articulated, including contributions to both theory and practice, as well as efforts to bridge the gap between the two. Lastly, recommendations for future work are provided.

REFERENCES

Archer MS (1995) *Realist Social Theory: The Morphogenetic Approach.* Cambridge University Press, Cambridge, UK.

Archer MS (2005) Structure, culture and agency. In *The Blackwell Companion to the Sociology of Culture.* Blackwell Publishing, Oxford, UK, pp. 17–34.

Bhaskar R (1975) *A Realist Theory of Science.* Leeds Books Ltd, Leeds, UK.

Bryman A (2016) *Social Research Methods.* Oxford University Press, Oxford, UK.

Checkland P (1999) *Systems Thinking, Systems Practice: Includes a 30-Year Retrospective.* John Wiley and Sons Ltd, Chichester, UK.

Checkland P and Scholes J (1990) *Soft Systems Methodology in Action.* Wiley, Chichester, UK.

Chua WF (1986) Radical developments in accounting thought. *The Accounting Review* **61(4)**: 601–632.

Collier A (1994) *Critical Realism: An Introduction to Roy Bhaskar's Philosophy.* Verso, London, UK.

Creswell JW (2009) *Research Design: Qualitative, Quantitative, and Mixed Methods Approaches*, 3rd edn. SAGE Publications, Inc, Thousand Oaks, CA, USA.

Danermark B, Ekström M and Jakobsen L (2005) *Explaining Society: An Introduction to Critical Realism in the Social Sciences.* Routledge, London, UK.

Davidson EJ (2005) *Evaluation Methodology Basics: The Nuts and Bolts of Sound Evaluation.* SAGE Publications, Thousand Oaks, CA, USA. https://books.google.co.uk/books?id=ePfuba9tDbEC (accessed 11/08/2024).

Dresch A, Lacerda DP and Antunes Jr JAV (2015) *Design Science Research: A Method for Science and Technology Advancement.* Springer International Publishing, Cham. 10.1007/978-3-319-07374-3

Easton G (2010) Critical realism in case study research. *Case Study Research in Industrial Marketing* **39(1)**: 118–128. 10.1016/j.indmarman.2008.06.004

Gambetta D (1998) Concatenations of mechanisms. In *Social mechanisms: An analytical approach to social theory* (Hedström P and Swedberg R (ed.)). Cambridge University Press, Cambridge, UK, pp. 102–124.

Giddens A (1979) *Central Problems in Social Theory: Action, Structure, and Contradiction in Social Analysis*. University of California Press, Berkeley, CA, USA. https://books.google.co.uk/books?id=jpxkQ-1elyAC (accessed 11/08/2024).
Guba EG (1990) The alternative paradigm dialog. In *The Paradigm Dialog* (Guba EG (ed.)). Sage Publications, Newbury Park, CA, USA, pp. 17–30.
Hempel CG (1965) *Aspects of Scientific Explanation*. Free Press, New York, NY, USA.
Hume D (2000) *An Enquiry Concerning Human Understanding: A Critical Edition* (T. L. Beauchamp, Ed.). Clarendon Press. https://books.google.co.uk/books?id=3Vp-0Y3Yz_cC
Johnson P and Duberley J (2000) *Understanding Management Research*. Sage, London, UK.
Keat R and Urry J (1978) *Social Theory as Science*. Routledge and Kegan Paul, London, UK.
Kuechler W and Vaishnavi V (2011) Promoting relevance in IS research: an informing system for design science research. *Informing Science* **14**: 125.
Lawson T (1997) *Economics and Reality*. Routledge, London, UK. 10.4324/9780203195390
Lewin K (1951) *Field theory in social science: Selected theoretical papers (Edited by Dorwin Cartwright.)*. Harpers.
Manson NJ (2006) Is operations research really research? *ORiON* **22(2)**. https://orion.journals.ac.za/pub/article/view/255 (accessed 11/08/2024).
Mingers J (2006) *Realising Systems Thinking: Knowledge and Action in Management Science*. Springer US, Boston, MA, USA. https://books.google.co.uk/books?id=C-UcDQEACAAJ (accessed 11/08/2024).
Mingers J (2014) *Systems Thinking, Critical Realism and Philosophy: A Confluence of Ideas*. Routledge, Abingdon, UK.
Morse JM (1991) Approaches to qualitative-quantitative methodological triangulation. *Nursing Research* **40(2)**: 120–123.
Mouton J (1996) *Understanding Social Research*, Van Schaik Publishers, Pretoria. https://books.google.co.uk/books?id=yWwsAQAACAAJ (accessed 11/08/2024).
O'Brien R (2001) An overview of the methodological approach of action research. In *Theory and Practice of Action Research*, Ed. R. Richardson, Universidade Federal da Paraíba, João Pessoa, Brazil. http://www.web.ca/~robrien/papers/arfinal.html (accessed 11/08/2024).
Popper K (1957) *The Poverty of Historicism*. Routledge and Kegan Paul, London, UK.
Porpora DV (1989) Four Concepts of Social Structure. *Journal for the Theory of Social Behaviour* **19(2)**, 195–211. https://doi.org/10.1111/j.1468-5914.1989.tb00144.x
Porpora DV (2015). *Reconstructing Sociology: The Critical Realist Approach*. Cambridge University Press, Cambridge, UK. https://books.google.co.uk/books?id=oPhfCgAAQBAJ (accessed 11/08/2024).
Porpora DV (2019) A reflection on critical realism and ethics. *Journal of Critical Realism* **18(3)**: 274–284. 10.1080/14767430.2019.1618064
Reason P and Bradbury H (2007) *The SAGE Handbook of Action Research: Participative Inquiry and Practice*, 2nd edn. SAGE Publications, London, UK. https://books.google.co.uk/books?id=2fTlmcue2p0C (accessed 11/08/2024).
Sayer RA (1992) *Method in Social Science: A Realist Approach*, 2nd edn. Psychology Press, London, UK. https://books.google.co.uk/books?id=kr8QvOpM2RwC (accessed 11/08/2024).
Simon HA (1996) *The Sciences of the Artificial*, 3rd edn. MIT Press, Cambridge, MA.
Smith HW (1975) *Strategies of Social Research: The Methodological Imagination*. Prentice-Hall, Englewood Cliffs, NJ, USA. https://books.google.co.uk/books?id=Nhe25tTs-sgC (accessed 11/08/2024).
Smith ML (2006) Overcoming theory-practice inconsistencies: critical realism and information systems research. *Information and Organization* **16(3)**: 191–211. 10.1016/j.infoandorg.2005.10.003

Trigg R (1980) *Reality at Risk: A Defence of Realism in Philosophy and the Sciences*. Harvester Press, Brighton, UK. https://books.google.co.uk/books?id=8Y7pqn1MQBUC (accessed 11/08/2024).

Volkoff O, Strong DM and Elmes MB (2007) Technological embeddedness and organizational change. *Organization Science* **18(5)**: 832–848.

Walsham G (1993) *Interpreting Information Systems in Organizations*. Books on Demand, Chichester, UK. https://books.google.co.uk/books?id=IOQ3swEACAAJ (accessed 11/08/2024).

Ramy Elsehrawy and Bimal Kumar
ISBN 978-1-83662-235-2
https://doi.org/10.1108/978-1-83662-234-520251005
Copyright © 2025 by Ramy Elsehrawy and Bimal Kumar. Published under exclusive licence by Emerald Publishing Limited

Chapter 3
Navigating the landscape

3.1. Introduction

As explained in the previous chapter, this book aims to design and develop a theoretical artifact that supports the growth and maturity of the new paradigm, *DT for UM*. To design an effective artifact using the DSR methodology, explained in Chapter 1, the following are required

- a full understanding of the problem situation and current state
- a clear vision of the desired target state
- well-defined pathways 'that lead from the present to that desired future' (Simon, 1996: p. 148).

This chapter focuses on the first requirement. It aims to develop a deep understanding of the current situation and the problem under investigation. As Dresch *et al.* (2015: p. 129) emphasise, 'it is important that the researcher be aware of what was previously researched, how it was researched, what results were found, and, perhaps most importantly, what has not yet been researched.' Therefore, a thorough systematic literature review is conducted in this chapter to grasp the current state of the *DT for UM* paradigm, identify gaps in the relevant literature, and understand how DT is currently utilised to support UM research and practices.

Before delving into the systematic review of the current state of *DT for UM*, Section 3.2 explores the concepts of systems and systems thinking. The nature of the problem at hand influences the selection of the approach used to address it. While the scientific method, rooted in reductionism, is highly effective for addressing scientific puzzles within a 'closed' lab environment, the urban problems central to UM are complex, multidimensional and must be addressed in the 'open' urban environment. Therefore, a holistic approach based on systems thinking is essential in this context.

Following this, the two key terms in this book, UM and DT, are thoroughly explored and defined in Sections 3.3 and 3.4, respectively. This sets the stage for the systematic literature review of the specific paradigm *DT for UM*. As explained in Section 2.4, the review is conducted from a critical realist perspective, viewing the reviewed studies as stratified and composed of three analytical levels: methods, methodology and philosophy (Figure 2.4).

3.2. Systems thinking

The scientific method has demonstrated considerable power in uncovering the natural laws governing observed phenomena. Through the three foundational pillars of the scientific method – reductionism, repeatability, and refutability (Checkland, 1999) – scientists can dissect phenomena into smaller components, adhering to Descartes's second rule for 'properly conducting one's reason'. This approach helps manage the complexity of natural phenomena and reveals underlying regularities. However, as we move beyond these natural systems, the contemporary world increasingly presents unprecedented levels of messiness and interconnectedness within extremely complex socio-technical systems – systems that involve both social and artificial elements. Attempting to explore these systems scientifically by breaking them down into smaller parts to uncover hidden regularities or general laws poses significant challenges. These systems resist separation and dividing them into smaller components distorts the original system under examination, where the individual parts no longer resemble the whole. This whole, along with its behaviour, is argued to be irreducible (Checkland, 1999).

The rise of the systems thinking movement in the early twentieth century was a rational response to the limitations of traditional science, particularly the inability of reductionism to effectively address overly complex systems. A systems approach does not seek to oppose science but rather to complement it by examining the more complex systems of the modern world. The systems school of thought refrains from separating or breaking down a system before studying it. Instead, it advocates for a broad and holistic view of the system, aiming to understand the interdependencies between its parts. Before delving deeper into the fundamentals of systems thinking, it is crucial to first define what constitutes a 'system', its properties, and what elements make up one.

3.2.1 System definition

A widely accepted definition of a 'system' describes it as a collection of interconnected elements that form a purposive whole, exhibiting new emergent behaviours or properties that are not attributable to the individual components (Armson, 2011). Several key terms used in this definition, such as 'interconnected', 'whole' and 'purposive' are further elaborated as follows.

- Interconnectedness: This term refers to the multidirectional relationships between the constituent parts of a system, where the state of each part is influenced by, and in turn influences, the state of the others. The term 'interconnected' is preferred over 'connected' to emphasise a complex web of multiple connections involving all parts together, rather than a simple concatenation of connections, such as a chain of pairs or other types of connections that can be reduced to two-term relations.
- Whole: This concept implies the formation of a new, standalone entity – created by the specific arrangement, not just the aggregation, of its parts – that is complete and functional on its own. This whole can achieve its purpose independently, even though it includes constituent parts as a 'system' and may itself be part of a higher-level system. The idea of different levels of complexity, also known as 'hierarchy', is further explored in Section 3.2.3.
- Purposive: This suggests that any system appears to have a goal that it consistently strives to achieve. This goal can be either intentionally designed, in the case of an 'artificial' system, or it may emerge unintentionally. The concept of 'emergence' is also discussed in detail in Section 3.2.3.

3.2.2 System features

A system's features, largely derived from its definition, are well recognised and widely accepted among systems thinkers. First, a system should have a boundary – this boundary defines the system and establishes its identity within the broader complexity of the world. Second, the system interacts with its environment, which consists of external complexities outside the system's boundary that both influence and are influenced by the system. Within this boundary, the system comprises sub-systems that are interconnected in a web of relationships, as previously defined. These sub-systems are themselves systems with their own boundaries and sub-systems, collectively forming a structural hierarchy.

Figure 3.1 Typical system features. Source: Armson, R (2011)

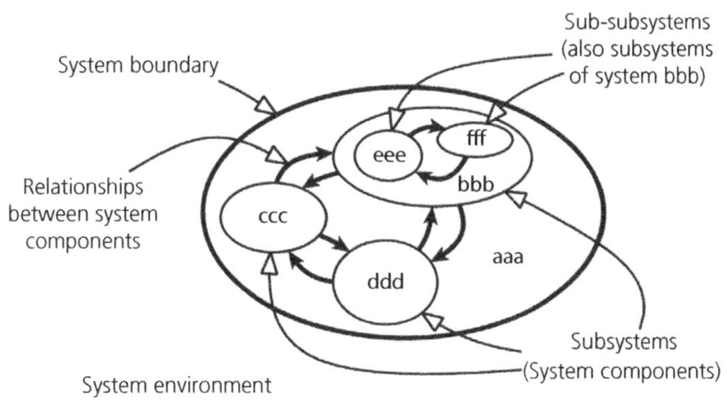

Figure 3.1 illustrates a typical system and its main features. Ackoff (1981) identified three crucial features of systems.

- Each sub-system impacts the functioning of the entire system.
- Each sub-system is influenced by at least one other sub-system.
- All possible subgroups of sub-systems exhibit the first two properties.

3.2.3 Hierarchy and emergence

A core principle of systems thinking, which forms the foundation of its modern movement, is the concept of hierarchy and emergence (Checkland, 1999). These two concepts are closely intertwined, often viewed as two sides of the same coin, and together they represent what is known as organised complexity. The idea of organised complexity is derived from the discovery of hierarchical structures in living organisms, such as molecules, cells, organs and the organism itself. This hierarchy suggests that each level of complexity is characterised by emergent properties that do not exist at lower levels and cannot be reduced to or explained in the language appropriate to those levels. As a result, hierarchy theory evolved, focusing on understanding how these levels of complexity are formed and how they relate to one another (Pattee, 1973). Checkland (1999: p. 78) eloquently captures this concept in a metaphor.

> *"The shape of an apple', although the result of processes which operate at the level of the cells, organelles, and organic molecules which comprise apple trees, and although, we hope, eventually explicable in terms of those processes, has no meaning at the lower levels of description. The processes at those levels result in an outcome which signals the existence of a new stable level of complexity – that the whole apple itself – which has emergent properties, one of them being the apple's shape.'*

In his exploration of the architecture of complexity, Simon (1996) argues that behaviours and properties at one hierarchical level of complexity are largely self-representative, meaning that understanding these behaviours does not usually require detailed knowledge of the lower levels. Simon describes how this hierarchical structure has been beneficial for the progress of natural sciences over the past three centuries. He explains that science has advanced by adopting a top-down approach, allowing us to understand higher levels before fully grasping the lower ones. Simon (1996: p. 16) elaborates, saying that the 'skyhook-skyscraper' construction of science – from the top down to the yet unconstructed foundations – was possible because the behaviour of systems at each level depended only on a simplified and abstracted characterisation of the system at the next lower level. In simpler terms, using Checkland's metaphor, one does not need to understand the cellular or molecular processes of an apple to recognise and learn about its shape. The underlying causalities and principles governing living matter are considered irrelevant in this context.

The invention of the microscope and the subsequent discovery of the cell provided scientific support for the concept of hierarchy, demonstrating that living organisms are structured hierarchically, from molecules to cells to organs and organisms. This discovery also paved the way for a deeper understanding of the lower levels of complexity, leading to insights into the causal mechanisms that drive behaviours at higher levels. It has been shown that these behaviours at higher levels of the 'skyhook-skyscraper' of science (e.g. biology) obey the laws governing the lower levels (e.g. physics and chemistry). Checkland (1999: p. 81) describes the phenomenon of emergent properties as the result of constraints acting on the 'degrees of freedom' of elements at the lower level. These emergent properties can then be described using a language appropriate to a meta-level above that of the lower-level elements – for example, in the language of chemistry, any arrangement of the bases in DNA follows the laws of physical chemistry. However, it is the constraints on ordinary chemistry that lead to the emergence of the specifically biological property of genetic coding, marking the transition from the level of chemistry to that of biology (Checkland, 1981: p. 81). This view has significantly influenced the thinking of artificial systems designers. Not only do artifacts inherently possess a hierarchical structure, but some researchers in design have also relied on the idea of imposing constraints on different elements within a level of complexity to create desired emergent properties at higher levels, thereby producing an artificial system capable of achieving a predefined goal (Stefik, 1981).

The increased accessibility to lower hierarchical levels of complexity and the consequent growth in knowledge at these levels have shaped the modern movement in biology. Crick (1966), in his work '*Of Molecules and Men*', claimed that the ultimate goal of biology is to explain it in terms of physics and chemistry. However, Grene (1974) argues that the idea of a one-level ontology – the belief that with enough knowledge of lower levels, all sciences can be explained in terms of atomic events – is inherently contradictory. To recognise such events as meaningful, one must establish a second, higher-level ontology. Checkland (1999) suggests that Crick failed to recognise this contradiction, as Crick himself implicitly acknowledged the existence of higher-level sciences that, in his view, require explanation in terms of lower ones.

To Simon, emergence is defined by the mutual relations between parts that do not exist for the parts in isolation (Simon, 1996: p. 183). These interactions arise from properties inherent in the interacting parts, which only function when these parts are brought together. In other words, the properties of the parts impose constraints on them at the hierarchical level that includes them, leading to the emergence of aggregate properties. These new properties are typically described using newly introduced terms to avoid referencing the detailed particulars of the lower-level parts and their mutual interactions. The greatest strength of this holistic perspective is its ability to provide a scientific explanation for 'why' a higher-level ontology behaves in a certain way, while simultaneously acknowledging that it cannot be reduced to the lower levels.

3.2.4 System boundary

Systems thinkers, unlike reductionists, consistently acknowledge and appreciate the complexity of the world. They emphasise the importance of considering the full picture, recognising that any picture derives its identity from how distinctly it is separated from its background, making it perceivable to the observer. In systems thinking, the boundary of a system is crucial as it encloses the interconnected significant variables and separates them structurally from the outer environment, which contains variables considered less relevant or significant by the observer. Mingers (2006: p. 73) notes that 'defining a system in terms of its components and their relations is effectively to delineate its boundary.'

Despite the foundational importance of the concept of a system boundary, it is rarely discussed in depth by systems thinkers, nor is there much guidance on how to define such boundaries. This reluctance seems to stem from the inherently complex nature of the topic, which is difficult to tackle. Mingers (2006) identifies this absence of thorough discussion about such a vital concept as an 'interesting paradox'. In response to this gap, he conducted a comprehensive review of the notion of system boundaries within the literature of systems thinking and provided a distinctive classification of all possible forms of boundaries. Mingers (2006) subsequently conceptualised four main types of system boundaries

- boundaries of physical systems
- boundaries of mathematical systems
- boundaries of conceptual and language systems
- boundaries of social systems.

While a detailed exploration of each boundary type is beyond the scope of this book, recognising these different types reveals that the answer to the question of system boundaries varies widely across a spectrum (Figure 3.2). At one end of this spectrum are physical systems, which have clear, well-defined edges, presenting an empirical and tangible system boundary. Systems near this end are typically concrete objects and physical artifacts. However, at the extreme opposite end are social systems, characterised by boundaries that are more notional and intangible, often subject to interpretation, value-laden and heavily influenced by the observer's perceptions. The idea that there could be a single, definitive boundary for such systems is highly problematic. Instead, these systems are flexible and open to being defined by multiple, potentially overlapping boundaries.

The most challenging aspect of boundaries arises when dealing with social systems, which occupy the extreme right-hand side of the spectrum in Figure 3.2. Although other types of systems may exhibit varying levels of ambiguity or dynamism, they are generally more straightforward compared to social systems. Social systems may sometimes display clear-cut boundaries, such as those found in formal groups or organisations with defined membership. However, this clarity is rare when addressing social or socio-technical 'wicked' problems. These systems are incredibly complex, characterised by intricate, interconnected causal loops that span both space and time, making it inherently difficult for the system to delineate itself from its environment. Determining where

Figure 3.2 The two extremes of the spectrum of types of system boundary (author's own)

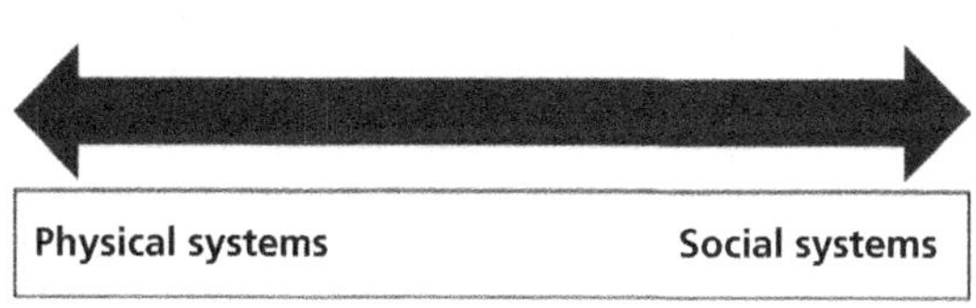

the boundary should extend – essentially identifying the limits of an 'organisationally closed' system (Varela, 1981) – is always a contentious issue. Giddens (1984, 1990) points out that the concept of a society's 'closure' over time and space is problematic, with many different possible interpretations.

Churchman (1968), influenced by systems thinking and holism, rejected the idea of experiment-like closed environments. This stance placed him in direct confrontation with the overwhelming complexity of social systems, which exhibit severe 'openness.' As Jackson (2019: p. xix) observed, 'the planner who works with open systems is caught up in the ambiguity of their causal webs.' This perspective, according to Mingers (2006), introduces several peculiar problems. First, it places the system's environment at the centre of attention, forcing the system to adapt to its environment. Consequently, the environment appears to dictate the structure of the system. This led Mingers (2006: p. 169) to pose several challenging questions.

> *'What exactly is the environment within which a social system, more especially a society, might exist? Is it the physical world, or other societies, or what? More generally, how would one draw a boundary to demarcate off some well-defined social system that then interacts with an equally well-specified environment? Second, what could possibly be the inputs and outputs of such a system? Does it really make sense to conceptualize a society, or part of it such as a family, as a processor of inputs into outputs?'*

These unanswered questions left both Mingers and Churchman before him curious and searching for the true 'frame' of an exceedingly 'rich picture.' In such a picture, countless interconnections between different parts can be traced and identified. However, reverting to reductionism by positing a boundary that consciously ignores these cross-boundary interconnections is not an option that any serious systems thinker could easily accept. The only other choice for those interested in making designed changes social systems is to acknowledge one's ignorance and inability to fully comprehend the sheer complexity of the system, and then proceed with a proposed intervention. To Churchman, this latter option is irrational. He argued that to be confident in any proposed design, solution, or intervention, one must first understand the entire system. He posed the question: 'How can we design improvement in large systems without understanding the whole system, and if the answer is that we cannot, how is it possible to understand the whole system?' (1968: p. 2).

Many authors have alluded to this intractable problem of boundaries. Muir (1911), through his famous words, gives us insight into the interconnected web of life, arguing that 'when we try to pick out anything by itself, we find it hitched to everything else in the universe.' Sunim (2018: p. 120) expands on this idea, describing how interconnected the entire universe is.

> *'The whole universe is contained in an apple wedge in a lunch box. Apple tree, sunlight, cloud, rain, earth, air, farmer's sweat are all in it. Delivery truck, gas, market, money, cashier's smile are all in it. Refrigerator, knife, cutting board, mother's love, are all in it. Everything in the whole universe depends on one another. Now think about what exists in you. The whole universe is in us.'*

Simon (1977: p. 258) also highlights the extent of this interconnectedness, noting that, 'to a Platonic mind, everything in the world is connected to everything else – and perhaps it is', although some things may be more connected than others. In such an interconnected world, examining a system and determining which elements to include in the analysis and which to classify as part of the system's environment – or even to ignore altogether – becomes a deeply problematic task.

3.2.5 A systems view of management problems

The contrasting perspectives of reductionists and systems thinkers offer valuable insights not only to scientists but also to managers and designers. While reductionism provides a detailed understanding of the components, systems thinking complements it by offering knowledge about irreducible 'wholes.' This holistic approach is crucial in fields like design, management, decision-making and problem-solving, particularly when dealing with complex, real-world issues. Numerous authors have attempted to classify or categorise the types of complex problems that arise in the real world.

Forrester (1968) characterised the structure of complex systems as a high-order system, involving multiple levels or states, controlled by non-linear relationships and comprising numerous interacting feedback loops. These loops include both positive feedback loops, which represent growth, and negative feedback loops, which denote purposiveness and self-control.

Boulton *et al.* (2015: p. 36) similarly view complex systems as being governed by underlying feedback loops. From their perspective, individual elements of a system interact in 'reflexive' ways, allowing for feedback while being influenced by the surrounding context, which is itself continuously evolving due to these interactions and feedbacks. They summarise the key features of a complex world as follows.

- Systemic and synergistic: Emerging from innumerable complex and interconnected causes; multiscalar, with interconnections occurring across multiple levels.
- Endlessly diverse and variant: Exhibiting variety, diversity, variation and fluctuations, which contribute to resilience and adaptability.
- Path-dependent: Relies on its local context and the sequence of events in a self-referential manner.
- Changing episodically: Occasionally undergoes radical changes that give rise to new regimes.
- Unpredictable future: Has multiple possible futures, making it impossible to predict future states based solely on the present.
- Self-regulative: Capable of self-organisation and self-regulation.

Rittel and Webber (1973) referred to the complex problems arising from the increasingly complex world as 'wicked problems.' These are problems that traditional reductionist science has struggled to address effectively, unlike the less complex or reducible 'tamed problems.' Rittel and Webber argued that 'wicked' problems, unlike 'tamed' ones, exhibit the following characteristics.

- No definitive formulation: Problem formulation and solution are intertwined. As one works on a solution, new insights into the problem emerge.
- No final solution: Because the problem is not fully understood, it is difficult to determine when it has been resolved.
- Solutions are not true or false, but good or bad: The ambiguity in the problem's definition means stakeholders will always propose different solutions.
- Implemented solutions cannot be immediately or ultimately tested: The problem involves numerous interconnected elements, leading to far-reaching consequences that are difficult to detect or predict.
- Every attempt counts: Once a solution is implemented, it triggers far-reaching consequences, making it impossible to revert to the initial state.
- A wicked problem is a symptom of another problem: Wicked problems are interconnected issues.
- Every wicked problem is unique: There are no predefined or off-the-shelf solutions for wicked problems.

Figure 3.3 Difficulties against messes. Source: Armson (2011)

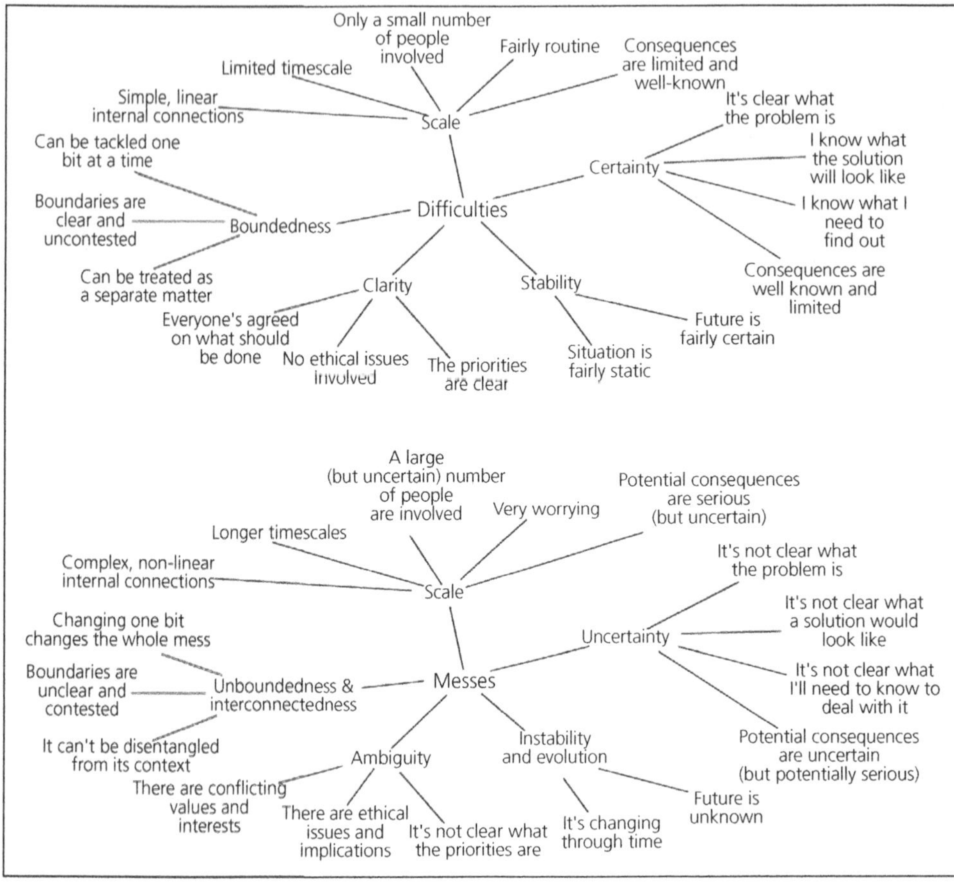

Building on Rittel and Webber's work, Armson (2011) viewed real-world problems as lying on a spectrum with two extreme ends: 'difficulties' at one end, which are more scientific-friendly and easier for reductionists to tackle, and 'messes' at the other end, representing the most perplexing types of problems. Figure 3.3, constructed by Armson and workshop participants, illustrates the differences between these types based on five criteria: scale, certainty, stability, clarity and boundedness. The concept of boundedness, in particular, underscores how challenging it is to define what is part of a perceived 'mess' and what is not. Consequently, Armson emphasises the need for thorough study of every 'mess,' noting that 'well-chosen, efficient and effective interventions will not cause unintended consequences' (Armson, 2011: p. 119). The issue of boundedness which involves understanding what constitutes a part of the system of a problem and what does not, is closely related to the problematic question of system boundaries discussed earlier in Section 3.2.4.

Kurtz and Snowden (2003) introduced the Cynefin framework, which categorises problem contexts into four main domains: simple, complicated, complex and chaotic (Figure 3.4). In the simple domain, strong central directions and weak connections between components lead to easily identifiable, linear cause-and-effect relationships with predictable outcomes. The complicated domain

Figure 3.4 Cynefin framework. Adapted from Kurtz and Snowden (2003)

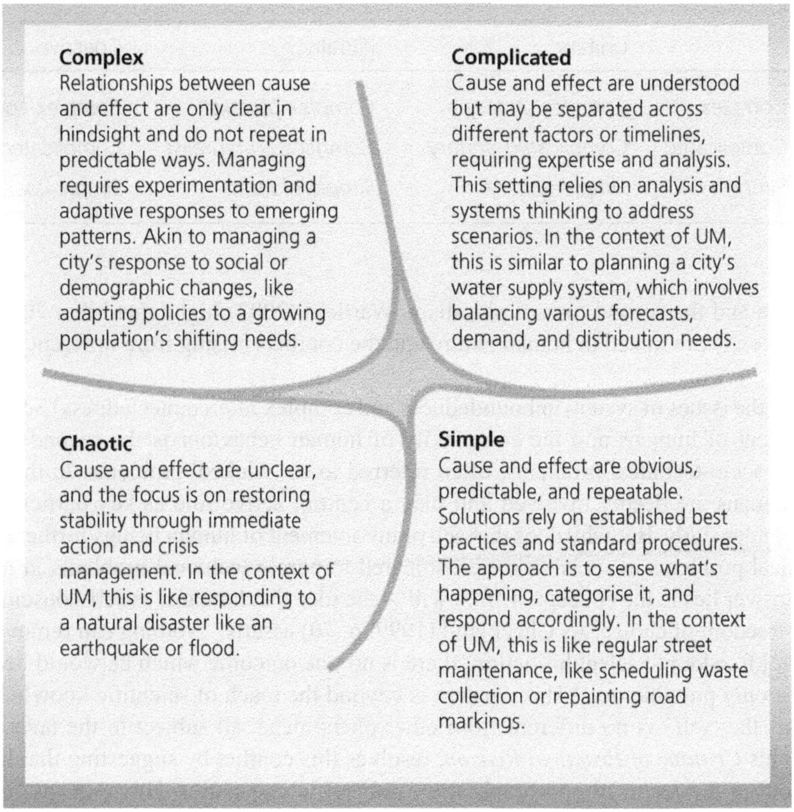

involves identifiable causal relationships that are separated by space and time, and are so intricately connected that understanding them becomes challenging. In the complex domain, predicting behavioural outcomes with high confidence is nearly impossible due to the myriad interconnected cause-and-effect sequences, though these can be traced retrospectively. The chaotic domain, however, shows no observable patterns or relationships between cause and effect.

The 'ideal-type'[1] grid of problem contexts, originally developed by Jackson and Keys (1984), has been refined over time, with the most recent version by Jackson (2019) shown in Table 3.1. This grid is defined by two dimensions: the first describes systems complexity, while the second pertains to stakeholders' complexity, which involves the complications arising from human interactions and power dynamics. This latter dimension reflects the complexities of human nature, where individuals participating in a problem context often have differing identities, values, beliefs, purposes and interests. These differences, coupled with power structures, can lead to coercive situations where some groups are oppressed or marginalised by more powerful ones. This framework aligns with Bawden's (1995) observation of two significant shifts in the history of systems thinking: the first

[1]As Jackson (2019, p. 157) clarifies, 'the concept of an 'ideal type' … establishes that the grid presents some abstract, logical classes of problem context. It does not seek to describe actual problem contexts, which … will look different to different observers and even to one observer taking a look for different purposes.'

Table 3.1 'Ideal-type' grid of problem contexts. Source: (Jackson, 2019)

		Stakeholders		
		Unitary	Pluralist	Coercive
Systems	Complex	*Complex-unitary*	*Complex-pluralist*	*Complex-coercive*
	Complicated	*Complicated-unitary*	*Complicated-pluralist*	*Complicated-coercive*
	Simple	*Simple-unitary*	*Simple-pluralist*	*Simple-coercive*

toward holism and the second toward pluralism. Warfield (2002) highlighted that 70% of his 20 laws of complexity are linked to human nature and the conflicts arising from different perceptions.

In addition to the issues of system unboundedness and complex interconnectedness (Section 3.2.4), the involvement of humans and the complexity of human behaviour is the second major issue in real-world socio-technical problems, often referred to as 'wicked' problems. In these types of problems, humans are deeply involved and play a central, active role as key participants in the phenomena under study. But why does this active involvement of human beings further complicate socio-technical problems in social science compared to purely technical problems in natural science? The answer lies in the concept of 'free will' – the idea that humans, as self-conscious beings, possess the freedom of choice. As Checkland (1999: p. 70) asserts, 'Nothing can remove from the agent his [sic] freedom to select his action; there is no one outcome which he would be correct to regard as the only possible one.' This concept is beyond the reach of scientific knowledge, where science treats the 'self' as no different from other phenomena, all subject to the laws of causality. Kant, in his *Critique of Practical Reason*, resolves this conflict by suggesting that humans, as 'phenomenon' in the realm of appearances, are subject to causalities. However, as 'noumenon' (things in themselves), they are beyond our knowledge, perception and accessibility – they are transcendental. Kant explains: 'we have in the world beings of but one kind whose causality is teleological, or directed to ends, and which at the same time are beings of such character that the law according to which they have to determine ends for themselves is represented by themselves as unconditioned and not dependent on anything in nature, but as necessary in itself. The being of this kind is man [sic], but man regarded as noumenon.' (Kant, quoted in Kemp, 1968: pp. 120–121).

This human nature and their direct involvement in the phenomena under study have profound implications for predictability. Unlike natural sciences, which evolved based on the idea of making accurate predictions and enabling the repeatability of results to generate public knowledge, the involvement of humans has severely reduced the ability to make accurate predictions about the behaviour and future state of social systems in two ways. First, humans can recognise, be aware of, and be conscious of predictions made. In response, they can react uniquely and modify the situation, thereby affecting the outcome and undermining the predictions. Popper, in *The Poverty of Historicism* (1957: p. 22), explains it this way.

> *'A prediction is a social happening that may interact with other social happenings, and among them with the one which it predicts. It may, as we have seen, help to precipitate this event; but it is easy to see that it may also influence it in other ways. It may, in an extreme case, even cause the happening it predicts: the happening might not have occurred at all if it had not been predicted. At the other extreme, the prediction of an impending event may lead to its prevention.'*

Second, the behaviour of natural and technical systems and their components is consistent – they obey natural laws consistently, which are hidden by the messiness of the world and waiting to be discovered, regardless of human knowledge. However, human agents, who are the core of social and socio-technical systems under examination, behave according to the knowledge they have, not based on a 'priori' set of laws. Since future human knowledge is unknown, predicting future human behaviour becomes impossible. Popper addresses this problem in his logical refutation of historicism (1957: p. 10).

'The course of human history is strongly influenced by the growth of human knowledge...We cannot predict, by rational or scientific methods, the future growth of our scientific knowledge...We cannot, therefore, predict the future course of human history.'

3.3. Urban management

This section aims to establish a clear definition and scope of the term urban management (UM), with further details provided in Chapter 4. UM is identified as 'an applied science' (Ding *et al.*, 2012: p. 1). However, the concept of UM has been consistently recognised as elusive (Engin *et al.*, 2020; Kearns and Paddison, 2000; Stren, 1993; Werna, 1995). Over the years, the term has been interpreted differently by various scholars. For instance, Davey (1993) perceived it as a loosely connected set of policies, plans and practices designed to ensure access to public services. Mattingly (1994: p. 202) understood it as 'taking sustained responsibility for actions to achieve particular objectives with regard to human settlements.' In contrast, Bačlija (2011) viewed UM as a reform of city administration, while Ding *et al.* (2012) conceptualised it as the integration of two 'seemingly disparate elements' – cities and management.

Urban managers and researchers have identified various objectives related to different dimensions of the urban environment that motivate UM and its associated activities – for example, McGill (1998: p. 463) argues that UM has two central and related objectives: 'first, to plan for, provide, and maintain a city's infrastructure and services; and second, to make sure that the city's government is in a fit state, organisationally and financially, to ensure that provision and maintenance.' The World Bank (1991) suggests that the objectives of UM include alleviating poverty, enhancing urban productivity and protecting the urban environment, which correspond to the social, economic and natural dimensions of urban space. In alignment with these three dimensions, Ding et al. (2012) view UM as efforts aimed at promoting social welfare for urban residents and ensuring sustainable urban growth. Others have emphasised the concept of optimisation or 'satisficing' (Simon, 1996), where effective UM involves balancing social and economic development (Bačlija, 2011). Additionally, some have highlighted the importance of UM in achieving urban resilience (McGill, 2020; Simone *et al.*, 2021).

SUD, as defined by Camagni (1998: p. 17), is 'a process of synergetic integration and co-evolution among the great subsystems making up a city (economic, social, physical and environmental), which guarantees the local population a non-decreasing level of well-being in the long term, without compromising the possibilities of development of surrounding areas and contributing by this towards reducing the harmful effects of development on the biosphere.' In a more recent and ambitious vision for the built environment, a group of key figures interested in the future of the built environment suggests that better management of the interconnected urban environment – which includes built systems, natural systems, cyber-physical systems, services and people – is fundamentally about 'improving outcomes for people and nature' and enabling them to 'flourish together for generations' (A collaboration of leading figures in the built environment, 2021: p. 5).

Aligning with these views, UM is defined in this book from a systemic perspective as an applied science that involves planning for and implementing interventions into the urban environment – that is, the system of systems comprising interconnected technical, social and natural systems – in ways that lead to the emergence of better conditions for people and nature.

3.4. Digital twin

This section provides a brief overview of the history and origin of DT and formulates a clear definition of the term. Further details on the implementation of DT in UM are discussed in Section 3.5.

3.4.1 Origin

The concept of DT has evolved over several decades. In 1991, David Gelernter, a specialist in massively parallel computation at Yale, introduced the notion of 'mirror worlds'. This concept refers to a city-scale model of reality continuously updated with massive amounts of data. The model enables users to zoom in or out to visualise desired levels of detail on a computer screen as 'a single dense, live, pulsing, swarming, moving, changing picture' (Gelernter, 1991: p. 30). Years later, in 2002, Grieves presented a conceptual model named 'conceptual ideal for PLM' that included real space, virtual space, virtual sub-spaces and data flow between these spaces for product lifecycle management (PLM). This concept was later embedded in the first PLM course at the University of Michigan in 2003, referred to as the 'mirrored spaces model', and published in a journal article in 2005 (Grieves, 2005). By 2006, Grieves had renamed this concept the 'information mirroring model' in his book *Product Lifecycle Management: Driving the Next Generation of Lean Thinking* (Grieves, 2006).

As this concept expanded and matured, Piascik and colleagues (2012) coined the term 'digital twin' in one of NASA's strategic technological roadmaps as a synonym for their long-term vision called 'virtual digital fleet leader.' Since then, the term digital twin has been widely adopted, incorporating earlier concepts like the information mirroring model (Grieves, 2011).

3.4.2 Definition

In recent years, interest in the term digital twin has surged (Figure 3.5). Gartner, Inc. (2019) predicted that by 2022, two-thirds of companies implementing an internet of things (IoT) would deploy at least one DT. This exponential growth in interest has significantly driven research into DT advancements. Gartner, Inc. (2019) also argued that DTs are entering mainstream use, despite the fact that the concept is still in its infancy and lacks a common definition across the literature (Kritzinger *et al.*, 2018).

Figure 3.5 Interest in 'digital twin' over 5 years (2014–2019). Source: courtesy of Google Trends https://trends.google.com/

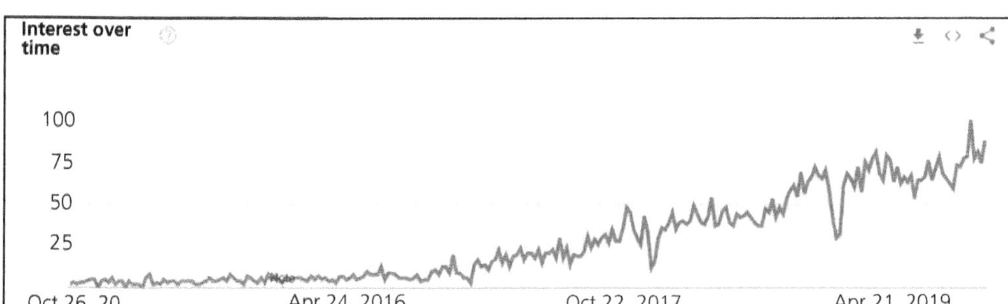

The Urban Innovation Labs (2019) noted that many have used the term DT without fully understanding its meaning or, in some cases, have even used misleading definitions (Tomko and Winter, 2019). Batty (2018) described the term DT as a 'cliché', a sentiment justified by the widespread lack of understanding that has led to unmet expectations and unrealised benefits (Hicks, 2019).

From the origins and history of DT, it is clear that the manufacturing and aerospace industries are the most mature and the earliest adopters of digital twinning. Therefore, before defining what a DT is, we first analyse the definitions provided by key manufacturing and aerospace studies (Table 3.2). Subsequently, definitions from researchers in UM or closely related disciplines such as architecture, engineering, construction and operation (AECO) were also analysed. The definitions were broken down and analysed based on three aspects

- the identity, nature or structure of the DT *(What is it?)*
- the aim, purpose or function of DT *(What is it for?)*
- the main constituents, elements or components of a DT *(What is it made of?)*.

The focus was on explicit definitions of the term DT provided by the cited authors without interpreting or analysing their further discussions. The analysis revealed that some less comprehensive definitions included only two of these aspects, while others addressed only one.

A closer examination of the first aspect (nature) shows that some researchers view DT as a concept, emphasising the underlying idea, while others see it as a model or a digital representation, focusing on the tangible product. Reviewing the second aspect (aim, purpose or function) provides insight into the potential of DT and the emerging set of aims and uses. One aim of DT, as suggested by some researchers, is the ability to predict the future state of a physical system. However, the general aim of DT, as mentioned by most researchers, is to provide insights that help make better decisions and unlock value. This creates several opportunities, including early detection of issues or anomalies, which can lead to early warnings (Parrott and Warshaw, 2017), improved performance during the operational phase of the physical system through proactive maintenance (Farsi *et al.*, 2020), detection of inefficient processes (Vachálek *et al.*, 2017), and operational optimisation (Tao *et al.*, 2018).

Considering the third aspect (main elements or components), it is evident that models and data are core constituents of a DT. Moreover, if the primary aim of a DT is to predict the future state, assist in strategic decision-making, provide feedback and recommend actions, then the need to simulate the behaviour and functions of the physical system with exceptionally high accuracy becomes crucial (DebRoy *et al.*, 2017; Schleich *et al.*, 2017). Achieving this requires *DT input*, including extensive data input and models representing all relevant attributes of the physical system's nature and context. To unlock the value of this input, 'new IT' (Tao *et al.*, 2019), such as big data analytics (BDA) and artificial intelligence (AI), are utilised to produce DT output, including high-fidelity virtual models, simulations and feedback.

Therefore, based on the reviewed definitions (Table 3.2), a DT can be described as

> *'The concept of connecting a physical system in the physical world to its virtual representation in the cyber world by way of bidirectional communication, with or without human involvement. This process involves transmitting data from the physical to the cyber world, followed by quantitative or qualitative data analysis and processing to unlock value. The value is realised through automatic control and intervention (human-out-of-the-loop) or by providing insights that support decision-making (human-in-the-loop).'*

Table 3.2 Key definitions of DT across different disciplines (continued on next page)

Author \ Aspect	Identity, nature or structure of a DT	Aim, purpose or function of a DT	Main constituents of a DT
Manufacturing and aerospace			
(Glaessgen and Stargel, 2012)	'An integrated multi-physics, multi-scale, probabilistic simulation of a complex product.'	'To mirror the life of its corresponding twin.'	'uses the best available physical models, sensor updates, etc.' 'digital twin consists of three parts: physical product, virtual product, and connected data that tie the physical and virtual product.'
(Hochhalter *et al.*, 2014)	'Digital twin is an emerging concept.'	'To enable high-fidelity modeling of individual vehicles throughout their service lives.'	'which employs modeling and simulation of the as-built vehicle state, as-experienced loads and environments, and other vehicle-specific history.'
(Boschert and Rosen, 2016)	NA	The simulation models making up the DT are specific for their intended use and apply the suitable fidelity for the problem to be solved. The DT evolves with the real system along the whole life cycle. The DT is not only used to describe the behaviour but also to derive solutions relevant for the real system.	the linked collection of the relevant digital artefacts including engineering data, operation data and behaviour descriptions by way of several simulation models. 'integrates the currently available knowledge about it [whole life cycle].'
(Grieves and Vickers, 2017)	'a digital informational construct about a physical system could be created as an entity on its own.'	'This digital information would be a 'twin' of the information that was embedded within the physical system itself and be linked with that physical system through the entire lifecycle of the system.'	NA

(Söderberg *et al.*, 2017)	'The concept of using a digital copy of the physical system.'	'to perform real-time Optimization'	NA
(Tao *et al.*, 2018)	NA	'… that [connected data] tie the physical and virtual product.'	'it [DT] consists of three parts: physical product, virtual product, and connected data…'
(Vachálek *et al.*, 2017)	'A DT is essentially a functional system.'	'…continuous process optimization.'	'…physical production lines [physical systems in manufacturing context] with a digital 'copy.'
(Negri *et al.*, 2017)	'a virtual representation of a production system that is able to run on different simulation disciplines that is characterized by the synchronization between the virtual and real system.'	'to forecast and optimize the behaviour of the production system at each life cycle phase in real time.'	'sensed data and connected smart devices, mathematical models and real time data elaboration.'
(Parrott and Warshaw, 2017)	'a near-real-time digital image of a physical object or process.'	'that helps optimize […] performance.'	Not applicable (NA)
(Rosen *et al.*, 2019)	'A comprehensive physical and functional description.'	NA	'all available operational data of a component, product or system, which includes more or less all information which could be useful in all - the current and subsequent - lifecycle phases.'
(Kritzinger *et al.*, 2018)	'It is the virtual and computerized counterpart of a physical system.'	'A Digital Twin can be used to simulate it for various purposes, … and is able to decide between a set of actions with the focus to orchestrate and execute the whole production system in an optimal way. This Results in a higher efficiency, accuracy and gains economic benefits in the production.'	'exploiting a real time synchronization of the sensed data originating from the field-level.' Fully integrated Bidirectional communication and flow of data.

Table 3.2 Continued

Aspect / Author	Identity, nature or structure of a DT	Aim, purpose or function of a DT	Main constituents of a DT
(Tao *et al.*, 2019)	'The models serve as a communication and recording mechanism.' 'The DT is another concept associated with cyber–physical integration.'	'to help interpret the behaviors of machines or systems and to predict their future state.'	'based on real-time data, historical data, experience, and knowledge, as well as on data from models.'
Urban management / AECO			
(National Infrastructure Commission, 2017)	'A digital model: a dynamic representation of a system…'	'mimics its [digital model] real-world Behavior (and, in some cases, the surrounding environment).'	'real-time, updated collection of data, models, algorithms or analysis.'
	'A virtual representation of a physical object or system across its lifecycle.'	'to enable understanding, learning and reasoning.'	
(Bolton *et al.*, 2018)	'a realistic digital representation of assets, processes or systems in the built or natural environment.'	'They unlock value by enabling improved insights that support better decisions, leading to better outcomes in the physical world.'	NA
(Dawkins *et al.*, 2018)	'The concept of 'Digital Twin' refers to the coupling of a physical system with its digital representation in a computer…'	'…such that any relevant change of state in the physical system is detected and triggers a flow of data that causes a corresponding change in the state of its digital counterpart.'	NA
(Batty, 2018)	'A mirror image of a physical process that is articulated alongside the process in question…'	'…usually matching exactly the operation of the physical process…'	'which takes place in real time.'

(Tomko and Winter, 2019)	'a cyber-physical-social system with coupled properties.'	'capable not only of reaction and prediction, but increasingly also of action (rather than being the passive reflection of a mirror).'	'properties of an organism: (a) it can sense the (physical or digital) environment and updates its counterpart accordingly, (b) it has agency and can change the (physical or digital) environment based on instructions, possibly from the counterpart environment and including engagement with people, and (c) it has a moderation system that attempts to preserve its operational inner state, here called an immune system.'
(Farsi *et al.*, 2020)	a concept or a workflow	to monitor its past and present operational state, detect issues, accurately forecast and predict its future state, maintenance as well as its prediction of remaining useful life (RUL).	Uses collected real-time using sensors and IoT, along with data-driven analytics and intelligence (e.g. machine learning algorithms) to create a continuously adaptable model representing the structure, behaviour and context of a physical asset.
		DT determines the best course of action.	Physical entity in real space, sensors and IoT for real-time data acquisition.

3.5. Systematic literature review

As interest in DT continues to grow (Figure 3.5) and the volume of studies related to their use in UM accumulates, conducting a systematic literature review becomes essential (Saunders *et al.*, 2009). This approach enables the extraction of relevant information from the ever-expanding body of publications (Seuring and Gold, 2012). Systematic literature reviews are defined as.

> *'Secondary studies used to map, find, critically evaluate, consolidate, and aggregate the results of relevant primary studies on an issue or specific research topic, as well as to identify gaps to be filled, resulting in a coherent report or synthesis. The term systematic means that the review should be performed according to an explicit, planned, responsible, and justifiable method, similarly to the expectations of primary studies'* (Dresch *et al.*, 2015: pp. 129–130).

3.5.1 Stratified view of reality

As outlined in Chapter 2, this book adopts a critical realist position. As briefly explained in Section 2.4, the corpus of *DT for UM* studies, retrieved through a systematic literature review, is analysed at three distinct levels. At the most abstract level are the structures and mechanisms within the domain of the *real*. In this context, these represent the *philosophical* worldviews underpinning the reviewed studies. The philosophical assumptions endorsed by researchers – 'spanning ontology, epistemology and methodology' (Healy and Perry, 2000: p. 121) – define 'the nature of possible research and intervention' (Mingers, 1997: p. 429), or the *methodological* approaches adopted at the higher level of analysis, where events occur in the *actual* domain. In other words, based on the philosophical assumptions DT practitioners endorse – whether implicitly or unconsciously – they tend to develop and implement DTs in particular ways that reflect their underlying worldviews and beliefs. Consequently, the methodological approach or form of practice adopted by the DT practitioner makes them more inclined to use specific *methods*, including observable tools, techniques and technologies manifested at the most concrete level of analysis – the experiences in the *empirical* domain. Thus, the conducted systematic literature review, fully illustrated in Figure 3.6, is guided by the three levels of analysis explained above, encompassing philosophies, methodologies and methods of DT-based interventions.

3.5.2 Review questions

'The objective and the reasons for conducting a systematic review … should be made explicit at the start of the process' (Smith *et al.*, 2011: p. 2). In line with this principle, four research questions, addressing all three levels of analysis – philosophy, methodology and methods – are put forward (Table 3.3). These questions play a crucial role in defining the boundaries and scope of this book. While additional questions could have been raised to explore each level of analysis in greater detail, they would fall beyond the scope of this book as presented in Chapter 1.

3.5.3 Review strategy

The nature of the questions posed (Table 3.3) influenced the choice of an appropriate research strategy tailored to the objectives of this systematic review. According to Dresch *et al.* (2015), research strategies span a continuum from configurative to aggregative approaches (Figure 3.7). The four questions are inherently open-ended, each to varying degrees, indicating a need for a broader exploration of the literature. Such questions are best addressed through a configurative literature

[2] 'Methods' in this review only involve uses and functions of DT. Details on specific technologies, software and tools enabling the technical solutions for DT applications are beyond the scope of this book.

Figure 3.6 Systematic literature review steps (author's own)

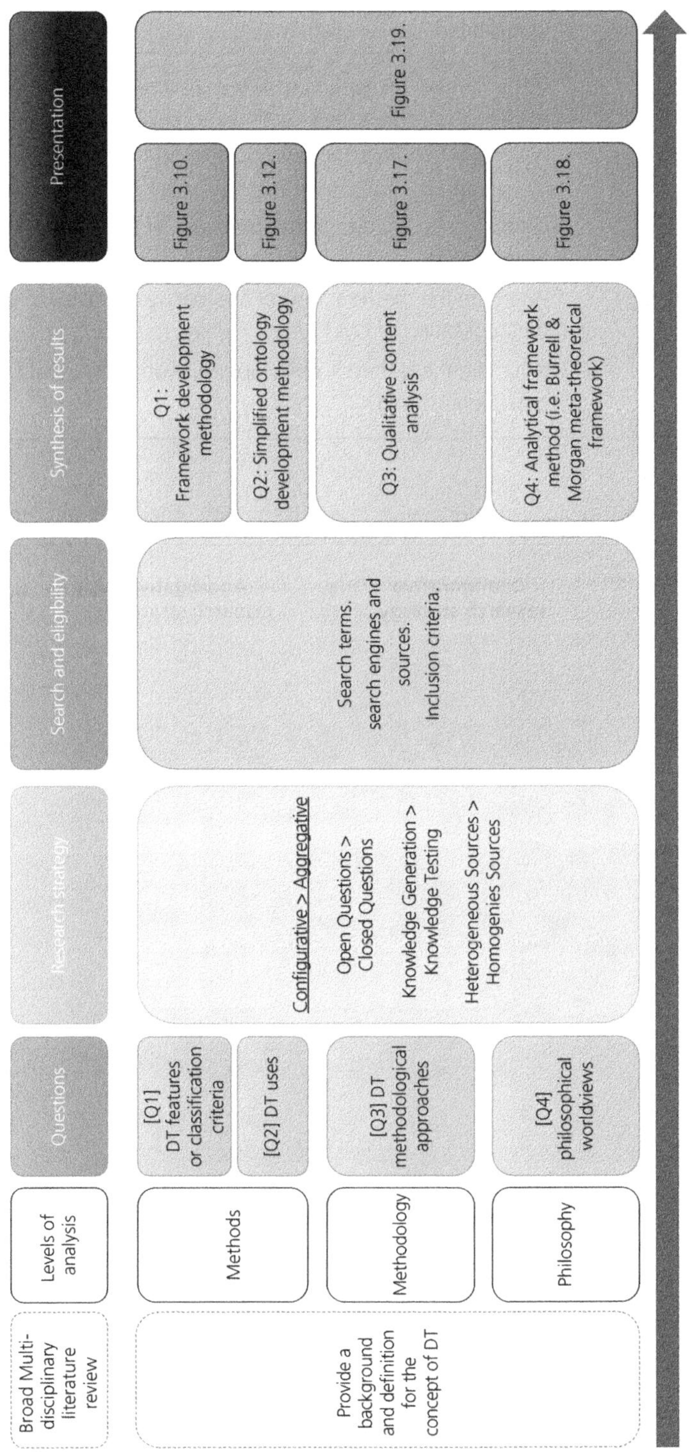

Table 3.3 Systematic literature review questions

No.	Level of analysis	Question
[Q1]	Methods	What are the different features of a DT that distinguish DT use cases from one another? Here, a DT use case refers to a unique situation in which a DT could potentially be applied to create value.
[Q2]	Methods	What are the various 'DT uses' identified across the literature? In this context, DT uses refer to the functions or technical actions a DT performs or executes.[2]
[Q3]	Methodology	What are the methodological approaches or styles of practice adopted in the application of DTs?
[Q4]	Philosophy	What are the different philosophical worldviews that DT researchers and practitioners hold?

Figure 3.7 Configurative against aggregative research strategy. Adapted from Dresch *et al.* (2015)

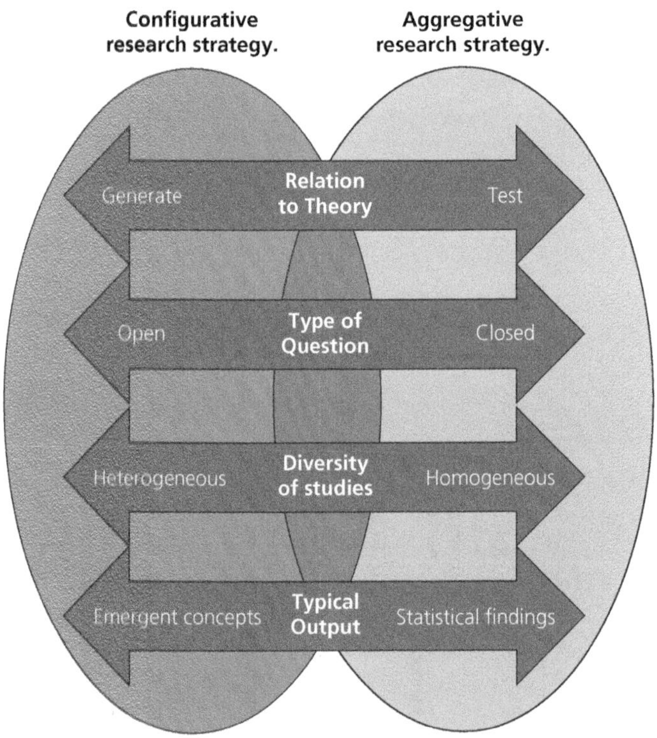

review (Figure 3.7), which contrasts with aggregative reviews. In this approach, answers to the research questions are derived from qualitative data gathered from diverse sources, ultimately leading to the generation of new conceptualisations.

3.5.4 Search keywords and eligibility criteria

A range of search keywords was categorised based on broad research themes (Table 3.4), which were deemed relevant to the research questions and the definitions of UM and DT as outlined in Sections 3.3 and 3.4, respectively. These keywords were refined in tandem with the review and analysis of retrieved results – for example, 'smart city' initiatives were included as many of them 'bore direct resemblance to digital twin conceptualisations' (Nochta *et al.*, 2019: p. 14).

Various concepts or systems such as virtual engineering, building management systems (BMS), and urban management systems were proposed across the broader literature. Although some of these solutions might overlap with or share common features with DT, DT, as defined in Section 3.4.2, remains the focal point throughout this book – for instance, the BMS referenced in the action research described in Section 8.4 is considered solely in relation to supporting the implementation of a DT.

Search strings were developed iteratively, refined using the keywords in Table 3.4, and combined using 'AND', 'OR', and 'NEAR/#' Boolean operators to ensure comprehensive coverage and minimise blind spots. These strings were employed to retrieve results from a variety of databases and search engines, encompassing both academic and grey literature (Table 3.5). This approach aimed to ensure a diversity of studies and capture DT solutions and innovative projects led by industry, not just academia. The search process began in 2019, with string-based alerts set on key database search engines to notify the researchers of any newly published relevant studies throughout the course of the research.

The inclusion criteria for the studies retrieved during the systematic literature search were established to ensure the relevance and quality of the research. Studies were considered eligible if they met the following conditions.

- Language: The study must be published in English.
- Publication Date: Only studies published from 2017 onwards were included. This time frame is significant because 2017 marks the release of the 'Data for Public Good' report by the National Infrastructure Commission [NIC], which is widely regarded as a catalyst for the DT movement in urban planning and city infrastructure management. Additionally, Min *et al.* (2019) highlighted that 2017 saw a significant surge in research related to smart cities within the field of urban planning.

Table 3.4 Keywords used in systematic literature search

Urban	Infrastructure	Data	Management	Digital Twin	Big Data
city	'built environment'	representation	'decision making'	smart city	data-driven
'city scale'	systems	collection	planning	future city	analytics
civic	'system of systems'	analysis	strategy		informatics
town	assets	interpretation	regulation		computing
		insights	maintenance		
			governance		

Table 3.5 Sources and search engines used in systematic literature search

Academic Literature Databases / Search Engines:	Grey Literature Databases / Search Engines:
• BASE • Emerald Journals* • Google Scholar* • IEEE Xplore* • Science Direct* • SCOPUS • Web of Science* • Zetoc	• CABE • CORE* • Google* • IEEE Xplore* • OpenDOAR • OpenGrey • Royal Town Planning Institute • Semantic Scholar • BSI Standards*

*Resources produced the most successful searches

- Type of Study: The study must be either an empirical case study demonstrating a DT in action, or a theoretical or conceptual study proposing how a DT should be developed or implemented in the context of UM.

The initial screening process involved reviewing the titles and abstracts (or introductions) of the returned documents to identify those that were potentially eligible for inclusion. Studies that did not meet the criteria, including duplicates and full-text dissertations, were immediately excluded. The studies that passed this initial screening were then subjected to a full-text analysis to confirm their eligibility against the established criteria (Figure 3.8).

Figure 3.8 Systematic literature review search and screening (author's own)

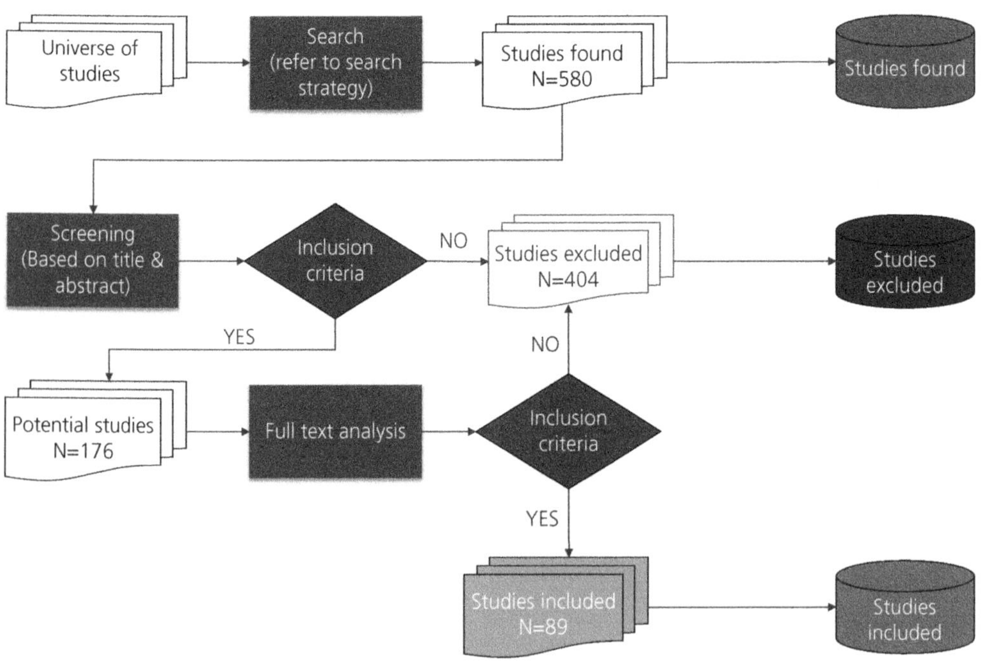

3.6. Q1 – Synthesis of results

3.6.1 Introduction

This section focuses on analysing and synthesising the results to answer the first research question [Q1]: What are the different features of a DT that distinguish DT use cases from one another? This question pertains to the 'methodical' level of analysis.

The literature presents various approaches to thematising or classifying DT general use cases (GUCs) – the specific purposes for which a DT is designed. Some authors classify GUCs based on broad application purposes (Evans *et al.*, 2019) as shown in Figure 3.9-a. Others use a framework based on the three management levels where DTs can contribute: strategic, tactical and operational (Xie *et al.*, 2018). Another classification method allocates DT use cases to different temporal scales: long-term, short-term or real-time/near-real-time (Bibri, 2018) – for instance, Figure 3.9-b displays a framework by Brilakis *et al.* (2019), highlighting four significant areas of DT application. Additionally, in a workshop organised by the Digital Twin Hub[3], participants identified 28 different GUCs, which were subsequently grouped into a three-themed framework based on the pillars of sustainability: social, economic and environmental (Figure 3.9-c). Similarly, Figure 3.9-d illustrates the DT use cases framework developed by the National Digital Twin Programme (2021), which categorises use cases into three groups representing distinct DT purposes or application areas.

While the diversity of these frameworks showcases the broad and extensive potential of DTs, it also highlights a lack of consistency and comprehensibility across the literature. Moreover, these DT use case classification frameworks have significant limitations. None of them comprehensively capture the various DT features and characteristics in sufficient detail. Therefore, the value of the synthesised answer to [Q1] is twofold: First, it thoroughly captures the key DT features identified in the literature concerning *DT for UM*. Second, it establishes a standard classification system that can be applied universally to categorise all DTs consistently and uniformly.

3.6.2 Methodology

To create a comprehensive DT use case classification framework[4] that effectively captures the various features of DTs developed for UM purposes, a three-stage framework development methodology was employed, as outlined by McMeekin *et al.* (2020). This methodology was designed to address the research question [Q1] (Table 3.3) and involves the following stages.

- Data extraction: This initial stage involved a thorough full-text analysis of all relevant studies identified through the systematic literature review, specifically focusing on answering [Q1].
- Qualitative content analysis: In this stage, the focus was on identifying the key features of DTs as described in the reviewed studies. This involved analysing the qualitative data to extract the essential characteristics and attributes of the DTs.
- Grouping and amalgamation: The final stage involved combining and organising the identified features into multiple dimensions, which collectively form the proposed DT use case classification system. This stage, along with the second stage, was conducted in an evolving and iterative manner to refine and enhance the framework as new insights emerged.

[3]https://digitaltwinhub.co.uk

[4]This classification framework is part of the artefact DATUM, developed in Chapter 5. However, the process of developing it is so intrinsic to examining the literature and synthesis of results that it seemed appropriate to cover it in this chapter.

Figure 3.9 Frameworks for classifying DT use cases in literature. Sources: (a) Evans *et al.* (2019); (b) Brilakis *et al.* (2019); (c) Digital Twin Hub https://digitaltwinhub.co.uk; (d) National Digital Twin Programme (2021)

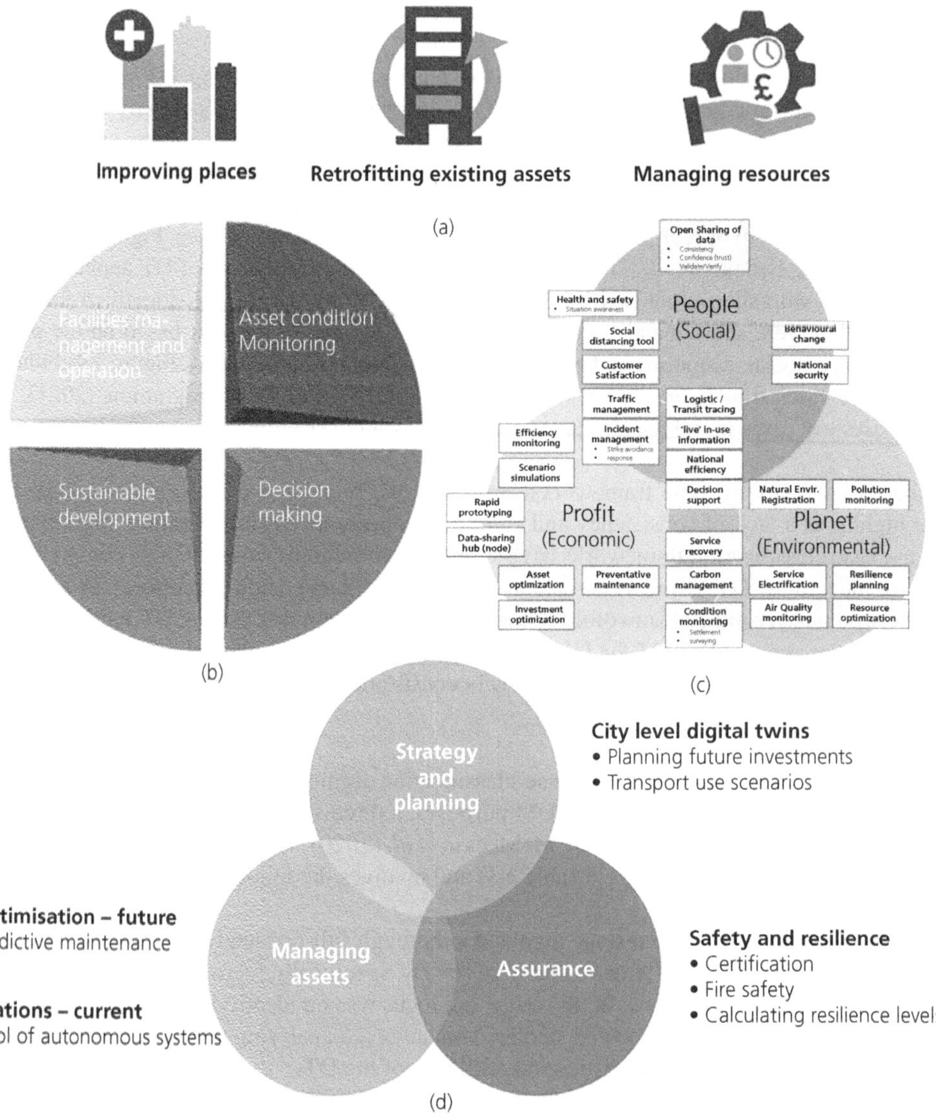

This structured methodology ensured that the developed framework could adequately capture the diverse features of DTs and provide a robust and comprehensive classification system.

3.6.3 DT use case multi-dimensional classification framework

3.6.3.1 Structure of the classification framework

Brilakis *et al.* (2019: p. 29) emphasise that 'digital twins should be driven by purpose [therefore] different use cases will require different update methods and different levels of detail.' In response

Figure 3.10 DT use case multi-dimensional classification framework (author's own)

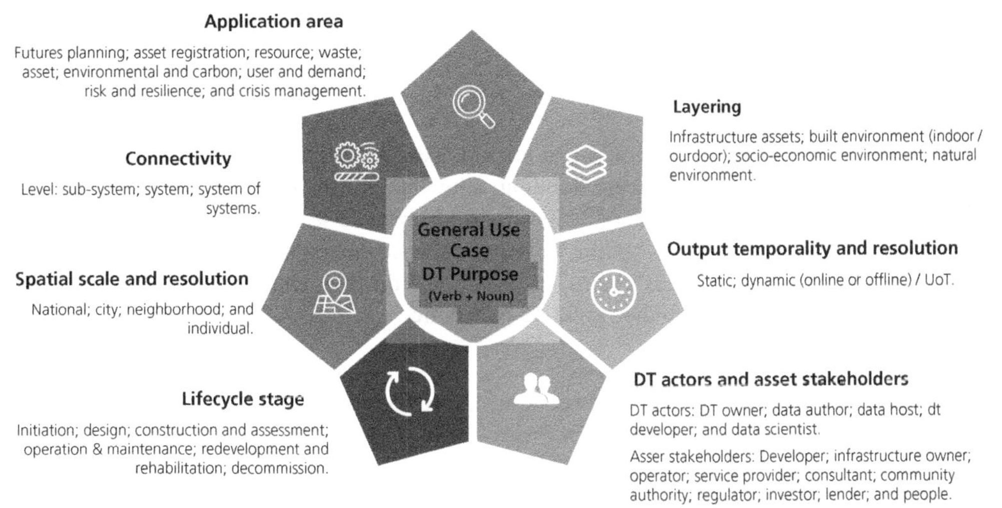

to this, the classification framework depicted in Figure 3.10 places the GUC at its core. A GUC is typically described using a 'verb' followed by a 'noun,' such as 'optimise traffic.' Surrounding the GUC are seven dimensions that represent various features which can be adjusted and tailored to align with the central purpose of the DT. This framework is designed to differentiate and categorise various DT use cases based on these features, facilitating a structured approach to classifying different applications of DTs.

3.6.3.2 D1: Application area

Although the application of *DT for UM* is still in its early stages, they exhibit significant potential for a wide range of uses. The following is a summary of different application areas, with examples from the literature demonstrating how DTs have contributed to each.

3.6.3.2.1 FUTURES PLANNING

This area focuses on long-term planning, future studies and strategic foresight – for instance, Anejionu *et al.* (2019) used a DT to identify urban areas with low liveability, guiding future infrastructure development; Kourtit and Nijkamp (2018) developed urban performance indicators to aid strategic decision-making; Pettit *et al.* (2018) used a DT for planning residential land allocation for 2051; while Nochta *et al.* (2021) created a city DT to forecast patterns of private car use in 2031 to promote sustainable growth.

3.6.3.2.2 ASSET MANAGEMENT

This application focuses on the operation and maintenance of assets to ensure service levels. Nallaperuma *et al.* (2019) integrated diverse datasets to differentiate traffic incidents and forecast traffic flow, optimising operational decisions. Witteborg (2021) demonstrated how DTs could support the smart operation of complex wastewater facilities.

3.6.3.2.3 RISK AND RESILIENCE MANAGEMENT

DTs are valuable for managing infrastructure risks and enhancing urban resilience. White *et al.* (2021) used a DT to simulate flood scenarios, while Bartos and Kerkez (2020) modelled urban stormwater networks in real-time. Wang *et al.* (2020) leveraged smart rail card data to improve public transportation safety, and Mavrokapnidis *et al.* (2021) used computer vision to predict heat stress risks for pedestrians.

3.6.3.2.4 CRISIS MANAGEMENT

DTs aid in decision-making during crises and natural disasters. White *et al.* (2021) used a DT to identify safe routes and locations during flooding, allowing citizens to signal for help, thus enabling the DT to identify the most vulnerable locations during the disaster. Lwin *et al.* (2018) developed an hourly-updated DT to show traffic flow and enhance emergency preparedness. Yabe and Ukkusuri (2019) predicted evacuees' return behaviour post-disaster, and Pang *et al.* (2021) used a DT for managing city responses to the Covid-19 pandemic.

3.6.3.2.5 USER AND DEMAND MANAGEMENT

DTs can manage user behaviours and usage patterns. Leleux and Webster (2018) created a gamified platform to encourage energy conservation. Projects like 'Connecting Bristol'[5] aimed to modify citizens' transportation choices (Kırdar and Ardıç, 2020). Balletto *et al.* (2021) sought to increase awareness and improve public behaviour by utilising abandoned city assets to promote walkability. Orellana and Guerrero (2019) analysed crowdsourced urban data to better understand how street network configurations influence cyclists' behaviour.

3.6.3.2.6 ENVIRONMENTAL AND CARBON MANAGEMENT

DTs offer solutions for environmental and carbon management. Honarvar and Sami (2019) used DTs to predict air pollution based on traffic data. Francisco *et al.* (2020) monitored and benchmarked city building energy consumption in real-time, aiding in environmental performance improvements.

3.6.3.2.7 WASTE MANAGEMENT

DTs introduce innovative approaches to waste management. Recent studies used IoT and computer vision to monitor waste levels in smart bins, alerting collection teams and optimising collection routes (Aktemur *et al.*, 2020; Jadli and Hain, 2020; Rao *et al.*, 2020).

3.6.3.2.8 RESOURCE MANAGEMENT

This application involves optimising the use of scarce resources, whether monetary or physical. McHugh and Thakuriah (2018: p. 4) raised questions about infrastructure investments and service dissatisfaction, such as: '*where would new infrastructure or transportation service investment deliver greatest benefits? Where is the evidence of dissatisfaction with existing services and resources?*'

Examples include managing human waste for energy generation or using DTs capturing data sensed from wind turbines, analysed with facts about landscape and current wind dynamics to optimise wind turbine configurations in wind farms[6]. Evans *et al.* (2019) highlighted the role of DTs and running 'what-if' scenario simulations in supporting sustainable natural resource allocation and managing the food-water-energy nexus.

3.6.3.2.9 ASSET REGISTRATION

DTs excel in asset registration by accurately capturing physical realities. An example is the 'National Underground Asset Register' project by the Geospatial Commission in the UK, which aims to

[5] www.youtube.com/@connectingbristol

[6] https://www.ge.com/renewableenergy/digital-solutions/lifespan/

improve underground infrastructure mapping for safer construction (Al-Sehrawy *et al.*, 2021). Pilot projects have been conducted in London and northeastern England (Brammall and Kessler, 2020).

3.6.3.3 D2: Connectivity

A DT can function using a single dataset, multiple datasets, or by integrating a collection of distinct DTs to achieve its purpose. These scenarios align with three levels of connectivity: sub-system, system, and system of systems (SoS).

3.6.3.3.1 SUB-SYSTEM LEVEL

At the sub-system level, DTs typically rely on one primary dataset. Several studies exemplify this approach, such as the use of sensor data from basins within a stormwater network (Bartos and Kerkez, 2020), pedestrian routes, road and rail networks, or energy consumption data from buildings (Barmpounakis and Geroliminis, 2020; Honarvar and Sami, 2019; Mavrokapnidis *et al.*, 2021; Nallaperuma *et al.*, 2019; Wang *et al.*, 2020).

3.6.3.3.2 SYSTEM LEVEL

At the system level, DTs incorporate multiple datasets within a single DT framework – for example, Kourtit and Nijkamp (2018) and Anejionu *et al.* (2019) developed DTs that considered all modes of transportation. Castelli *et al.* (2019) included data from multiple infrastructure systems in one DT. Similarly, Aktemur *et al.* (2020) and Jadli and Hain (2020) integrated data on waste quantities in smart bins with the bin locations and city road network data to optimise waste collection routes.

3.6.3.3.3 SYSTEM OF SYSTEMS (SOS) LEVEL

At the SoS level, the focus is on understanding and managing the interdependencies between various urban systems, which have traditionally been treated as independent. These interdependencies can be geospatial, cyber, physical or logical (ISO, 2020; Whyte *et al.*, 2019). The SoS level represents the pinnacle of systemic thinking, breaking down organisational silos and integrating independently functioning DTs to create synergistic value on a larger scale. An example of this is a recent study by the Centre for Digital Built Britain (CDBB), which explored how DTs from different infrastructure systems can be used together to identify, prioritise, and manage relationships and interdependencies between these systems (Whyte *et al.*, 2019).

3.6.3.4 D3: Layering

A key characteristic of a DT use case is the extent to which it spans different urban layers. The interdisciplinary nature of most smart city research and urban DTs, which involve interdependent urban data and models (Ma *et al.*, 2019), underscores the complexity and multiplicity of layers within urban environments. Researchers have made various attempts to disentangle and categorise these layers. White *et al.* (2021) identified four layers: terrain, buildings, infrastructure and mobility. Ibrahim *et al.* (2020) expanded this to five layers: built environment, human interactions, transportation and traffic, infrastructure, and natural environment. Similarly, Ma *et al.* (2019) defined five city domains: transportation, energy, emergency and public safety, social sensing, and natural environment. In a review of urban planning needs and urban sensing technologies, Cunningham and Verbraeck (2018) noted three conceptual perspectives of the city: physical and infrastructural, natural resources, and political economy. Drawing from these studies, this book considers four distinct urban layers: *infrastructure*; *built environment* (including both indoor and outdoor spaces); *socio-economic environment*; and *natural environment*.

Depending on its GUC, a DT may focus on capturing relevant datasets or models from a single layer. For instance, Rao *et al.* (2020) and Aktemur *et al.* (2020) focused exclusively on the infrastructure layer, optimising public smart bins and supporting city waste management plans. However, more often, a DT must span multiple urban layers to achieve its objectives. Bartos and Kerkez (2020) developed a DT encompassing both infrastructure and natural environments to manage urban drainage networks. Other studies explored the interactions between infrastructure and socio-economic layers (Barmpounakis and Geroliminis, 2020; Jadli and Hain, 2020; Nallaperuma *et al.*, 2019; Y. Wang *et al.*, 2020). Some researchers considered combinations of three urban layers when developing DTs (Anejionu *et al.*, 2019; Barkham *et al.*, 2022; Francisco *et al.*, 2020; Honarvar and Sami, 2019; Kourtit and Nijkamp, 2018; Mavrokapnidis *et al.*, 2021; Mayaud *et al.*, 2019), while others included all four layers to meet specific application needs (Pettit *et al.*, 2018) – for example, Yabe and Ukkusuri (2019) integrated data from diverse sources across all urban layers to predict post-disaster return behaviour.

3.6.3.5 D4: Spatial scale and resolution

Depending on the defined GUC, a DT can vary significantly in terms of spatial scale – corresponding to spatial coverage – and resolution (Gardner and Hespanhol, 2018; Kontokosta, 2021). The spatial scale and resolution can range from *national*, *city* and *neighbourhood* levels to *individual* elements. Two key points need clarification here. First, it is essential to distinguish between scale and resolution – for example, Anejionu *et al.* (2019) developed a DT with national-scale spatial coverage across the UK but offered visualisation of 'liveability' at a finer neighbourhood resolution. Second, the individual level does not necessarily refer to individual people but can denote any singular element within a neighbourhood, such as a building, infrastructural unit, natural entity or specific location. This concept of an individual element is similar to the 'intelligent planning unit' (IPU) described by Hastak and Koo (2017: p. 3) as a 'well-defined planning unit that can be initiated to achieve any specific purpose.'

Some DTs are designed with a focus solely on neighbourhood scale and resolution – for instance, Panagoulia (2017) created a DT concentrated on a specific district, predicting heat exposure for citizens within that area. However, other DTs spatially encompass entire cities, albeit with varying levels of resolution – for example, Kourtit and Nijkamp (2018) provided aggregate urban key performance indicators (KPIs) at the city level without finer resolution, while Honarvar and Sami (2019) addressed air quality issues at a neighbourhood level within the city. Similarly, Mayaud *et al.* (2019) developed a city-scale DT with neighbourhood-level resolution to assess accessibility to healthcare facilities.

Some DTs with higher resolution capture details beyond city neighbourhoods – for example, Barmpounakis and Geroliminis (2020) and Nallaperuma *et al.* (2019) provided detailed information about individual vehicles. Other studies introduced DTs offering data on individual smart bins across a city (Aktemur *et al.*, 2020; Jadli and Hain, 2020; Rao *et al.*, 2020), while others captured details of individual city buildings (Francisco *et al.*, 2020) or even specific 50m x 50m grids (Kim, 2020). On a national scale, Pang *et al.* (2021) integrated city DTs across the nation to improve predictions of pandemic infection patterns. One of the most ambitious DTs under development at a national scale is the UK's National Digital Twin (NDT)[7].

3.6.3.6 D5: Temporality and resolution

Similar to spatial scale and resolution, DTs also vary in terms of temporality, specifically regarding temporal scale and resolution, which correspond to the extent or horizon and granularity of time, respectively (Li *et al.*, 2018). It is important to distinguish between three types of DTs based on temporality. The first type is the *static DT*, which at best generates static output without involving

[7]https://www.gov.uk/government/collections/the-national-digital-twin-programme-ndtp

any temporal information (e.g. As-built Asset Information Model). The second type is the *offline dynamic DT*, which utilises historical temporal data to generate dynamic behaviour but is not connected to the current situation, meaning it could become outdated. This type of DT is more suitable for slowly evolving systems, such as urban spatial configurations, where significant changes occur over long periods. For example, Kourtit and Nijkamp (2018) developed a DT that computed urban performance indices and KPIs based on a 4-year temporal scale, using data collected from 2012 to 2016. Similarly, Wang *et al.* (2020) used a DT to infer patterns of railway passenger flow from one-year temporal scale smart card ticket data collected in 2017.

The third type is the *online dynamic DT*, considered more mature by the DT maturity spectrum developed by Evans *et al.* (2019). This type maintains a live connection with its physical counterpart, receiving a continuous influx of up-to-date data, which means it never becomes obsolete as long as this connection persists – for instance, Francisco *et al.* (2020) used IoT technology and smart meters to monitor the energy consumption of buildings in real-time. Similar examples include Aktemur *et al.* (2020), Jadli and Hain (2020), Nallaperuma *et al.* (2019), and Rao *et al.* (2020) who also relied on real-time data to feed their DTs.

Dynamic DTs, whether online or offline, require a certain level of temporal resolution, which refers to the temporal steps or increments by which the DT input changes. Researchers have attempted to define objective levels of temporal resolution – for example, Kontokosta (2021) identified four distinct levels of temporal resolution: real-time, daily, annual and decennial. However, it is acknowledged that DT output may fall anywhere between these thresholds. The concept of real-time is flexible and is often defined by the GUC driving the creation of the DT, meaning that the frequency of data capture or updates depends on the specific objective. Therefore, the concept of 'real-time' can overlap with 'right-time', depending on the context. Consequently, this book supports a flexible classification of temporal resolution and proposes that the temporal resolution of DTs be measured by the *unit of time* (UoT).

3.6.3.7 D6: Asset lifecycle stage

Similar to how DTs vary by spatial and temporal scale, they also differ in their applicability across the lifecycle phases of an asset. As Al-Sehrawy and Kumar (2021) recommend, a vertically integrated DT can connect various lifecycle phases in a circular manner, allowing knowledge and insights from one phase to inform others. However, most DT use cases in the literature primarily target specific lifecycle phases, even though other phases might benefit from the generated output. Drawing from the ISO (2020) suggestions for lifecycle phases of smart communities' infrastructure assets, the following discussion outlines how DTs can be utilised across different stages.

3.6.3.7.1 INITIATION PHASE

In this phase, DTs are often used to identify and crystallise the needs and motivations for interventions in the urban environment, particularly through infrastructure development. Various studies have focused on DT-based collaborative initiatives, such as geo-participation and geo-discussion platforms that capture the city's status quo and encourage public contributions – for example, Haklay *et al.* (2018) and Hasegawa *et al.* (2019) utilised such platforms, while Afzalan and Sanchez (2017) developed an interactive GIS interface for citizens to propose their views on bike-share infrastructure planning. Similarly, Dembski *et al.* (2020) engaged citizens in evaluating traffic development scenarios using VR technology, allowing intrinsic needs to emerge during the process. Kovacs-Györi *et al.* (2020) used social media data and spatial information to understand urban dynamics, helping to infer public needs and preferences.

3.6.3.7.2 DESIGN PHASE

At the design stage, DTs increase confidence in proposed infrastructure development plans by revealing how new interventions might unfold in the urban environment. White *et al.* (2021) suggested using data on sunlight, wind and seismic activity to evaluate the impact of new buildings on city features and assess the effects of existing challenges and risks. DTs can ensure that urban planning decisions do not negatively impact citizens or ecosystems, compare alternative design options (e.g. land use allocations for economic value and job creation), and assess whether new infrastructure developments may hinder current operations (McHugh and Thakuriah, 2018; Oléron-Evans and Salhab, 2021). Additionally, Barmpounakis and Geroliminis (2020) used DTs to investigate traffic congestion, generating insights to support the design of new roads.

3.6.3.7.3 CONSTRUCTION PHASE

Although relatively few studies focus on DT use cases during construction, significant benefits are promised. One application is the underground asset register, which has the potential to deliver strike-less construction and a safer working environment (Brammall and Kessler, 2020). Another valuable use case involves monitoring and controlling construction progress – for example, Tang *et al.* (2019) implemented a clustering method to assess urban development progress by evaluating the conformity between planned urban clusters and the actual current state.

3.6.3.7.4 OPERATION AND MAINTENANCE PHASE

Several DT case studies have been advanced to provide asset managers with a better understanding of urban assets' behaviour and state, supporting efficient operations and high-quality service delivery – for instance, Wang *et al.* (2020) developed a DT to predict short-term rail passenger flow, aiding operational decision-making. Similarly, Francisco *et al.* (2020) used real-time monitoring of energy consumption in buildings to detect operational faults.

3.6.3.7.5 REDEVELOPMENT AND REHABILITATION PHASE

In this phase, DT insights are used to reflect on the current state of the urban environment and guide its redevelopment to offer improved services and adapt to changing urban dynamics. Kourtit and Nijkamp (2018) used city-scale DT indicators and KPIs to guide city redevelopment strategies. Continuous monitoring of pedestrian heat stress led to decisions on redevelopments, such as adding building shades to mitigate the issue (Mavrokapnidis *et al.*, 2021). Another approach involved using DTs to guide the redevelopment of road networks to reduce air pollution (Honarvar and Sami, 2019). This phase highlights the importance of integrating lifecycle phases, where knowledge gained from one phase, such as operation and maintenance, can benefit subsequent phases like redevelopment and rehabilitation (Al-Sehrawy and Kumar, 2021).

3.6.3.7.6 DECOMMISSION PHASE

This phase has received the least attention from DT researchers and practitioners. However, DTs still hold potential for real-life applications and case studies. Al-Sehrawy *et al.* (2022) discussed how DTs can support knowledge transfer from old to new assets, guide end-of-life procedures (e.g. disposal or decommissioning) towards a circular asset lifecycle, and offer more sustainable solutions.

3.6.3.8. D7: DT actors and stakeholders

While many DT use cases in the literature do not explicitly address the dimension of stakeholder involvement, it is expected that any DT use case will inevitably involve multiple stakeholders. The ISO (2020) provides a comprehensive list of potential stakeholders in the development of smart communities, including developers, infrastructure owners, operators, service providers, consultants, community authorities, regulators, investors, lenders and people. Although this list is based on the perspective of physical assets, it is also useful to consider these groups from the DT perspective.

Several studies have proposed smart city frameworks and theoretical constructs for DT development (Bibri and Krogstie, 2018; Kent *et al.*, 2019; Mamta and Nagpal, 2018). From these frameworks, key roles, responsibilities and actors in the DT development process can be deduced. These roles include.

- DT owner: This is the client or entity that defines the purpose of the DT and the outcomes it aims to achieve. The DT owner is responsible for setting the objectives and ensuring the DT aligns with the strategic goals.
- Data author: The creator and issuer of data, responsible for generating and maintaining the datasets used in the DT. This role is crucial for ensuring data quality and relevance.
- Data host: The entity that provides the infrastructure for storing large datasets, often a cloud storage service provider. The data host ensures data accessibility and security.
- DT developer: The consultant or technical expert responsible for designing and building the DT. This role involves developing the system architecture, specifications, and functions of the DT, ensuring it meets the requirements set by the DT Owner.
- Data scientist: The professional responsible for cleaning, standardising, re-formatting, analysing, processing and visualising data in alignment with the DT Owner's requirements. The data scientist plays a key role in transforming raw data into actionable insights.

In the context of UM, a DT use case would include both infrastructure asset stakeholders and the DT actors listed above. These roles are critical to the successful implementation and operation of the DT, and in many cases, they may overlap – for instance, future city officials and leaders are expected to enhance their knowledge in data science (Kontokosta, 2017). As a result, an infrastructure asset operator could also serve as the DT owner for an operation and maintenance DT, with expert personnel fulfilling the responsibilities typically associated with a data scientist. This integration of roles highlights the evolving nature of stakeholder involvement in DT use cases, where expertise in data science becomes increasingly valuable across various functions.

3.7. Q2 – Synthesis of results

3.7.1 Introduction

This section provides an analysis and synthesis of the studies reviewed to answer question [Q2] ('What are the various 'DT uses' identified across the literature?') at the methodological level. DT uses refer to the functions or technical actions a DT performs to achieve its intended purpose. These uses are the fundamental building blocks that can be combined in customised ways to fulfil a GUC.

The reviewed studies present a wide array of DT uses, showcasing the diverse capabilities and functions that a DT can execute to achieve a defined GUC. However, capturing and categorising these DT uses posed a significant challenge due to the inconsistency in terminology across the literature and the absence of a standardised vocabulary. Even when the same DT use is implemented in different studies, authors often refer to it by different names or terms. As a result, the synthesis of the answer to [Q2] serves a dual purpose.

First, it captures and groups the various DT uses identified in the literature related to DTs for UM. This categorisation helps to clarify and organise the different functions that DTs can perform, making it easier to understand and apply these uses in practical contexts.

Second, it establishes a standardised vocabulary for the various DT uses. Standardising the terminology used by DT stakeholders is crucial for the effective dissemination of experiences and practical knowledge. When case studies are described using ad-hoc or localised technical language that is familiar only to the authors, there is a higher likelihood of misinterpretation by other stakeholders. By adopting a common language, DT stakeholders can ensure that the knowledge gained from different DT use cases is accurately understood and applied across various contexts. This standardisation is key to enhancing communication, collaboration and the overall success of DT implementations in urban management.

3.7.2 Methodology

The various DT uses identified in the literature have been organised into a taxonomy to reflect the hierarchical nature of their relationships[8]. This taxonomy was developed using a simplified ontology development methodology. An ontology, as defined by Guarino (1998), is 'a hierarchy of concepts related by subsumption relationships.' The methodology used to create this taxonomy involves the following five general steps (Figure 3.11).

(a) Defining the domain and scope: The first step involves establishing the domain and scope of the intended standard ontology. Key questions addressed at this stage include: Which domain will the ontology cover? What is the purpose of the ontology? What types of questions should the knowledge represented in the ontology be able to answer? These competence questions help to clearly define the objectives and boundaries of the ontology.

(b) Reviewing existing ontologies: The second step involves determining whether any similar ontologies have previously been developed within the same domain. Adapting or building upon existing and well-established ontologies offers several advantages. It saves time and effort, as the groundwork has already been laid. Additionally, building on familiar ontologies or systems makes it easier for practitioners in the domain to understand and adopt the new system. This approach also ensures consistency with established practices and provides an additional layer of validation, as existing ontologies have already undergone multiple checks and refinements through practical applications. One relevant example identified is the 'BIM uses' classification system (Kreider and Messner, 2013).

(c) Enumerating terms: The third step involves developing a pool of terms. These terms were gradually gathered from the studies reviewed in the systematic literature review, supplemented by insights from previously developed systems identified in the second step. This process ensures that the taxonomy is comprehensive and grounded in existing research and practices.

(d) Grouping terms: In the fourth step, the listed terms, which refer to various DT uses, were grouped based on shared and mutual properties inherent to those terms and the defined classes to which they belong. A mixed method approach – incorporating top-down, bottom-up, and middle-out techniques – was used, working iteratively between higher-level classes and the

Figure 3.11 The simplified ontology development methodology (author's own)

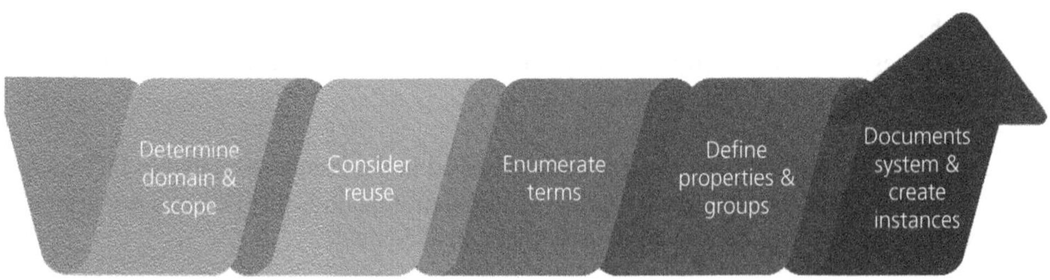

[8]The 'taxonomy of DT uses' is part of the artefact DATUM, developed in Chapter 5. However, the process of developing it is so intrinsic to examining the literature and synthesis of results that it seemed appropriate to cover it in this chapter.

lowest-level terms. This approach allowed for a thorough and flexible organisation of the terms within the taxonomy.

(e) Documenting the taxonomy: The fifth and final step focused on documenting the standard classes and terms in a taxonomic form. This documentation process included providing examples of common DT practices to help map them onto the taxonomy. These examples also demonstrate how the standards developed can address the competency questions raised in step one, thereby adding an extra degree of validation to the taxonomy.

This structured approach ensures that the taxonomy not only categorises DT uses effectively but also aligns with established knowledge and practices, providing a robust framework for future DT developments.

3.7.3 DT uses taxonomy

3.7.3.1 Structure

Figure 3.12 presents the developed taxonomy of DT uses. The taxonomy is structured hierarchically beneath the GUC and comprises three distinct levels, organised in a parent-child-grandchild relationship. These levels, from the highest to the lowest, are as follows.

Included uses: This is the topmost level of the taxonomy and consists of four high-level cornerstone uses – *mirror*, *analyse*, *communicate* and *control* – are outlined in Sections 3.7.3.2 to 3.7.3.5. These broad categories capture the essential functions that DTs perform across various applications.

Specialised uses: The second level of the hierarchy, specialised uses, includes more specific forms of the included uses. Each specialised use is characterised by unique strengths and is tailored to meet specific purposes within DT projects. This level of the taxonomy allows for the identification of DT uses that go beyond the basic functions, offering more targeted and refined capabilities that cater to particular needs within a project.

Specialised sub-uses: At the lowest level of the taxonomy are the specialised sub-uses. These further differentiate between different types within the same specialised use, based on very fine inherent variations. These subtle distinctions enhance the practical adequacy of the DT, enabling it to better address alternative contexts and user-defined purposes. This granularity ensures that the taxonomy can accommodate the diverse and nuanced requirements that may arise in different DT applications.

This hierarchical structure ensures that the taxonomy is both comprehensive and flexible, allowing it to capture a wide range of DT uses while also providing the specificity needed to address particular contexts and objectives. The parent-child-grandchild relationship among the levels facilitates a clear and organised categorisation of DT uses, supporting both broad and detailed analyses of DT capabilities.

3.7.3.2 Mirror

The concept of a DT fundamentally revolves around the idea of mirroring or duplicating a physical system that operates in the real world into a virtual counterpart within the cyber world (Al-Sehrawy and Kumar, 2021). This included use, known as *mirror*, does not initially focus on deriving meanings or gaining insights from the collected data or understanding its implications for decision-making or future interventions. Instead, the primary objective of the *mirror* use is to sense and collect raw data produced by the physical system, as required by the GUC.

The captured data must represent a certain level of abstraction of reality, which is determined according to the specific goal in mind. Therefore, *mirror* involves the simple creation of a shadow

Figure 3.12 DT uses taxonomy (author's own)

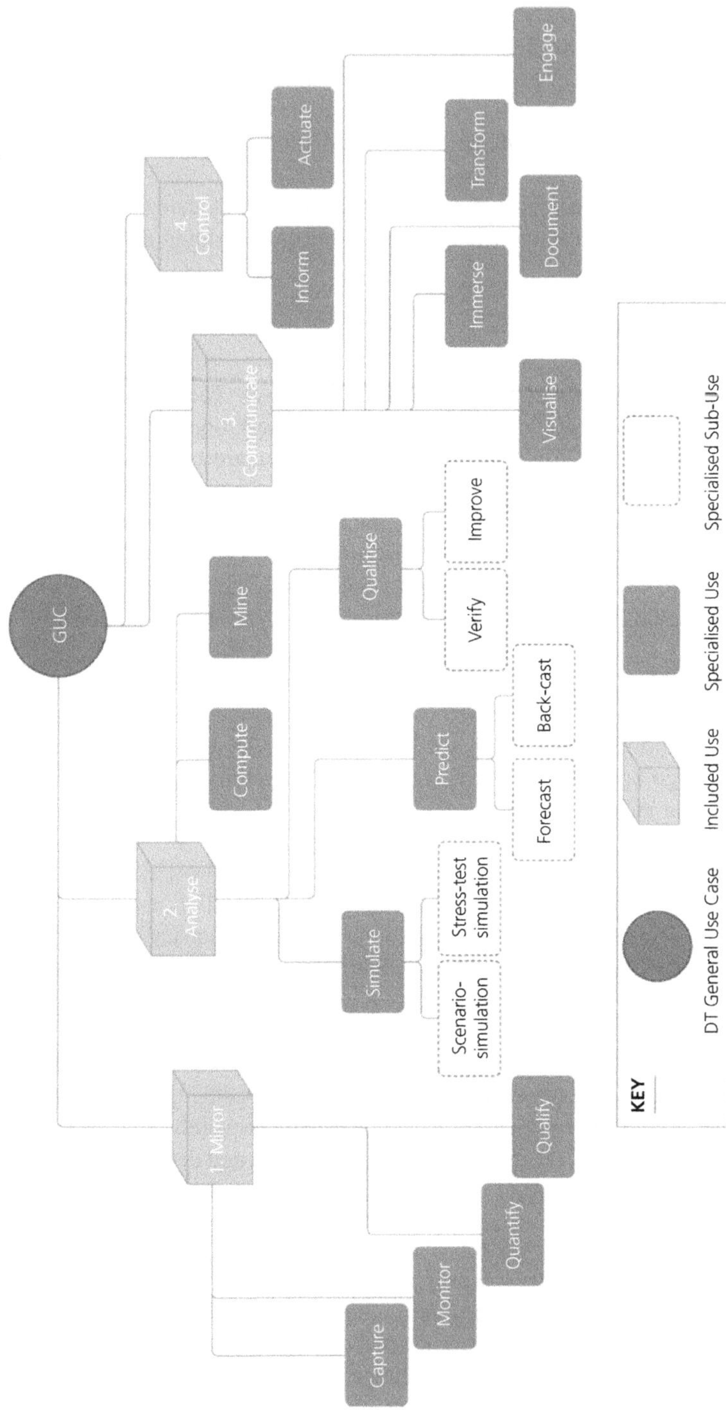

or replica of the observed real entity, as described by Kritzinger *et al.* (2018). Typically, this DT use serves as the initial step in the roadmap toward the actual delivery and realisation of a DT GUC.

In the context of the urban environment, which is viewed as a socio-technical system, the datasets that make up a DT can be classified as either primary or secondary. Primary data is directly, deliberately, and consciously created with the intention of feeding the DT to achieve a pre-defined purpose or GUC. In contrast, secondary data is naturally generated through common daily urban operations and dynamics. Although secondary data is not specifically created for the DT, it is valuable and relevant, and, thus, it is gathered to supply the DT with useful datasets.

The *mirror* DT use is further divided into four *specialised uses*: *capture*, *quantify*, *monitor* and *qualify*. These *specialised uses* reflect different aspects of the mirroring process, each tailored to a specific function within the overall goal of creating an accurate and useful digital replica of the physical system (Table 3.6).

Capture: DTs are employed to capture the structural and physical attributes of real-world systems, primarily focusing on spatial data related to a specific geographic location at a particular point in time. This process involves various techniques and datasets such as laser scanning, photogrammetry, morphological structure analysis, demographic studies, geographic surveys, censuses, spatial information models, infrastructure network topologies and geographic information system (GIS) data (Panagoulia, 2017). The key characteristic of this specialised use is that the data collected is typically historical, static or discrete, involving straightforward recording of existing spatial information. Essentially, the *capture* use provides a 'static base layer', akin to a snapshot of reality at a specific moment, often serving as the backbone of an urban structure (Penn and Al Sayed, 2017) – for instance, Boeing (2021) captured extensive data on urban forms and morphological structures from several major cities as part of a DT application. This type of data is generally more feasibly captured by way of a *static DT* (for more on *static DTs*, see Section 3.6.3.6). Further examples of the *capture* use can be found in studies by Lu *et al.* (2017), Lwin *et al.* (2018), Pettit *et al.* (2018) and Yabe and Ukkusuri (2019).

Table 3.6 Included use *mirror* and its specialised uses

DT Use			Definition	Synonyms
01	Mirror		To duplicate a physical system in the real world in the form of a virtual system in the cyber world.	Replicate, represent, twin, model, shadow, mimic
	1.1	Capture	Express in a digital format within the virtual world the status of a physical system at a point of time. (Usually, offline DT)	collect, scan, survey, digitalize
	1.2	Monitor	Collecting information related to the performance of a physical system. (Online or Offline DT)	Sense, observe, measure
	1.3	Quantify	Measure quantity of a physical system's particulars, instances or incidents. (Online or Offline DT)	Numerate, takeoff, count
	1.4	Qualify	Track the ongoing status of a physical system (Online or Offline DT)	Diagnose, follow, track

Monitor: This DT use involves dynamically mirroring the behaviour or performance of a physical system over time through spatiotemporal data communication, projecting it onto a virtual system, whether an *online* or *offline DT* (for more on *online* and *offline DTs*, see Section 3.6.3.6). The digital output of this use is dynamic, reflecting temporal changes in the real system. *Offline DTs* are typically better suited for long-term strategic planning of slowly evolving systems, relying on historical records and stored longitudinal data with temporal information that displays gradual changes in reality, such as tracking demographic changes in a city over a decade. Conversely, *online DTs* depend on a live influx of data streams with appropriate spatial and temporal resolutions. The *monitor* use facilitates the collection of longitudinal data related to the state, performance, and behaviour of dynamic physical systems over time.

Secondary data for monitoring is often retrieved through 'passive' means, such as environmental sensors, IoT devices, crowdsensing, social media (Cerrone *et al.*, 2018; Yabe and Ukkusuri, 2019), cellular phone activity (Kim, 2020), computer vision (Ibrahim *et al.*, 2020), cyber-physical sensing (Celes *et al.*, 2019), drones (Barmpounakis and Geroliminis, 2020), or lidar systems (Shirowzhan *et al.*, 2020) to monitor traffic congestion. Primary data, however, requires 'active' monitoring, which involves direct human input through methods like GIS-web tools, collaborative platforms, e-participation systems, or social media, such as gathering citizens' feedback on proposed urban development plans.

Quantify: DTs can also be used to count and numerically quantify elements, instances or incidents within a real-world system. These measured quantities are defined according to the elements or units identified in the physical reality, based on the DT's scale and resolution. This may involve using static, spatial-only data with no temporal dimension, such as population or local pedestrian count data showing pedestrian volumes relative to land use, which can provide insights into a city's socio-spatial structure (Lai and Kontokosta, 2018). Alternatively, it might involve utilising dynamic data at a defined temporal resolution, such as the daily number of commuters. Other examples include counting traffic vehicles (Honarvar and Sami, 2019; Nallaperuma *et al.*, 2019) or train passengers (Wang *et al.*, 2020) over time. When these quantities are supplemented by a temporal dimension, they can support the inference of spatiotemporal urban patterns.

Qualify: This DT use is focused on tracking the status, condition or mode of a physical system, resulting in a 'qualified' or 'unqualified' type of output. A simple example of this could be checking whether traffic lights are operational. A more dynamic example might involve qualifying the safety of a bridge by continuously monitoring its structural health throughout its lifespan (Ye *et al.*, 2019) or detecting whether a trash bin is full (Aktemur *et al.*, 2020; Jadli and Hain, 2020; Rao *et al.*, 2020). During disasters, qualification could include user-tagging to mark whether a citizen is safe or unsafe (White *et al.*, 2021).

3.7.3.3 Analyse

This DT included use is arguably where the greatest value of a DT can be realised. It enables DT owners, users and stakeholders to leverage the data gathered through *mirroring* to generate new, insightful information that can expand knowledge boundaries and deepen the understanding of a physical entity, or a complex sociocultural-economic system, such as an urban environment. This understanding supports planning and decision-making processes. This value can be unlocked through one or more of the following five specialised uses: *compute*, *mine*, *simulate*, *predict*, and *qualitise*, as shown in Table 3.7.

Table 3.7 Included use *analyse*, its specialised uses and specialised sub-uses

DT Use				Definition	Synonyms
02	Analyse			To create new knowledge and provide insights for users and stakeholders about a physical system.	Examine, manage
	2.1	Compute		To perform conventional arithmetical calculations, traditional mathematical operations and functions and simple statistical techniques like correlations.	Calculate, add, subtract, multiply, divide
	2.2	Mine		To uncover, identify and recognise the web of interdependencies, interconnected mechanisms, complex processes, interwoven feedback loops, masked classes, clusters or typologies, hidden trends and patterns within the physical system.	Learn, recognise, identify, detect, AI, ML, BDA
	2.3	Simulate		To explore and discover the implications and possible emerging behaviours of a complex web of interacting set of variables.	
		2.3.1	Scenario	To find out the implications, impacts or consequences of implementing pre-defined scenarios (akin to non-destructive tests).	What-if, evaluate, assess
		2.3.2	Stress-Test	To identify the scenarios that may lead to failure or breakdown of physical system (akin to destructive tests).	Test, inspect, investigate
	2.4	Predict		Concerned with futures studies.	
		2.4.1	Forecast	to predict the most likely state of a real system in the future, by projecting the known current trends forward over a specified time horizon.	foresee
		2.4.2	Back-cast	To question or prove in a prospective manner, how the physical system is operating towards achieving the pre-set aims and goals.	manage, confirm
	2.5	Qualitise		Enhance and improve the quality of the outcomes or deliverables produced by an intervention in real world.	
		2.5.1	Verify	Verify conformance and compliance of physical system with standards, specifications and best practice.	Validate, check, comply, conform
		2.5.2	Improve	Inform the future updating, modifying or enhancing the current standards to be in better coherence and harmony with the actual operational and usage behaviours and patterns.	Update, upgrade, revise

Compute: This specialised use involves conventional arithmetic calculations, traditional mathematical operations and statistical techniques. These can include calculating spatial correlations to identify cycling patterns within a city (Lieske *et al.*, 2021) or computing KPIs using formulae comprising independent variables (Anejionu *et al.*, 2019; Kourtit and Nijkamp, 2018). The *compute* use offers a basic yet essential layer of data processing that supports further analysis and decision-making by translating raw data into quantifiable metrics.

Mine: Rooted in the concept of data-driven research, this specialised use is central to what Kitchin (2014) refers to as the '4th paradigm of science'. It heavily relies on AI techniques, such as machine learning (ML) algorithms, data mining (Bibri and Krogstie, 2018; Ghaemi *et al.*, 2017), and innovative BDA, to uncover hidden patterns and underlying regularities in large datasets. This powerful analysis provides unprecedented insights into real-world phenomena. While theoretically validated knowledge may guide this data-driven research, the focus is on uncovering new information that can inform decisions. Examples include using machine learning (ML) techniques to support urban management and planning, such as selecting appropriate locations for public events and optimising land use (Sideris *et al.*, 2019), deploying clustering methods to detect distinct urban clusters (Tang *et al.*, 2019), or quantitatively analysing social media data to understand citizens' feelings and activities in public parks to inform future developments (Plunz *et al.*, 2019).

Simulate: This specialised use involves employing digital simulation techniques to explore and discover the implications and potential emerging behaviours of a complex web of interconnected variables recognised from real-world data. At a lower hierarchical level, two distinct approaches or specialised sub-uses of *simulation* are identified:

1. Scenario-simulation: This approach primarily involves raising 'what-if' questions based on predefined scenarios, whether derived from a proposed plan or expected risks and hazards. It helps understand how an infrastructure system might react to expected shocks and stresses – for instance, White *et al.* (2021) carried out flooding and skyline simulations to evaluate their impacts on urban environments, while Pettit *et al.* (2018) simulated what-if scenarios for land use futures planning to balance economic and environmental goals.
2. Stress-test simulation: This approach focuses on questioning the vulnerabilities of the system itself, rather than the impacts of a posited plan or expected scenario. It aims to determine the 'breaking point' of the physical counterpart of the virtual system, akin to 'destructive' testing. The key question addressed is: 'What does it take for this urban arrangement or infrastructure system to fail or break down?'

Predict: A DT can be used to predict the future state of a mirrored physical system, which can be pursued through two alternative specialised sub-uses:

1. Forecasting: This method predicts the most likely future state of a real system by projecting current trends forward over a specified time horizon. Examples include using federated learning to predict the spread of COVID-19 and its impact on infrastructure services (Pang *et al.*, 2021), using neural networks to predict transportation carbon emissions (Lu *et al.*, 2017), or computing future accessibility levels to healthcare facilities in different neighbourhoods considering current population trends and demographic changes (Mayaud *et al.*, 2019).
2. Backcasting: Unlike forecasting, which projects trends to predict future outcomes, backcasting is concerned with answering the question of 'how' a desirable future can be attained.

It is typically used as part of an active intervention aimed at achieving a desirable state, determining the most effective scenarios or plans of action to meet pre-set aims and objectives.

Qualitise: This specialised use is concerned with baseline plans, ideal states, benchmarks, thresholds, or standards such as building codes, environmental assessment methods, or resilience standards. It is further divided into two specialised sub-uses:

1. Verify: Similar to quality assurance and control activities, this use involves using a DT to verify or validate a plan, design, operational practice, or ongoing intervention against existing standards and best practices – for example, attempts to satisfy indoor air quality (IAQ) standards under different operational circumstances (Rogage *et al.*, 2019) or using clustering techniques to verify the conformity of ongoing urban development with set plans (Tang *et al.*, 2019).
2. Improve: This sub-use goes beyond mere validation to challenge existing standards based on real-life operations, as opposed to theoretical assumptions. For instance, a DT could inform future planning and design based on actual usage rather than standards, offering the potential to set more realistic and fit-for-purpose benchmarks. An example is provided by Francisco *et al.* (2020), where daily energy consumption benchmarks for buildings were computed across different seasons, enabling the development of standards inferred from actual operations rather than theoretical models.

3.7.3.4 Communicate

A fundamental use of DTs is to facilitate communication by conveying either the mirrored raw data (Section 3.7.3.2) or the more insightful results of analysis (Section 3.7.3.3) to various stakeholders. This communication is tailored according to the stakeholders' interests and the initial purpose for which the DT was deployed. This use allows DT users to interpret, share and exchange information, thereby enhancing understanding and supporting decision-making. The *communicate* included use is typically executed either after, in the case of *offline DTs*, or concurrently with, in the case of *online DTs*, the implementation of the *mirror* and *analyse* included uses. The *communicate* included use is divided into five specialised uses: *visualise*, *immerse*, *document*, *transform* and *engage*, as summarised in Table 3.8.

Visualise: Visualisation 'is the process of representing data in a visual and meaningful way to facilitate its understanding' (García and Montané-Jiménez, 2020: p. 200). DTs enhance information exchange and sharing through various visualisation tools and techniques. Visualisation helps bring diverse perspectives to the forefront, allowing different stakeholders, even those unfamiliar with technical languages, to contribute insights and ideas. This can be achieved through tools such as realistic city-scale models (e.g. Virtual Singapore[9]), walkthroughs, geo-visualisation maps (Anejionu *et al.*, 2019), dashboards (Kourtit and Nijkamp, 2018), platforms, heatmaps (Barmpounakis and Geroliminis, 2020), 2D or 3D figures (Ghaemi *et al.*, 2017; Nallaperuma *et al.*, 2019), charts, scatterplots (García and Montané-Jiménez, 2020), or user-friendly interfaces (Hasegawa *et al.*, 2019). For instance, Boeing (2021) visualised the street orientations of various cities using simple rose-diagrams. Other studies employed geo-visualisation techniques to represent traffic patterns and dynamics of vehicles, cyclists and pedestrians across the city.

[9]https://www.nrf.gov.sg/programmes/virtual-singapore

Table 3.8 Included use *communicate* and its specialised uses

DT Use			Definition	Synonyms
03	Communicate		To exchange collected and analysed information amongst stakeholders.	Interact
	3.1	Visualise	To form and vision a realistic representation or model of current or predicted physical system.	View, visioning
	3.2	Immerse	To involve interested stakeholders in real-like experiences using immersive technologies such as VR, AR and MR.	Involve
	3.3	Document	Document and represent gathered and/or analysed data in a professional manner and technical language, forms or symbols.	Present
	3.4	Transform	To modify, process or standardise information to be published and received by other DT(s) or other DT users (e.g. a National DT) or overcome interoperability issues.	Translate, map
	3.5	Engage	To involve citizens and large groups of people including marginalised groups in policy and decision-making processes.	Empower, include

Immerse: To complement the communicational means and collaborative environment where stakeholders appreciate and interpret gathered and analysed data, DTs can enhance real-life perception through immersive technologies like virtual reality (VR), augmented reality (AR), mixed reality (MR) or the metaverse. These techniques deliver information in an easy-to-interpret manner, enabling wider participation and involvement, and ensuring equal understanding among stakeholders with different social, cultural and language backgrounds – for example, in Herrenberg, Germany, a DT was used to develop an immersive environment to display simulations of different traffic planning scenarios in a realistic and interactive experience (Dembski *et al.*, 2020). Similarly, Kent *et al.* (2019) recommended using VR for city planning, Lock *et al.* (2019) utilised AR cityscapes to enhance shared understanding of big urban data, and Allam *et al.* (2022) explored the opportunities and challenges of the metaverse for smart cities and urban sustainability.

Document: This specialised use involves exporting collected or analysed data into a representable and often printable form for documentation purposes, such as conducting studies, preparing reports or obtaining organisational approvals. Unlike the outcomes of *visualise* or *immerse*, the products of this use are less comprehensible to non-technical audiences. Examples of documentation may include producing a business case to justify a proposed strategic intervention or creating monthly operation and maintenance reports – for instance, Afzalan and Sanchez (2017) used a DT to inform the feasibility study of bike-share infrastructure planning. The documentation of a DT use case scenario, including the modelling of DT-actor interactions (as explained later in Chapter 5), is another form of this specialised use.

Transform: This specialised use involves converting a DT's output into a global or standard format, allowing for integration with other DTs using a common shared language or data format. This use is particularly relevant in constructing a web of digital twins – for example, in the vision of the NDT for the UK (Hetherington and West, 2020), a DT must be transformed from its original form (e.g. application data model) to be compliant with a predefined common standard language (e.g. integration data model). This transformation is part of proper information management, such as the information management framework envisaged to support the NDT. This framework is based on a foundation data model (FDM), equivalent to high level core concepts of the standard language, and a reference data library (RDL) equivalent to sub-classes and vocabularies.

Engage: Considering the socio-political aspects of the urban environment, DTs offer novel opportunities for people empowerment, civic engagement, and public participation in decision-making and planning processes for future cities. As Kontokosta (2017: p. 4) notes, this DT use focuses on enhancing participation by a broader range of stakeholders in planning strategies such as visioning, goal-setting and value definition, rather than on analytical methods to solve problems. While other specialised uses, like *compute* or *mine*, provide quantitative analysis of mirrored data, the *engage* use equips DT users with tools to qualitatively interpret citizens' input, convey their views and feedback, and foster a more human- or citizen-centric DT. This approach can revolutionise city governance and support public consultation and involvement (Nochta *et al.*, 2019). The new generation of IT has endowed the heterogeneous DT toolkit with both 'top-down' and 'bottom-up' methods, including social media, open-source platforms, the internet of people (IoP), gamified engagement tools and more. Unlike passive crowdsensing, the *engage* use involves active citizen participation in real-world interventions, such as in the 'Future City Glasgow' project[10]. Here, citizens are not merely monitored but actively contribute to decision-making processes. Examples include the 'FixMyStreet' initiative (Gardner and Hespanhol, 2018), public participation in city traffic planning (Dembski *et al.*, 2020), online models for citizen feedback on urban policies (White *et al.*, 2021), and crowdsourced information gathering by way of web-GIS tools for bike-share feasibility studies (Afzalan and Sanchez, 2017).

3.7.3.5 Control

The *control* included use illustrates how a DT intervenes back into the real-world. This use firmly establishes the concept of bidirectional communication between the physical and virtual systems, which is fundamental to the DT concept. As shown in Table 3.9, this included use is divided into two specialised uses: *inform* and *actuate*, which represent the 'passive' (i.e. human-in-the-loop) and 'active' (i.e. human-out-of-the-loop) forms of bidirectional communication, respectively, as defined by Al-Sehrawy and Kumar (2021). This dichotomy is analogous to the distinction between 'programmed' and 'non-programmed' types of decisions, as explained by Simon (1960). It is important to note that for a DT to *inform* or *actuate*, it may need to rely on the output produced by the successful execution of other DT included uses, such as *mirror*, *analyse* and *communicate*. This output is then purposefully accumulated to enable the DT to *control* the problem situation, taking into account the current circumstances, context, and the general purpose of the DT as outlined by the GUC.

Inform: Most DT applications in urban management conclude by enabling this specialised use. Whether the DT is *mirroring* a real system, *analysing* captured information or *communicating* results to stakeholders, these steps ultimately support and *inform* interventions in the real world. This specialised use is akin to human-in-the-loop bidirectional communication, where decision-makers interpret analysed data and decide on the appropriate intervention. This use is equally prevalent in both *online* and *offline DTs*. Through this specialised use, humans take DT-informed

[10]https://futurecity.glasgow.gov.uk/

Table 3.9 Included use *control* and its specialised uses

DT Use			Definition	Synonyms
04	Control		To leverage the collected and analysed information to intervene back into the real world to achieve a desirable state.	Implement, execute
	4.1	Inform	To support human decision making throughout the implementation of interventions in the real world.	Support, aid
	4.2	Actuate	Using Cyber Physical Systems and actuators to implement changes to physical system.	Regulate, manipulate, direct, automate, self-govern

actions to implement interventions in the real world (Ma *et al.*, 2019). The necessity of this use reflects the complexity and dynamism of the urban environment, as well as our limited understanding of its intrinsically interconnected and ever-changing nature. These challenges require collective human judgement to make decisions that, in Simon's (1960) terms, satisfice rather than optimise.

Actuate: *Actuate* is another specialised use of DT control. While actuation may not be as beneficial or appropriate for strategic interventions requiring human judgement and collective wisdom, it is particularly useful for repetitive operational and short-term tasks within relatively closed environments. Such routine tasks are better automated to increase productivity and provide more efficient solutions. *Actuation* does not necessarily represent a more intelligent use or approach compared to *informing*; rather, it focuses on removing humans from the loop of bidirectional communication between physical and virtual systems (Akanmu *et al.*, 2013). This specialised use is predominantly associated with *online* DTs, which are expected to enable the efficient execution of such tasks.

3.7.3.6 Relations between DT uses

The different uses and sub-uses of a DT, as explained above and illustrated in Figure 3.12, can interrelate in two primary ways. First, DT uses can serve as enablers for one another, creating valuable synergistic effects when combined – for instance, after *mirroring* an existing physical system through *monitoring*, a DT can perform effective *mining* to uncover hidden patterns and regularities. Similarly, a DT can better *inform* decision-makers by leveraging its *communication* uses and capabilities. Second, DT uses can overlap at certain levels. Distinct DT uses may employ the same tools, techniques, or technologies to achieve different functions – for example, the computational and analytical technologies used for *prediction* are often similar to those employed for *simulation*, despite the different purposes and sub-uses associated with these two functions.

3.7.4 DT capabilities

This section builds upon the taxonomy above and the various DT uses to highlight four key unique capabilities of a DT. The DT uses – *mirror, analyse, communicate* and *control* (as detailed in Section 3.7 and illustrated in Figure 3.12) – are rooted in the DT's greatest capabilities, which become particularly evident when handling big data. The rise of sensor devices and advancements in information and communication technology (ICT) have led to the emergence of big data, characterised by the '5Vs' as outlined by Ishwarappa and Anuradha (2015). These are.

Volume: This is perhaps the most recognised feature of big data, referring to the vast amounts of data that are continuously being collected.

Velocity: This aspect pertains to the unprecedented speed at which data is generated.

Variety: Big data is gathered from various sources in a multitude of formats.

Veracity: With the high volume, velocity and variety of data, there is an increased likelihood of capturing inaccurate or low-quality data, leading to varying levels of data reliability.

Value: Considered 'the most important aspect of big data' (Ishwarappa and Anuradha, 2015: p. 321), value is realised through the powerful computational and processing capabilities, and big data analytics, which can be integrated into a DT to unlock the full potential of big data.

3.7.4.1 Mirroring capability

Regarding the first three Vs of big data – volume, velocity and variety – it is beyond human capability to fully capture the flow of data due to our inherently 'unique' and 'limited' observations of the world, as noted by Armson (2011). The uniqueness of our perspective stems from individual biases, subjectivity, personal interests and experiences, which influence how we perceive and interpret phenomena. As Armson (2011: p. 41) suggests, 'no one has access to the view-from-here that I have.' Using multiple methods and a variety of data sources can help neutralise the influence of such biases. Moreover, the limited nature of human perception reflects our sensory limitations, which force us to rely on expectations, interests and past experiences to fill in gaps – for instance, Armson (2011: p. 41) describes how our brains can only process a minuscule fraction of the data our senses receive, forcing us to create a mental picture based on incomplete information.

> *'As I walk around my neighbourhood, I encounter an overwhelming amount of data. My senses receive 400 thousand million bits of data every second. My brain only deals with 2,000 bits per second so I only notice a very small fraction – a half a millionth of one percent – of what I see, hear and smell. More extraordinary still is the observation that the 100 bits per second that trigger my visual perception are not enough to form any image of what is going on around me. My brain fills in the deficiency. My expectations and previous experience create my sense of the outside world.'*

A DT, however, can mirror a vast amount of data, addressing both the uniqueness and limitedness of human observations, thus offering a more comprehensive view.

3.7.4.2 Analysing capability

Human senses are not only incapable of capturing the immense volume of data around us but also lack the capacity to process and comprehend it. The sheer scale of big data presents a significant challenge in unlocking its potential value. However, a DT can enhance this process through its powerful analysis capability, which supports two main approaches.

Mining: A DT can uncover hidden patterns, regularities, or demi-regs that might be invisible to human observers. Simon (1996) argues that natural science seeks to simplify complexity, revealing patterns hidden within apparent chaos. In a complex environment like an urban setting, systems are often irreducible and open or 'quasi-closed', making it difficult to detect regularities. A DT, through uses like *mining* or *computing* (Section 3.7.3.3) can help identify correlations between variables or use advanced analytical techniques such as AI or ML to assess the relative importance, strength and frequency of key events.

Simulation: A DT can provide insights into the consequences of complex interactions within a system. While simulations are limited by the assumptions built into them, they are invaluable for understanding the implications of these interactions, especially when numerous variables and complex initial conditions are involved. A DT can *simulate* (Section 3.7.3.3) such scenarios, helping practitioners understand the potential outcomes of various interventions.

3.7.4.3 Communication capability

Communication is one of the most powerful capabilities of a DT, making big data and its analysis outcomes accessible to our senses and cognition. As Pavlovskaya (2006) emphasises data must be visualised – whether quantitative or qualitative – to assess their quality, suitability or completeness. This process of communication allows DT practitioners to gain insights, acquire new knowledge and develop a deeper understanding of the phenomena they are studying. By enabling us to see 'beyond what is currently seen' (National Infrastructure Commission, 2017: p. 63), a DT enhances our ability to interpret complex data and make informed decisions.

3.7.4.4 Control capability

In certain scenarios, particularly within closed or quasi-closed systems, a DT can operate autonomously and with a teleological focus. Autonomy refers to the DT's ability to intervene in the real world without human involvement in the bi-directional communication loop between the digital and physical realms. Teleology implies that the DT can execute these interventions with a predefined goal and a desired outcome in mind. Tomko and Winter (2019: p.395) describe this capability as the DT's 'agency' that 'can change the (physical or digital) environment based on instructions, possibly from the counterpart environment and including engagement with people'. A DT can, therefore, react, predict and act independently, making it a powerful tool for managing complex systems.

3.8. Q3 and Q4 – Synthesis of results

3.8.1 Introduction

This section delves into the analysis and synthesis of results to address two key questions: [Q3] What are the methodological approaches or styles of practice adopted in the application of DTs? and [Q4] What are the different philosophical worldviews that DT researchers and practitioners hold? The discussion is conducted at both the 'methodological' and 'philosophical' levels of analysis, respectively.

A review of the collected studies reveals the presence of a general framework that appears to underpin the majority of DT projects, guiding their development and implementation in a broader context. This overarching methodological workflow, as identified by Pregnolato *et al.* (2022), or framework, subtly influences most studies and, in some cases, is discussed more explicitly, as noted by Council and Lamb (2022). This framework reflects the prevailing methodology by which DT researchers and practitioners operate, particularly in the context of the paradigm *DT for UM.*

Figure 3.13 illustrates the key elements and processes that shape this general methodological framework, as observed across most of the reviewed studies. The framework comprises three distinct, yet interconnected worlds: the physical, the social and the digital. These three worlds are continuously linked throughout any DT project through the exchange of data and the execution of actions from or by one to another. This interconnectedness underscores how DT projects are not merely technical endeavours but are deeply integrated with social and physical realities. The framework serves as a methodological guide, implicitly shaping the approach to DT development and implementation.

Figure 3.13 Existing general methodological framework for the implementation of *DT for UM* (author's own)

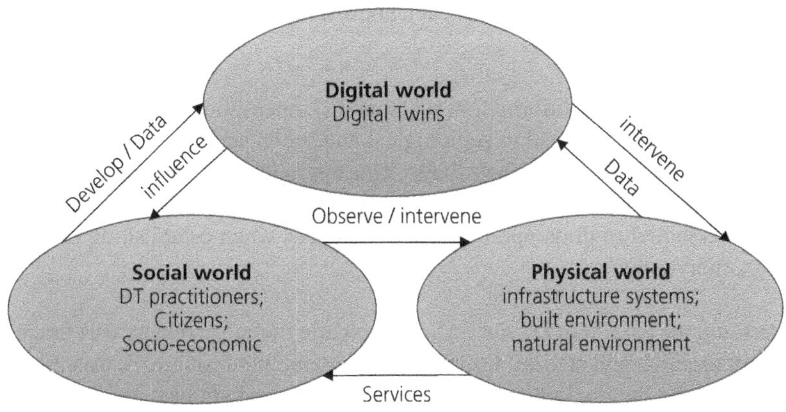

After a thorough review of the literature, it becomes evident that authors demonstrate different styles or forms of practice when implementing the general framework (Figure 3.13). These variations in methodological approaches necessitate a deeper analysis of the philosophical worldviews that underpin them. Creswell (2009: p. 6) defines a worldview as a 'general orientation about the world and the nature of the research that the researcher holds.' To elaborate further, the manner in which a DT is constructed and utilised represents a specific 'DT approach' or form of practice, adopted by the responsible DT practitioner.

Each DT approach inherently carries a set of underlying philosophical assumptions that reveal how the practitioner views the urban world. These assumptions may include ontological and epistemological beliefs about the urban environment and its constitutive elements – for instance, they may encompass questions such as what is considered real, how knowledge about these elements can be acquired, and how human beings behave, interact and make decisions. Given the intertwined nature of methodological approaches and philosophical worldviews, it is essential to address both questions [Q3] and [Q4] simultaneously, as detailed in the following sections.

3.8.2 Methodology

The approach to answering [Q3] and [Q4] involves two main steps.

1. Conceptualisation of DT methodological approaches: This involves identifying and defining the various *ideal* methodological approaches to DT. The term *ideal* here refers not to worlds of perfection, morality or exemplars but to that of ideas and mental constructs. It only represents a group of abstract characteristics or features rather than a realistic comprehensive reflection of a phenomenon or a detailed representation of any actual concrete case. As such, DT ideal approaches help bring theoretical order to the messy world of practice. Thus, a single real-world event like a DT project may exhibit a mix of several ideal approaches. The implications of mixing different ideal approaches in the same intervention are discussed below.
2. Linking *ideal* DT approaches to philosophical worldviews: Each identified ideal DT approach is explicitly connected to its underlying philosophical worldview.

To achieve these objectives, an analytical framework methodology is employed. This involves conducting a qualitative analysis of the reviewed studies using a meta-framework. This framework facilitates the rigorous conceptualisation of the various ideal DT approaches while clearly articulating the philosophical position underlying each approach.

3.8.2.1 Selection of a meta-framework

In various disciplines, including information systems (IS), operational research (OR), and systems dynamics (SD), scholars have turned to sociological and social science meta-frameworks to delve into the philosophical foundations of their fields. This approach is particularly relevant within the context of UM, where human beings are undeniably central to the urban landscape. Given their integral role, it is essential to draw upon the social sciences when establishing theoretical frameworks for urban interventions.

To better understand the paradigm *DT for UM*, a systematic literature review was first conducted to identify pertinent research and studies. Following this, a suitable framework was selected to serve as a theoretical lens for the qualitative analysis of the retrieved studies, aiming to uncover their implicit philosophical assumptions. Various meta-frameworks and meta-theories, which encapsulate the expansive domain of sociology and its foundational theories, were considered to provide a comprehensive understanding of social science. Influential works, such as Habermas's exploration of human cognitive interests (1978) and Perspectives in Sociology by Cuff *et al.* (1990), were instrumental in this process.

Ultimately, the seminal Burrell and Morgan framework (BMF) (Burrell and Morgan, 1979) was chosen for its effectiveness in conceptualising practices through different philosophical perspectives. The BMF's established use in previous studies across IS, OR, and SD literature (Hirschheim and Klein, 1989; Lane, 1999; Olaisen, 1991; Walsham, 1995) further validated its applicability in exploring the philosophical underpinnings of *DT for UM*.

3.8.3 Burrell and Morgan framework [BMF]

3.8.3.1 Objective–subjective (dimension 1 [D1])

This dimension, referred to as [D1], encompasses four critical strands of theory – ontology, epistemology, human nature and methodology – through which Burrell and Morgan conceptualise social science. Combining the extreme ends of these strands, they give rise to two polarised intellectual traditions. Together, these traditions form one of the principal dimensions of the BMF, known as the 'objective–subjective' dimension, as depicted in Figure 3.14.

Ontology, the first of these strands, deals with the fundamental nature of reality and existence. Ontological perspectives can vary significantly. At one extreme, reality is perceived as 'objective,' existing independently of individuals, regardless of whether they are aware of it or not. At the opposite extreme, reality is viewed as 'subjective,' shaped by the consciousness and perceptions of the observer, constructed through concepts, names and labels.

Epistemology, another crucial strand, concerns the nature of knowledge and the means by which it is acquired. A 'positivist' stance regards knowledge as real, concrete and measurable, deriving from the discovery of regularities and laws governing the components of a system. In contrast, 'anti-positivists' view knowledge as more subjective, intangible and even transcendental, acquired through social interactions. They reject the binary of 'true' or 'false' knowledge, seeing it instead as a relative construct, shaped by individual experiences.

Figure 3.14 1st dimension (D1) of BMF: objective–subjective. Adapted from Burrell and Morgan (1979)

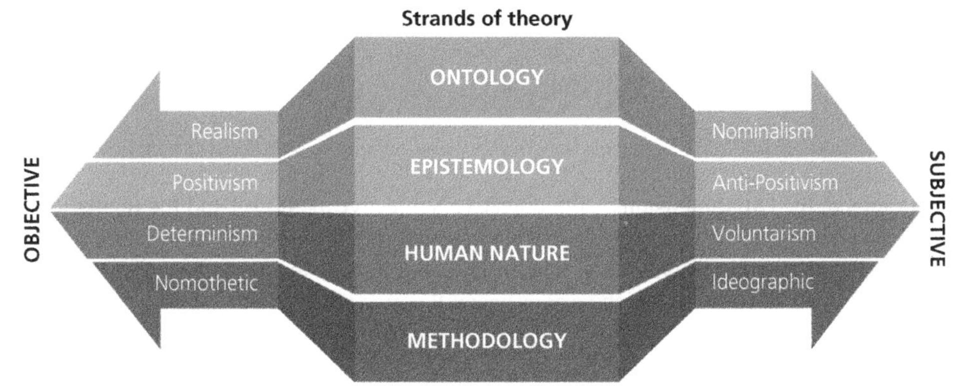

Human involvement is the defining feature of social science, with human beings at the core of inquiry. The strand of human nature addresses how humans, as the central subjects of study, are perceived. From a 'determinist' viewpoint, human behaviour is mechanistic, dictated by external circumstances and environmental factors. On the other hand, a 'voluntarist' perspective sees humans as autonomous agents with free will, shaping and influencing their environment rather than being shaped by it.

The varying stances on ontology, epistemology and human nature have given rise to a diverse array of methodologies in social science. At one extreme, a social scientist may adopt a 'nomothetic' methodology, which treats the social world similarly to the natural world – real, tangible, and governed by universal laws despite its apparent complexity. This approach relies on rigorous 'scientific' processes, including hypothesis formulation, testing and refutation, to generate knowledge. Opposing this is the 'ideographic' methodology, which places the individual at the centre of analysis. This approach, which contrasts with the methods of natural sciences, eschews generalisations in favour of exploring the unique role of the individual within a social phenomenon, challenging the notion of an external reality and instead advocating for a relativistic view where reality is seen as a construct of the individual.

The dichotomies within these four strands – ontology, epistemology, human nature and methodology – have led to the emergence of two polarised perspectives: subjective and objective. These extremes define the first dimension of the BMF.

3.8.3.2 Regulation–radical change (dimension 2 [D2])

On one side, a group of social theorists focuses on understanding how society maintains its stability and cohesion. They are concerned with the mechanisms that ensure societal unity, believing that regulation and the preservation of the status quo are essential for a stable and cohesive social entity. Their work often revolves around exploring the ways in which social order is sustained and how societies hold together.

In contrast, another group of theorists is more concerned with the forces of radical change within societies. They examine how deep-seated social conflicts and structural contradictions drive societal transformation. This group envisions more utopian scenarios, where individuals are liberated from the constraints imposed by social structures or from their own misconceptions and lack of awareness, which hinder the realisation of their full potential. These theorists are driven by the pursuit of new possibilities and alternatives, rejecting the notion of simply maintaining the status quo.

These two perspectives represent the opposing ends of the second dimension of the BMF, as illustrated in Figure 3.15.

3.8.3.3 D1 and D2 combined

When combined, the two dimensions – the objective-subjective axis (D1) and the regulation-radical change axis (D2) – create a framework that is divided into four quadrants. These quadrants represent the four paradigms: *functionalism, radical structuralism, interpretivism* and *radical humanism* (Figure 3.16), within which various social theories are situated.

Functionalism: This paradigm is firmly grounded in objectivism and aligns closely with the principles of regulatory sociology. Functionalists are the most pragmatic among social theorists, striving to provide scientific explanations for social phenomena using methods akin to those employed in the natural sciences. This approach allows for the creation of a more generalisable body of public knowledge that can be applied broadly within a social engineering framework to address social issues. The underlying assumption of this paradigm is that, despite any apparent chaos, the social world is composed of concrete systems, characterised by objective artefacts and interconnections that shape the human experiences we observe.

Interpretivism: Perspectives within this paradigm are aligned with the sociology of regulation but are deeply rooted in subjectivism, focusing on the intangible and implicit aspects of the social world. Interpretivism centres on understanding social phenomena at the individual level, emphasising the unique worldview and subjective experiences of each person. It fully acknowledges the role of the observer, who is seen as the creator of their own social reality.

Radical structuralism: While maintaining a commitment to objectivism, this paradigm resonates with the sociology of radical change. Both radical structuralists and radical humanists aim to liberate

Figure 3.15 2nd dimension (D2) of BMF: regulation–radical change. Adapted from Burrell and Morgan (1979)

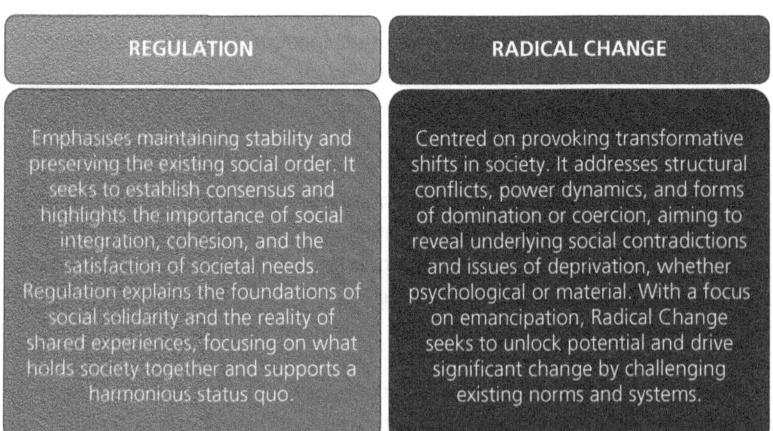

Figure 3.16 Four paradigms of social science. Adapted from Burrell and Morgan (1979)

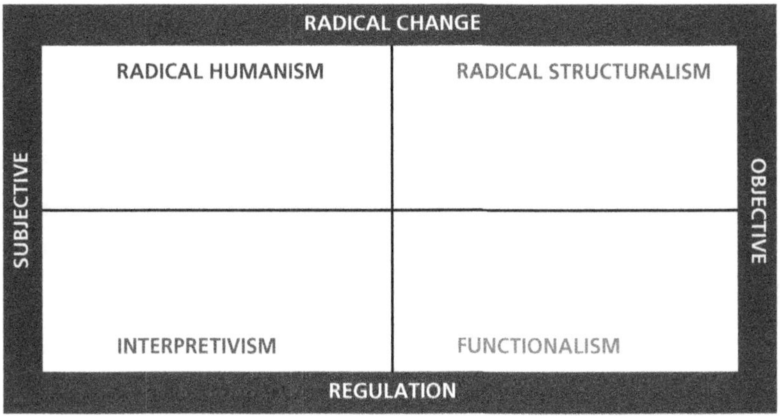

individuals from societal domination, relying on a radical critique of society. However, radical structuralists concentrate on the structural realities and interrelationships within society rather than individual consciousness. They seek to explain and justify social changes through the underlying, deep-seated structures, contradictions and feedback loops that shape society. Radical structuralists believe that societal changes are driven by these fundamental conflicts, which, when understood, can facilitate the process of emancipation. While this paradigm shares a belief in the existence of underlying laws with regulatory sociology, it uses science not just to explore the social world but to comprehend how the inherent conflicts within contemporary social structures lead to radical change.

Radical humanism: Social scientists within this paradigm adopt a subjectivist approach, similar to interpretivism, but with a foundation in German idealism, they engage in a radical critique of society based on individual consciousness. Proponents of this paradigm view society as a constraint that limits individual potential, advocating for human emancipation as a means of realising this potential. Burrell and Morgan describe the core of this paradigm as the belief that human consciousness is dominated by ideological superstructures, creating a cognitive wedge of alienation or false consciousness that hinders true human fulfilment (Burrell and Morgan, 1979: p. 32). Since society is seen as inherently anti-human, this paradigm's ultimate goal is to free individuals so they can achieve their full potential.

3.8.4 DT ideal approaches

As the concept of DT has gained increasing prominence within the realm of UM, its perception and implementation have undergone significant transformations, leading to the emergence of various DT approaches or practices. To rigorously conceptualise these approaches, the studies reviewed were analysed through the lens of the BMF explained above. This analysis led to the identification of four *ideal* types of DT practice – tech-driven, disruptive, cognitive and humanistic – each of which corresponds to one of the four paradigms of the BMF: functionalism, radical structuralism, interpretivism and radical humanism, respectively (Figure 3.17).

3.8.4.1 Tech-driven DT approach

The *tech-driven* approach to digital twinning represents a clear manifestation of the 'convergence of technology and the city' (Yigitcanlar *et al.*, 2018: p. 145). This approach focuses on leveraging

Figure 3.17 Four *ideal* DT approaches (author's own)

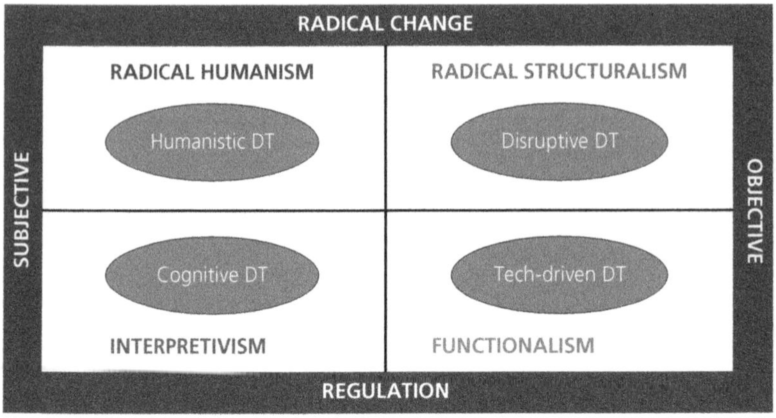

cutting-edge digital technologies to replicate urban physical systems with high fidelity, detect patterns, uncover urban dynamics,and predict future events with objectivity and precision. Practitioners of this approach assess the success of DT outcomes based on their efficiency and effectiveness.

DTs have already proven their worth in early-adopting industries such as manufacturing and aerospace (Al-Sehrawy and Kumar, 2021). These sectors have utilised DTs – built with sensors, bi-directional communication, analytics and controllers – to optimise performance and increase efficiency across the lifecycle of technical artefacts like aircraft and space vehicles (Piascik *et al.*, 2012; Ríos *et al.*, 2015). Urban managers who seek to emulate these successes in UM are likely to adopt a *tech-driven* DT approach – for instance, the initial proposal for a National Digital Twin in the UK frequently referenced successful DT projects in other industries, such as real ships and data centres, to highlight the potential for similar achievements within urban environments (National Infrastructure Commission, 2017). Likewise, other studies have drawn on early adopters' literature to define the key concepts and principles of DTs in the context of the built environment.

The *tech-driven* approach is characterised by several distinct features. First, from this perspective, the higher the fidelity of a DT, the more reliable and valuable its outputs are perceived to be. This concept dates back to 1991 when David Gelernter introduced the notion of 'mirror worlds,' envisioning a city-scale model of reality that processes vast amounts of real-time data, allowing urban managers to view detailed and dynamic representations of the city on a computer screen (Gelernter, 1991: p. 30). Al-Sehrawy and Kumar (2021) extended this idea by proposing a three-dimensional method for assessing the visual, reflective and performance fidelity of DTs within the built environment, drawing inspiration from early adopters like Grieves (2006).

Tech-driven DT practitioners view cities as organised complexities (Batty, 2013). This worldview significantly influences the *tech-driven* approach, positing that cities are complex systems governed by urban laws and exhibiting consistent patterns. Consequently, *tech-driven* researchers aim to use big urban data and advanced analytics to quantify and scientifically formulate mathematical expressions of observed urban phenomena and citizen behaviours (Hirschheim and Klein, 1989; Hu *et al.*, 2021). These formulations are then used to predict future urban dynamics, support and automate decision-making processes in UM (Kontokosta, 2021) – for example, a *tech-driven* approach might use urban big data to forecast transportation carbon emissions (Lu *et al.*, 2017), apply machine learning to estimate the suitability of buildings for specific urban uses (Sideris *et al.*, 2019) or leverage spatiotemporal data to detect urban clusters (Tang *et al.*, 2019).

This school of DT practice aligns with a broader movement within urban science, often referred to by terms such as data-driven urban management (Engin *et al.*, 2020), urban analytics (Bibri and Krogstie, 2018; Engin *et al.*, 2020; Ibrahim *et al.*, 2020), urban computing (Nourian *et al.*, 2018; Zheng *et al.*, 2014), urban informatics (Barns, 2017; Kontokosta, 2017; Nourian *et al.*, 2018; Thakuriah *et al.*, 2017), urban intelligence (Castelli *et al.*, 2019; de Castro Neto and de Melo Cartaxo, 2019), civic analytics (Kontokosta, 2017) and the new science of cities (Batty, 2013). The *tech-driven* approach, therefore, equates scientific progress, rigour and objectivity with quantification. Since big data analytics and advanced computational techniques like AI and ML are inherently quantitative (Bettencourt, 2014; Townsend, 2013; Zheng *et al.*, 2014), there is a strong emphasis on computer and data sciences. This focus is reflected in the use of 'algorithms … computational methods … etc.' (Kontokosta, 2017: p. 55) to evaluate the success of a DT. Similarly, the maturity of a DT is often measured by its level of technological sophistication (National Digital Twin programme, 2021; The Institution of Engineering and Technology, 2019).

In light of these characteristics, the tech-driven approach is situated within the functionalist paradigm (Figure 3.17), as detailed in Tables 3.10 and 3.11 below.

3.8.4.2 Disruptive DT approach

The *tech-driven* approach focuses on innovation through the use of advanced technologies. In contrast, the *disruptive* approach seeks to leverage these technologies to expose and challenge the

Table 3.10 Positioning of tech-driven DT approach in the objective–subjective dimension [D1] of BMF

Objective – Subjective (D1)	Description	Position
Ontology	Collect objective data, 'out there', using senses and empirical observations. What is observed, measured or sensed is real.	Realism
Epistemology	Consequently, knowledge about cities is 'scientifically' produced in the form of objective and reproducible law-like mathematical expressions (Batty, 2013).	Positivism
Human nature	For the discovered universal urban laws to be valid and for urban dynamics to be predictable, aggregate citizens' behaviour, giving rise to observed urban regularities, must be necessarily determined and controlled by these laws. Computer simulation techniques, like agent-based modelling, bring this perspective into practice (Barnes *et al.*, 2021; Crooks *et al.*, 2021).	Deterministic
Methodology	Quantitative analytics used to formally express regularities and scaling relationships manifested within cities (Batty, 2013; Bettencourt, 2014).	Nomothetic

Table 3.11 Positioning of tech-driven DT approach in the Regulation-Radical change dimension [D2] of BMF

Regulation–radical change (D2)	Description	Position
Regulation against radical change	While digital innovation may eventually reshape the urban environment, the *tech-driven* approach tends to quantify physiological conditions and subjective emotions using solutions like crowdsourcing or 'people-as-sensors' (Resch *et al.*, 2015), or quantify sentiment levels based on X (formerly Twitter) data (Plunz *et al.*, 2019). This, however, may result in losing much of the richness that could have been expressed by way of qualitative data and individualistic approaches. It also ignores issues like digital divide where, for example, sentiments of offline citizens are ignored. Therefore, the *tech-driven* approach is argued to stabilise the status quo and hinders potential attempts to bring about radical change.	Regulation

injustices and inequalities embedded within existing urban structures. Practitioners of the *disruptive* approach typically adopt it when they believe that current urban systems are asymmetrical, unjust and restrictive, particularly towards marginalised groups.

Central to the *disruptive* approach is the examination of urban structures, both physical and non-physical. Physical structures include the natural environment (such as green-blue infrastructure and air quality), and the built environment and infrastructure (such as homes, highways and grey infrastructure). Non-physical structures encompass socio-economic, organisational or power dynamics. *Disruptivists* understand that contradictions within one type of structure are often mirrored in the other – for example, socio-economic inequalities in a city can manifest as unequal access to infrastructure services (Lin *et al.*, 2021) or the development of slums (Khan *et al.*, 2022). Similarly, inadequate communication infrastructure in certain communities can diminish their influence over urban management decisions.

Practitioners of the *disruptive* approach view disadvantaged citizens as powerless, lacking access to essential urban services and facilities. To address this, a DT could be developed to assess urban health justice by comparing different neighborhoods' access to medical facilities (Xia *et al.*, 2019). Another DT might analyse how road network configurations affect impoverished communities (Shuster, 2021). By highlighting these disparities, the *disruptive* approach can support interventions aimed at reducing income inequality and combating poverty (Wahba, 2021).

The *disruptive* approach employs advanced quantitative methods and big data analytics to uncover the contradictions and inequalities rooted in asymmetric urban structures – for instance, it might analyse areas affected by the digital divide to empower residents to oppose harmful interventions (Shuster, 2021).

Disruptivists are also aware of how siloed organisational and sectoral urban structures can lead to systemic issues (Kontokosta, 2017; Nochta *et al.*, 2021). They advocate viewing the urban environment as a system of systems, akin to a complex machine (Schooling *et al.*, 2021), which should be managed with systems-based policies and strategies (A collaboration of leading figures in the built environment, 2021). This approach aims to break down sectoral silos, connect different organisations and authorities, and facilitate the transparent flow of big urban data across city infrastructure assets (de Castro Neto and de Melo Cartaxo, 2019).

From the *disruptivist* perspective, the primary goal of DT outcomes is to transform existing urban structures, both physical and non-physical, to remove unnecessary constraints and create new opportunities for empowerment. Given these characteristics, the *disruptive* approach is aligned with the radical structuralism paradigm (Figure 3.17), as detailed in Tables 3.12 and 3.13 below.

3.8.4.3 Cognitive DT approach

This approach is conceptualised as a form of practice that rejects the excessive emphasis on technology. Instead, several authors have placed human beings, rather than technology, at the centre of digital twinning (Almeida *et al.*, 2018; Bouzguenda *et al.*, 2019; Carvalho, 2015). A DT practitioner is likely

Table 3.12 Positioning of disruptive DT approach in the objective–subjective dimension [D1] of BMF

Objective–subjective (D1)	Description	Position
Ontology	Acknowledging the existence of real urban physical and non-physical structures external to and independent of humans is central to the *disruptive* approach.	Realism
Epistemology	Draws on inherently positivist DT technologies including quantitative and statistical analytics to prove that urban inequalities exist and emerge because of existing urban structures (Xia *et al.*, 2019).	Positivism
Human nature	For the *disruptive* approach, people's actions and capabilities are largely determined by the urban structures constituting the environment within which they live.	Deterministic
Methodology	Quantitative analytics are used to unveil the biases and asymmetricity in existing urban physical and non-physical structural configurations (Shuster, 2021).	Nomothetic

Table 3.13 Positioning of disruptive DT approach in the regulation–radical change dimension [D2] of BMF

Regulation–radical Change (D2)	Description	Position
Regulation against radical change	*Disruptive* practices are more concerned with unveiling contradictions inherent to the deep-seated coercive urban structures. A DT is seen as an innovative idea that enables interventions needed to reconstruct the existing a-symmetric urban structures for the sake of citizen emancipation and human flourishing (A collaboration of leading figures in the built environment, 2021).	Radical change

to adopt a *cognitive* approach upon realising that the phenomena explored by early DT adopters in industries such as manufacturing and aerospace (Piascik *et al.*, 2012; Ríos *et al.*, 2015) bear little resemblance to the unique challenges of UM (Jiang, 2021; Verrest and Pfeffer, 2019). UM issues involve human subjects, whose relevance extends beyond their physical presence to their capacity for thought and conscious experiences. The principles underpinning the *cognitive* DT approach are numerous.

First, cities are viewed as the 'setting for human economic and social life' (Alberti, 2017: p. 2), rather than merely as 'complex machines' composed of interrelated infrastructure systems (Schooling *et al.*, 2021). This perspective adds an additional layer of complexity to DTs developed for UM purposes, defining them as 'cyber-physical-social ecosystems' that not only focus on the convergence of physical and digital systems (Tomko and Winter, 2019) but also incorporate unique human characteristics as integral parts of the system (Wang, 2010). The approach emphasises the 'coupling across the digital, physical, and social spheres' (Nochta *et al.*, 2021: p. 268) and the complexities that emerge from this triadic interconnection. As a result, social actors are seen as deeply embedded within the urban environment, playing a central and active role as key participants in the creation, appreciation and interpretation of various urban phenomena.

Stakeholders, recognised by *cognitive* DT practitioners as subjects with unique and sometimes conflicting consciousnesses, feelings, opinions, interests and perceptions, are unlikely to perceive a complex urban problem in the same way or propose identical solutions. Different perceptions of a problem may suggest varying boundaries for the system under investigation. In this context, *cognitivists* argue that one 'ought not to treat models as if they were perceived reality' (Checkland, 1988: p. 236). For them, a DT can, at best, be a twin not of an objective independent reality, but of the thoughts, perceptions, and interests of the stakeholders involved.

Moreover, for a *cognitive* practitioner, the development and use of a DT is a social endeavour that is inherently value laden, far removed from the notion of value-free scientific activity. In this sense, 'big data and new data analytics answer only the questions that we are able to formulate' (Alberti, 2017: p. 2). Similarly, all datasets are 'socially constructed and different forms of data allow for competing representations of place' (Shelton *et al.*, 2015: p. 18). A DT might, for example, support the monitoring and evaluation of sustainability within a city based on urban KPIs agreed upon by multiple stakeholders with competing interests (Kourtit and Nijkamp, 2018). Collaborative methods, such as learning and action alliances (Maskrey *et al.*, 2020), participatory systems dynamics (Pluchinotta *et al.*, 2021), or workshops (National Infrastructure Commission, 2020), are prevalent when exploring different urban phenomena. The subjectivity of the people involved is viewed as a strength, not a weakness, justifying the fostering of participation and collaboration. Participants' worldviews are considered the key source of understanding and explaining urban phenomena. In this sense, the focus of a DT is 'not on analytical methods to solve problems … it is to enhance substantive participation by a wider range of stakeholders in typical planning strategies of visioning, goal-setting, and value definition' (Kontokosta, 2017: p. 55).

The *cognitive* approach heavily relies on insights gained from different human subjects. As Mohammadi and Taylor (2020: p. 1657) put it, a DT 'draws on insights from human intuition and cognition to facilitate the collective discovery of knowledge from both social and sensor data.' Interventions designed using the *cognitive* DT approach are primarily evaluated in terms of effectiveness, the level of acceptance across various stakeholders, and the ability to preserve cultural identity and social norms. As detailed in Tables 3.14 and 3.15 below, the *cognitive* approach is situated at the heart of interpretivism (Figure 3.17).

Table 3.14 Positioning of cognitive DT approach in the objective–subjective dimension [D1] of BMF

Objective–subjective (D1)	Description	Position
Ontology	Reality is a mere construction by the observer and a projection of their consciousness. To this end, the greatest value of a DT lies in the insights every observer might gain 'beyond what is currently seen' (National Infrastructure Commission, 2017: p. 61) through observing DT outputs and not in the latter in itself.	Nominalism
Epistemology	The *cognitive* approach adopts the 'consensus' rather than the 'correspondence' theory of truth – an epistemic definition of truth. Hence, truth about the right interventions into the urban environment is sought through collaborative methods (Pluchinotta *et al.*, 2021).	Anti-positivism
Human nature	In the *cognitive* approach, humans are not mechanistic but autonomous agents and have free-will. Therefore, a *cognitive* practitioner would rather adopt participatory and collaborative methods to encourage people to articulate their own beliefs and insights instead of trying to predict and control their behaviours in a mechanistic way.	Voluntarism
Methodology	Relies on qualitative visualisations to help gain insights based on *cognitive* abilities and facilitate the involvement of several stakeholders. Based on how the *cognitive* approach is conceptualised, its advocates would be at unease with the use of quantitative techniques like correlation analysis, data mining or predictive analytics for they imply a mechanistic view of the urban environment and huma beings at its core.	Ideographic

Table 3.15 Positioning of cognitive DT approach in the regulation–radical change dimension [D2] of BMF

Regulation–radical change (D2)	Description	Position
Regulation against radical change	Fostering of collaboration and generation of insights paves the way for creativity and the molding of new unconventional worldviews. This indeed keeps the cognitivists away from the extreme end of regulation. However, because of its ontological nominalist stance, the *cognitive* approach finds it hard to acknowledge objective real power structures and thus, it practically has little to offer in face of possible political tension and the overpowering of less powerful groups like citizens. Hence, the outcomes of a *cognitive*-based DT would risk becoming a 'product of conformity to the uncontested authority … rather than the outcome of participation in democratic debate.' (Johnson and Duberley, 2000: p. 73). Like, for example, involving citizens only as means to 'validate the planning process by showing that they [decision makers] have conducted a participatory process' (Afzalan and Sanchez, 2017: p. 40).	Regulation

3.8.4.4 Humanistic DT approach

The *humanistic* DT approach is a self-critical form of DT practice that seeks to address issues of alienation, marginalisation and the depletion of natural resources within the urban environment. This approach uses DTs not merely as technological tools but as a means for empowering people, particularly by involving citizens and marginalised groups in the planning and management of their urban environments. It is also deeply conscious of the environmental impacts associated with the DT implementation process itself.

Practitioners likely to adopt the *humanistic* approach are critical of how datafication, automation, instrumentalism and functionalism are reshaping cities and societies in a top-down, dehumanising manner (Barnes *et al.*, 2021; Barns, 2017: p. 20; Leorke, 2020). While this criticism shares some overlap with the *disruptive* school of thought – that is, both seek the emancipation of coerced groups – the *humanistic* approach differs significantly. The *disruptive* approach advocates deconstructing the asymmetric urban structures that perpetuate coercion. In contrast, the *humanistic* approach posits that true emancipation must emerge from within humans themselves, by overcoming the false consciousness responsible for creating and sustaining oppressive structures. The ultimate aim is to 'set human consciousness or spirit free, thus facilitating the growth and development of human potentialities' (Burrell and Morgan, 1979: p. 306).

Humanistic practitioners are likely to reject the dehumanising classification that groups "people' under the same 'bucket' with other 'connected things', making no explicit distinction between the two' (Kamel Boulos *et al.*, 2015: p. 3). Instead, they would advocate for less dehumanising terminology, such as 'internet of things and people' (Kamel Boulos *et al.*, 2015). Moreover, they would aim to replace an extractive approach to human participation with an empowering one (UN-Habitat, 2012), viewing citizens and businesses not as passive recipients of services but as active owners and participants in the creation and delivery of city services (British Standards Institution, 2014). Digitally-inclusive citizen engagement initiatives under this approach may include the development of gamified platforms (Smith and Martín, 2021), participatory innovation platforms (Anttiroiko, 2016), or the use of games and play as tools to 'counter the dehumanising effects of smart city technologies' (Leorke, 2020). Consequently, people become 'active in shaping their urban environment' (De Lange and De Waal, 2017). Examples include DTs based on participatory design approaches (Panagoulia, 2020), public participation GIS (Hasegawa *et al.*, 2019), or Geo-citizen participation and Geo-discussion (Haklay *et al.*, 2018), which have been adopted to raise citizens' awareness of their urban environments and help marginalised groups participate in evaluating city infrastructure planning scenarios (Dembski *et al.*, 2020; White *et al.*, 2021). Other DTs enable citizens to report street damage issues (Gardner and Hespanhol, 2018) or use a web-GIS tool to involve people in developing a bike-share feasibility study (Afzalan and Sanchez, 2017).

The *humanistic* outlook is also acutely aware of the natural environment (Michalec *et al.*, 2019), either directly through developing DTs for ecological purposes or indirectly by considering the environmental impact of DT implementation, including associated communications technologies like 5G (Gandotra and Jha, 2017) and energy-intensive computation (Anthony *et al.*, 2020). *Humanistic* practitioners are inherently self-critical, continuously questioning the societal, individual and environmental impacts of the DTs they develop. Consequently, they give serious consideration to issues like equality, inclusivity, privacy, security and other ethical concerns related to DTs. Examples include addressing technological bias and transparency issues through concepts like 'participatory AI' (Falco, 2019) and adopting 'bias-aware data-driven processes' (Kontokosta and Hong, 2021) to overcome biases inherent in training data, algorithmic development and the contexts in which these technologies are applied. Additionally, they are mindful of the emotional responses participants may have when engaging with DT visualisations (Kennedy and Hill, 2018).

The outcomes of a *humanistic* DT are designed to empower alienated groups, raise awareness and establish harmony with the natural environment. The *humanistic* approach, as conceptualised above and detailed in Tables 3.16 and 3.17 below, is situated within the paradigm of radical humanism (Figure 3.17).

3.8.5. The dilemma of pluralism

3.8.5.1 Practice

The analysis of retrieved studies highlights the intrinsically pluralistic nature of DT approaches, revealing that many DT-based projects embody traits from multiple ideal types within a single intervention. For instance, some scholars advocate for urban models that can only be realised through a multi-approach strategy, such as Cities 4.0 (Miles, 2021) and Smart Sustainable Cities (Bibri, 2018). Nübel *et al.* (2021) propose using digital platforms, including DTs, to develop a

Table 3.16 Positioning of humanistic DT approach in the objective–subjective dimension [D1] of BMF

Objective–subjective (D1)	Description	Position
Ontology	Citizens are seen as agents who are capable of creating their own realities and have the right to shape the future of their urban environment.	Realism
Epistemology	The right interventions into the urban environment are the ones defined and shaped by citizen by way of engagement and participation.	Anti-positivism
Human nature	All individuals possess free-will, reflexivity and creativity to appreciate the urban phenomena and envisage a desirable urban reality.	Voluntarism
Methodology	*Humanistic* DT practitioners argue that relying on quantitative analytics is not the one and only method that can support urban DT practices (Bouzguenda *et al.*, 2019; Cohen, 2015; Miles, 2021), but attention must be paid to qualitative and unstructured methods that allow every individual to creatively express their own unique views.	Ideographic

Table 3.17 Positioning of humanistic DT approach in the regulation–radical change dimension [D2] of BMF

Regulation–radical Change (D2)	Description	Position
Regulation against radical change	Seeks to radically change the status quo through the empowerment and engagement of marginalised groups, emancipation of oppressed groups and endowing them with sense of ownership of their city (De Lange and De Waal, 2017), and protection of natural environment that has always been exploited.	Radical change

holistic infrastructure model that integrates several approaches: system-thinking in governance structures (*disruptive* approach), technology as an enabler (*tech-driven* approach), shared values and cultural norms (*cognitive* approach) and citizen interests (*humanistic* approach). This underscores the necessity for urban managers to employ a diverse set of methods when addressing complex and multi-dimensional urban problems.

This argument aligns with the principles of multimethodology and cybernetics research. Ross Ashby, a significant figure in cybernetics, emphasised the importance of variety in his law of 'requisite variety'. He argued that for a system to remain viable and survive, it must possess a range of responses equal to or greater than the variety of perturbations in its environment (Ashby, 1961). This principle is particularly relevant when dealing with the intricate challenges of urban management, where a single approach may not suffice.

3.8.5.2 Theory

Zhu (2011: p. 795) observes that operational research (OR) practitioners often adopt a pluralistic stance, freely mixing and matching various approaches – similar to the four DT ideal approaches discussed here – without necessarily addressing the potential paradigm incommensurability. This flexibility allows for practical solutions but can create theoretical tensions. Eden (1990: p. 91) contended that 'when different methods reflect different 'theories-in-use', it is unlikely that they will sit happily together in practice.' Ignoring these theoretical inconsistencies could undermine the call for a more theoretically aware discipline, as emphasised in this book.

One potential resolution to these theoretical challenges is 'isolationism' (Jackson, 2019), where only one DT approach is deemed sufficient and practically adequate to address all aspects of an urban problem. This strategy could maintain theoretical consistency and protect DT practices from accusations of oxymoronicity, ensuring alignment with the principles of paradigm incommensurability. However, this approach may sacrifice the benefits of pluralism and fail to provide the 'requisite variety' necessary to tackle wicked urban problems effectively. Thus, isolationism may prove to be a theoretically appealing but practically inadequate stance, risking the disintegration of the discipline into separate, isolated strands. Each strand, rooted in a specific philosophical paradigm, could become independent and disconnected from other strands, undermining the holistic integration necessary for comprehensive urban management.

Therefore, Figure 3.18 presents a more realistic depiction of how these paradigms intersect and interact in the messy world of practice, contrasting with the neat and 'ideal' representation found in Figure 3.17, which belongs to the theoretical realm. This messy interaction reflects the complex and multifaceted nature of urban challenges, where a pluralistic approach is often necessary to achieve effective solutions.

3.9. Summary

Figure 3.19 below encapsulates the key findings from the systematic literature review, highlighting the research gaps and issues identified within the paradigm *DT for UM*, across three distinct analytical levels: philosophy, methodology and methods. This figure serves as a comprehensive summary, mapping out the areas where current research falls short and where further investigation is needed. By distinguishing between these levels, the figure clarifies the philosophical foundations, methodological approaches and specific methods that are either underdeveloped or inadequately addressed in the existing literature, thereby providing a clear direction for future research in this field.

Figure 3.18 Four paradigms of social science as per the current pluralistic DT practices (author's own)

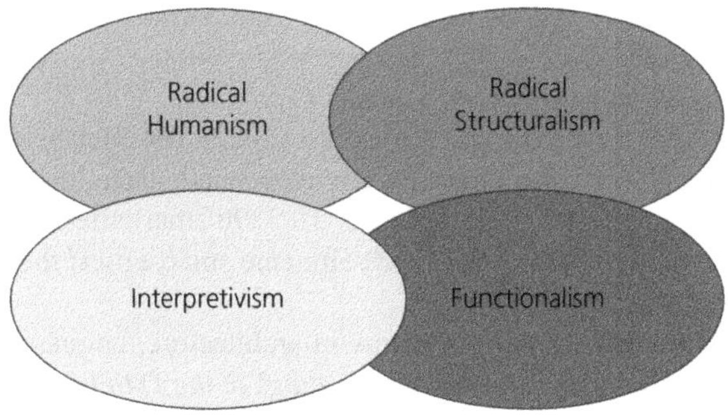

Figure 3.19 Current state of the new paradigm *DT for UM* (author's own)

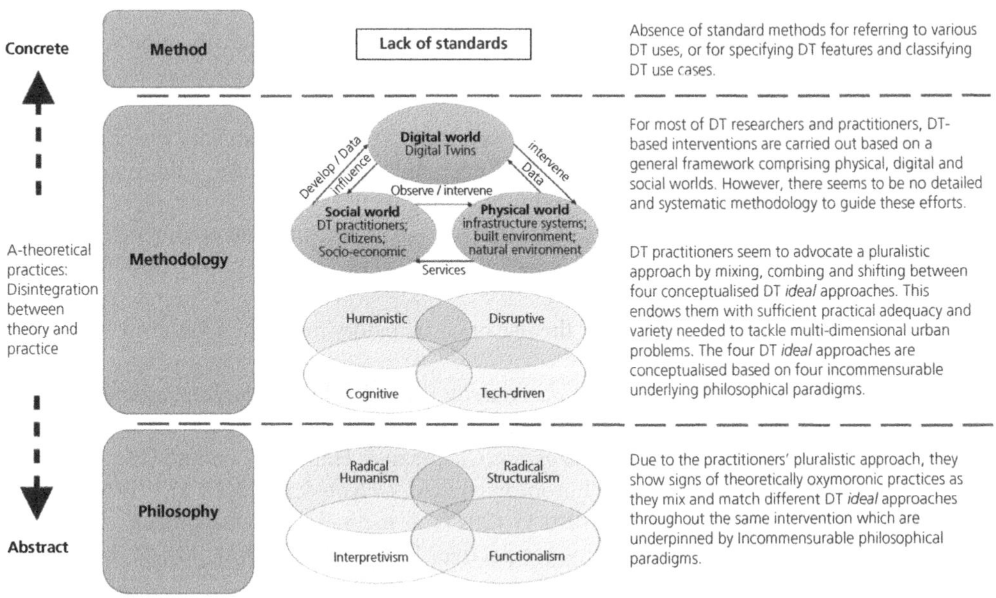

REFERENCES

A collaboration of leading figures in the built environment (2021) *Our Vision for the Built Environment.* CDBB. https://www.cdbb.cam.ac.uk/news/industry-unites-behind-vision-built-environment

Ackoff RL (1981) *Creating the Corporate Future*, Wiley, New York, USA.

Afzalan N and Sanchez T (2017) Testing the use of crowdsourced information: case study of bike-share infrastructure planning in Cincinnati, Ohio. *Urban Planning* **2(3)**: 33–44. 10.17645/up.v2i3.1013

Akanmu A, Anumba C and Messner J (2013) Scenarios for cyber-physical systems integration in construction. *Journal of Information Technology in Construction (ITcon)* **18(12)**: 240–260.

Aktemur I, Erensoy K and Kocyigit E (2020) Optimization of waste collection in smart cities with the use of evolutionary algorithms. In *2020 International Congress on Human-Computer Interaction, Optimization and Robotic Applications (HORA)*, pp. 1–8. 10.1109/HORA49412.2020.9152865

Alberti M (2017) Grand challenges in urban science. *Frontiers in Built Environment* **3**: 6.

Allam Z, Sharifi A, Bibri SE, Jones DS and Krogstie J (2022) The Metaverse as a virtual form of smart cities: opportunities and challenges for environmental, economic, and social sustainability in urban futures. *Smart Cities* **5(3)**: Article 3. 10.3390/smartcities5030040

Almeida VA, Doneda D and da Costa EM (2018) Humane smart cities: the need for governance. *IEEE Internet Computing* **22(2)**: 91–95.

Al-Sehrawy R and Kumar B (2021) Digital twins in architecture, engineering, construction and operations: a brief review and analysis. In *Proceedings of the 18th International Conference on Computing in Civil and Building Engineering* (Toledo Santos E and Scheer S (eds)). Springer International Publishing, Cham, pp. 924–939. 10.1007/978-3-030-51295-8_64

Al-Sehrawy R, Doukari O, Kumar B and Watson R (2022) A knowledge management strategy for urban digital twins. In *Proceedings of the 2022 European Conference on Computing in Construction* (Tagliabue LC, Hall DM, Chassiakos A, Nikolic D and Soman R (eds)), pp. 529–538. 10.35490/EC3.2022

Al-Sehrawy R, Kumar B and Watson R (2021) A digital twin uses classification system for urban planning and city infrastructure management. *Journal of Information Technology in Construction (ITCON)* **26**: 832–862.

Anejionu OCD, Thakuriah P (Vonu), McHugh A *et al.* (2019) Spatial urban data system: a cloud-enabled big data infrastructure for social and economic urban analytics. *Future Generation Computer Systems* **98**: 456–473. 10.1016/j.future.2019.03.052

Anthony LFW, Kanding B and Selvan R (2020) Carbontracker: tracking and predicting the carbon footprint of training deep learning models. *ArXiv:2007.03051*. https://arxiv.org/abs/2007.03051 (accessed 1/09/2024).

Anttiroiko A-V (2016) City-as-a-platform: the rise of participatory innovation platforms in Finnish cities. *Sustainability* **8(9)**: 922.

Armson R (2011) *Growing Wings on the Way*. Triarchy Press, Bridport, UK. https://books.google.co.uk/books?id=CvwlDwAAQBAJ (accessed 1/09/2024).

Ashby WR (1961) *An Introduction to Cybernetics*. Chapman & Hall Ltd, London, UK.

Bačlija I (2011) Urban management in a European context. *Urbani Izziv* **22(2)**: 137–146.

Balletto G, Ladu M, Milesi A and Borruso G (2021) A methodological approach on disused public properties in the 15-minute city perspective. *Sustainability* **13(2)**, Article 2. 10.3390/su13020593

Barkham R, Bokhari S and Saiz A (2022) Urban big data: city management and real estate markets. In *Artificial Intelligence, Machine Learning, and Optimization Tools for Smart Cities: Designing for Sustainability* (Pardalos PM, Rassia S.Th. and Tsokas A (eds)). Springer International Publishing, Cham, pp. 177–209. 10.1007/978-3-030-84459-2_10

Barmpounakis E and Geroliminis N (2020) On the new era of urban traffic monitoring with massive drone data: the pNEUMA large-scale field experiment. *Transportation Research Part C: Emerging Technologies* **111**: 50–71. 10.1016/j.trc.2019.11.023

Barnes B, Dunn S, Pearson C and Wilkinson S (2021) Improving human behaviour in macroscale city evacuation agent-based simulation. *International Journal of Disaster Risk Reduction* **60**: 102289.

Barns S (2017) Visions of urban informatics: from proximate futures to data-driven urbanism. *Fibreculture Journal* **29**: Article 29. https://eprints.qut.edu.au/209224/ (accessed 1/09/2024).

Bartos M and Kerkez B (2020) pipedream: an interactive digital twin model for urban drainage networks. *EarthArXiv*. https://eartharxiv.org/repository/view/74/ (accessed 1/09/2024).

Batty M (2013) *The New Science of Cities*. MIT Press, Cambridge, MA, USA.

Batty M (2018) Digital twins. *Environment and Planning B: Urban Analytics and City Science* **45(5)**: 817–820. 10.1177/2399808318796416

Bawden R (1995) *Systemic Development: A Learning Approach to Change*. University of Western Sydney. https://scholar.google.com/scholar_lookup?title=Systemic+development%3A+a+learning+approach+to+change&author=Bawden%2C+R.&publication_year=1995 (accessed 1/09/2024).

Bettencourt LM (2014) The uses of big data in cities. *Big Data* **2(1)**: 12–22.

Bibri SE and Krogstie J (2018) The big data deluge for transforming the knowledge of smart sustainable cities: a data mining framework for urban analytics. In *Proceedings of the 3rd International Conference on Smart City Applications*, pp. 1–10. 10.1145/3286606.3286788

Bibri SE (2018) *Smart Sustainable Cities of the Future: The Untapped Potential of Big Data Analytics and Context–Aware Computing for Advancing Sustainability*. Springer, Cham.

Boeing G (2021) Spatial information and the legibility of urban form: big data in urban morphology. *International Journal of Information Management* **56**: 102013. 10.1016/j.ijinfomgt.2019.09.009

Bolton A, Butler L, Dabson I *et al.* (2018). *Gemini Principles* [Report]. CDBB. 10.17863/CAM.32260

Boschert S and Rosen R (2016). Digital twin—the simulation aspect. In *Mechatronic Futures: Challenges and Solutions for Mechatronic Systems and their Designers* (Hehenberger P and Bradley D (eds)). Springer International Publishing, Cham, pp. 59–74. 10.1007/978-3-319-32156-1_5

Boulton JG, Allen PM and Bowman C (2015) *Embracing Complexity: Strategic Perspectives for an Age of Turbulence*. Oxford University Press, Oxford. https://books.google.co.uk/books?id=YlUbCgAAQBAJ (accessed 1/09/2024).

Bouzguenda I, Alalouch C and Fava N (2019) Towards smart sustainable cities: a review of the role digital citizen participation could play in advancing social sustainability. *Sustainable Cities and Society* **50**: 101627.

Brammall N and Kessler H (2020) An update on the UK government's plans for a national underground asset register. *Proceedings of the Institution of Civil Engineers - Civil Engineering* **173(2)**: 56–56. 10.1680/jcien.2020.173.2.56

Brilakis I, Pan Y, Borrmann A, Mayer H-G, Rhein F, Vos C, Pettinato E and Wagner S (2019) Built environment digital twining [Report]. *International Workshop on Built Environment Digital Twinning*, presented by TUM Institute for Advanced Study and Siemens AG. 10.17863/CAM.65445

British Standards Institution (2014) *Smart City Framework–Guide to Establishing Strategies for Smart Cities and Communities*. BSI London, UK.

Burrell G and Morgan G (1979) *Sociological Paradigms and Organisational Analysis*. Routledge, London, UK.

Camagni R (1998) Sustainable urban development: definition and reasons for a research programme. *International Journal of Environment and Pollution* **10(1)**: 6–27. 10.1504/IJEP.1998.002228

Carvalho L (2015) Smart cities from scratch? A socio-technical perspective. *Cambridge Journal of Regions, Economy and Society* **8(1)**: 43–60.

Castelli G, Cesta A, Diez M *et al.* (2019) Urban intelligence: a modular, fully integrated, and evolving model for cities digital twinning. In *2019 IEEE 16th International Conference on Smart Cities: Improving Quality of Life Using ICT & IoT and AI (HONET-ICT)*, pp. 033–037. 10.1109/HONET.2019.8907962

Celes C, Boukerche A and Loureiro AAF (2019) Crowd management: a new challenge for urban big data analytics. *IEEE Communications Magazine* **57(4)**: 20–25. 10.1109/MCOM.2019.1800640.

Cerrone D, López Baeza J and Lehtovouri P (2018) Integrative urbanism: using social media to map activity patterns for decision-making assessment. In *Proceedings of the 13th International Forum for Knowledge Asset Dynamics (IFKAD)*, Delft, pp. 1094–1107.

Checkland P (1981) *Systems Thinking, Systems Practice*. John Wiley & Sons, Chichester, UK.

Checkland P (1988) The case for 'holon.' *Systems Practice* **1(3)**: 235–238.

Checkland P (1999) *Systems Thinking, Systems Practice: Includes a 30-Year Retrospective*. John Wiley & Sons Ltd, Chichester, UK.

Churchman CW (1968) *Challenge to Reason*. McGraw-Hill, New York. https://books.google.co.uk/books?id=TXFEAAAAIAAJ (accessed 1/09/2024).

Cohen B (2015) The 3 generations of smart cities. *FASTCOMPANY*. https://www.fastcompany.com/3047795/the-3-generations-of-smart-cities (accessed 1/09/2024).

Council G and Lamb K (2022) *Gemini papers: How to enable an ecosystem of connected digital twins* [Report]. Centre for Digital Built Britain, Cambridge. 10.17863/CAM.82193

Creswell JW (2009) *Research Design: Qualitative, Quantitative, and Mixed Methods Approaches*. SAGE Publications, Inc, Los Angeles.

Crick F (1966) *Of molecules and men*. University of Washington Press, Seattle.

Crooks A, Heppenstall A, Malleson N and Manley E (2021) Agent-based modeling and the city: a gallery of applications. In *Urban Informatics* (Shi W, Goodchild MF, Batty M, Kwan M-P, and Zhang A (eds.)). Springer, Cham, pp. 885–910.

Cuff EC, Sharrock WW and Francis DW (1990) *Perspectives in Sociology*, 3rd edn, Routledge and Kegan Paul, London, UK.

Cunningham SW and Verbraeck A (2018) Concepts and constructs of urban sensing. In *2018 IEEE International Conference on Technology Management, Operations and Decisions (ICTMOD)*, pp. 184–189. 10.1109/ITMC.2018.8691274

Davey KJ (1993) *Elements of Urban Management*. The World Bank, Washington, DC. https://elibrary.worldbank.org/doi/abs/10.1596/0-8213-2424-1 (accessed 1/09/2024).

Dawkins O, Dennett A, & Hudson-Smith A P (2018) *Living with a Digital Twin: Operational management and engagement using IoT and Mixed Realities at UCL's Here East Campus on the Queen Elizabeth Olympic Park* [Proceedings paper]. 26th annual GIScience Research UK conference: GISRUK 2018. Giscience and Remote Sensing; GIS Research UK (GISRUK). http://leicester.gisruk.org/programme/

de Castro Neto M and de Melo Cartaxo T (2019) Smart and collective urban intelligence. In *Security at a Crossroad. New Tools for New Challenges*. Nova Science Publishers, Inc, New York, USA, pp. 83–94.

De Lange M and De Waal M (2017) Owning the city: new media and citizen engagement in urban design. In *Urban Land Use*. Apple Academic Press, pp. 109–130.

DebRoy T, Zhang W, Turner J and Babu SS (2017) Building digital twins of 3D printing machines. *Scripta Materialia* **135**: 119–124. 10.1016/j.scriptamat.2016.12.005

Dembski F, Wössner U, Letzgus M, Ruddat M and Yamu C (2020) Urban digital twins for smart cities and citizens: the case study of Herrenberg, Germany. *Sustainability* **12(6)**, Article 6. 10.3390/su12062307

Ding C, Lai S-K and Wang M-S (2012) Global urbanization and urban management. *Journal of Urban Management* **1(1)**: 1–2.

Dresch A, Lacerda DP and Antunes Jr JAV (2015) *Design Science Research: A Method for Science and Technology Advancement*. Springer International Publishing, Cham. 10.1007/978-3-319-07374-3

Eden C (1990) Part III: Mixing methods - introduction. In *Tackling Strategic Problems: The Role of Group Decision Support* (Eden C and Radford J (eds.)). Sage, London, UK, pp. 90–91.

Engin Z, van Dijk J, Lan T *et al.* (2020) Data-driven urban management: mapping the landscape. *Journal of Urban Management* **9(2)**: 140–150.

Evans S, Savian C, Burns A and Cooper C (2019) *Digital twins for the built environment* [pdf, p.24]. The Institution of Engineering and Technology. https://www.theiet.org/media/8762/digital-twins-for-the-built-environment.pdf (accessed 1/09/2024).

Falco G (2019) Participatory AI: reducing AI bias and developing socially responsible AI in smart cities. In *2019 IEEE International Conference on Computational Science and Engineering (CSE) and IEEE International Conference on Embedded and Ubiquitous Computing (EUC)*, pp. 154–158.

Farsi M, Daneshkhah A, Hosseinian-Far A and Jahankhani H (2020) *Digital Twin Technologies and Smart Cities*, 1st edn. Springer, Cham. 10.1007/978-3-030-18732-3

Forrester JW (1968) *Principles of Systems*. Productivity Press, Portland, OR.

Francisco A, Mohammadi N and Taylor JE (2020) Smart city digital twin–enabled energy management: toward real-time urban building energy benchmarking. *Journal of Management in Engineering* **36(2)**. 10.1061/(ASCE)ME.1943-5479.0000741

Gandotra P and Jha RK (2017) A survey on green communication and security challenges in 5G wireless communication networks. *Journal of Network and Computer Applications* **96**: 39–61.

García MT and Montané-Jiménez LG (2020) Visualization to support decision-making in cities: advances, technology, challenges, and opportunities. In *2020 8th International Conference in Software Engineering Research and Innovation (CONISOFT)*, pp. 198–207. 10.1109/CONISOFT50191.2020.00037

Gardner N and Hespanhol L (2018) SMLXL: scaling the smart city, from metropolis to individual. *City, Culture and Society* **12**: 54–61. 10.1016/j.ccs.2017.06.006

Gartner, Inc (2019) Gartner survey reveals digital twins are entering mainstream use. *Gartner*. https://www.gartner.com/en/newsroom/press-releases/2019-02-20-gartner-survey-reveals-digital-twins-are-entering-mai [Accessed 1 Sep. 2024].

Gelernter DH (1991) *Mirror Worlds, or, The Day Software Puts the Universe in a Shoebox—: How It Will Happen and What It Will Mean*. Oxford University Press, Oxford, UK. https://books.google.co.uk/books?id=PmCoPwAACAAJ (accessed 1/09/2024).

Ghaemi MS, Agard B, Trépanier M and Partovi Nia V (2017) A visual segmentation method for temporal smart card data. *Transportmetrica A: Transport Science* **13(5)**: 381–404. 10.1080/23249935.2016.1273273

Giddens A (1984) *The Constitution of Society: Outline of the Theory of Structuration*. Polity, Cambridge, UK.

Giddens A (1990) Structuration theory and sociological analysis. In *Consensus and Controversy*, pp. 297–315.

Glaessgen E H and Stargel D S (2012). *The Digital Twin Paradigm for Future NASA and U.S. Air Force Vehicles*. 53rd AIAA/ASME/ASCE/AHS/ASC Structures, Structural Dynamics and Materials Conference - Special Session on the Digital Twin, Honolulu, HI. https://ntrs.nasa.gov/citations/20120008178

Grene M, (1974) *The Understanding of Nature: Essays in the Philosophy of Biology*. Springer Netherlands, Dordrecht. https://books.google.co.uk/books?id=ngT5fVPH0QIC (accessed 1/09/2024).

Grieves M and Vickers J (2017). Digital twin: mitigating unpredictable, undesirable emergent behavior in complex systems. In *Transdisciplinary Perspectives on Complex Systems: New Findings and Approaches* (Kahlen F-J, Flumerfelt S and Alves A (eds)). Springer International Publishing, Cham, pp. 85–113. 10.1007/978-3-319-38756-7_4

Grieves M (2005) Product lifecycle management: the new paradigm for enterprises. *International Journal of Product Development* **2(1–2)**: 71–84.

Grieves M (2006) *Product Lifecycle Management: Driving the Next Generation of Lean Thinking*. McGraw-Hill, New York.

Grieves M (2011) *Virtually Perfect: Driving Innovative and Lean Products Through Product Lifecycle Management*. Space Coast Press.

Guarino N (1998) *Formal Ontology in Information Systems: Proceedings of the First International Conference (FOIS'98), June 6-8, Trento, Italy*. IOS Press, Amsterdam.

Habermas J (1978) *Knowledge and Human Interests*. Heinemann, London.

Haklay M, Jankowski P and Zwoliński Z (2018) Selected modern methods and tools for public participation in urban planning – A review. *Quaestiones Geographicae* **37(3)**: 127–149. 10.2478/quageo-2018-0030.

Hasegawa Y, Sekimoto Y, Seto T, Fukushima Y and Maeda M (2019) My City Forecast: urban planning communication tool for citizen with national open data. *Computers, Environment and Urban Systems* **77**: 101255. 10.1016/j.compenvurbsys.2018.06.001

Hastak M and Koo C (2017) Theory of an intelligent planning unit for the complex built environment. *Journal of Management in Engineering* **33(3)**: 04016046. 10.1061/(ASCE)ME.1943-5479.0000486

Healy M and Perry C (2000) Comprehensive criteria to judge validity and reliability of qualitative research within the realism paradigm. *Qualitative Market Research: An International Journal* **3(3)**: 118–126. 10.1108/13522750010333861

Hetherington J and West M (2020) The pathway towards an Information Management Framework—A 'Commons' for Digital Built Britain [Report]. CDBB. 10.17863/CAM.52659

Hicks B (2019) Industry 4.0 and digital twins: key lessons from NASA. *The Future Factory®: Business Transformation Training*. https://www.thefuturefactory.com/blog/24 (accessed 1/09/2024).

Hirschheim R and Klein HK (1989) Four paradigms of information systems development. *Communications of the ACM* **32(10)**: 1199–1216.

Hochhalter J, Leser WP, Newman JA *et al.* (2014). *Coupling Damage-Sensing Particles to the Digital Twin Concept* (pp. 1–9). NASA Technical Reports Server.

Honarvar AR and Sami A (2019) Towards sustainable smart city by particulate matter prediction using urban big data, excluding expensive air pollution infrastructures. *Big Data Research* **17**: 56–65. 10.1016/j.bdr.2018.05.006

Hu C, Fan W, Zeng E *et al.* (2021) Digital twin-assisted real-time traffic data prediction method for 5G-enabled internet of vehicles. *IEEE Transactions on Industrial Informatics* **18(4)**: 2811–2819.

Ibrahim MR, Haworth J and Cheng T (2020) Understanding cities with machine eyes: a review of deep computer vision in urban analytics. *Cities* **96**: 102481. 10.1016/j.cities.2019.102481

Ishwarappa and Anuradha J (2015) A brief introduction on big data 5Vs characteristics and Hadoop technology. *Procedia Computer Science* **48**: 319–324. 10.1016/j.procs.2015.04.188

ISO (2020) *ISO 37155-1: (2020). Framework for Integration and Operation of Smart Community Infrastructures. Part 1: Recommendations for Considering Opportunities and Challenges from Interactions in Smart Community Infrastructures from Relevant Aspects Through the Life Cycle*. International Organization for Standardization. https://www.iso.org/standard/69241.html (accessed 1/09/2024).

Jackson MC and Keys P (1984) Towards a system of systems methodologies. *Journal of the Operational Research Society* **35(6)**: 473–486. 10.1057/jors.1984.101

Jackson MC (2019) *Critical Systems Thinking and the Management of Complexity*. Wiley. https://books.google.co.uk/books?id=1jmNDwAAQBAJ (accessed 1/09/2024).

Jadli A and Hain M (2020) Toward a deep smart waste management system based on pattern recognition and transfer learning. In *2020 3rd International Conference on Advanced Communication Technologies and Networking (CommNet)*, pp.1–5. 10.1109/CommNet49926.2020.9199615

Jiang H (2021) Smart urban governance in the 'smart' era: why is it urgently needed? *Cities* **111**: 103004.

Johnson P and Duberley J (2000) *Understanding Management Research*. Sage, Thousand Oaks.

Kamel Boulos MN, Tsouros AD and Holopainen A (2015) 'Social, innovative and smart cities are happy and resilient': insights from the WHO EURO 2014 International Healthy Cities Conference. *International Journal of Health Geographics* **14(1)**: 1–9.

Kearns A and Paddison R (2000) New challenges for urban governance. *Urban Studies* **37(5–6)**: 845–850.

Kemp J (1968) *The Philosophy of Kant*. Oxford University Press, Oxford.

Kennedy H and Hill RL (2018) The feeling of numbers: emotions in everyday engagements with data and their visualisation. *Sociology* **52(4)**: 830–848.

Kent L, Snider C and Hicks B (2019) Early stage digital-physical twinning to engage citizens with city planning and design. In *2019 IEEE Conference on Virtual Reality and 3D User Interfaces (VR)*: 1014–1015. 10.1109/VR.2019.8798250

Khan M, Wu Q, Yan S and Peng J (2022) Data-driven urban management and its impact on upgradation of slums in Islamabad, Pakistan: mediating role of privacy and security concerns. *Journal of Urban Planning and Development* **148(2)**: 04022009.

Kim Y-L (2020) Data-driven approach to characterize urban vitality: how spatiotemporal context dynamically defines Seoul's nighttime. *International Journal of Geographical Information Science* **34(6)**: 1235–1256. 10.1080/13658816.2019.1694680

Kırdar G and Ardıç Sİ (2020) A design proposal of integrated smart mobility application for travel behavior change towards sustainable mobility. *Civil Engineering and Architecture* **8(5)**: 1095–1106. 10.13189/cea.2020.080536

Kitchin R (2014) Big data, new epistemologies and paradigm shifts. *Big Data & Society*, April-June 2014: 1–12. 10.1177/2053951714528481

Kontokosta CE and Hong B (2021) Bias in smart city governance: how socio-spatial disparities in 311 complaint behavior impact the fairness of data-driven decisions. *Sustainable Cities and Society* **64**: 102503.

Kontokosta CE (2017) Urban informatics for social good: definitions, tensions, and challenges. In *Proceedings of the 2nd International Workshop on Science of Smart City Operations and Platforms Engineering*, pp. 52–56. 10.1145/3063386.3064888

Kontokosta CE (2021) Urban informatics in the science and practice of planning. *Journal of Planning Education and Research* **41(4)**: 382–395. 10.1177/0739456X18793716

Kourtit K and Nijkamp P (2018) Big data dashboards as smart decision support tools for i-cities – An experiment on Stockholm. *Land Use Policy* **71**: 24–35. 10.1016/j.landusepol.2017.10.019

Kovacs-Györi A, Ristea A, Havas C *et al.* (2020) Opportunities and challenges of geospatial analysis for promoting urban livability in the era of big data and machine learning. *ISPRS International Journal of Geo-Information* **9(12)**: Article 12. 10.3390/ijgi9120752

Kreider RG and Messner JI (2013) *The Uses of BIM: Classifying and Selecting BIM Uses*. Version 0.9. The Pennsylvania State University, University Park, PA, USA. http://bim.psu.edu (accessed 1/09/2024).

Kritzinger W, Karner M, Traar G, Henjes J and Sihn W (2018) Digital twin in manufacturing: a categorical literature review and classification. *IFAC-PapersOnLine* **51(11)**: 1016–1022. 10.1016/j.ifacol.2018.08.474

Kurtz CF and Snowden DJ (2003) The new dynamics of strategy: sense-making in a complex and complicated world. *IBM Systems Journal* **42(3)**: 462–483. IBM, UK. 10.1147/sj.423.0462

Lai Y and Kontokosta CE (2018) Quantifying place: analyzing the drivers of pedestrian activity in dense urban environments. *Landscape and Urban Planning* **180**: 166–178. 10.1016/j.landurbplan.2018.08.018

Lane DC (1999) Social theory and system dynamics practice. *European Journal of Operational Research* **113(3)**: 501–527.

Leleux C and Webster W (2018) Delivering smart governance in a future city: the case of Glasgow. *Media and Communication* **6(4)**: 163–174. 10.17645/mac.v6i4.1639

Leorke D (2020) Reappropriating, reconfiguring and augmenting the smart city through play. In *Making Smart Cities More Playable* (Nijholt A (ed.)). Springer, Cham, pp. 51–70.

Li M, Ye X, Zhang S, Tang X and Shen Z (2018) A framework of comparative urban trajectory analysis. *Environment and Planning B: Urban Analytics and City Science* **45(3)**: 489–507. 10.1177/2399808317710023

Lieske SN, Leao SZ, Conrow L and Pettit C (2021) Assessing geographical representativeness of crowdsourced urban mobility data: an empirical investigation of Australian bicycling. *Environment and Planning B: Urban Analytics and City Science* **48(4)**: 775–792. 10.1177/2399808319894334

Lin Y, Zhou Y, Lin M, Wu S and Li B (2021) Exploring the disparities in park accessibility through mobile phone data: evidence from Fuzhou of China. *Journal of Environmental Management* **281**: 111849.

Lock O, Bednarz T and Pettit C (2019) HoloCity – Exploring the use of augmented reality cityscapes for collaborative understanding of high-volume urban sensor data. In *The 17th International Conference on Virtual-Reality Continuum and Its Applications in Industry*: 1-2. 10.1145/3359997.3365734

Lu X, Ota K, Dong M, Yu C and Jin H (2017) Predicting transportation carbon emission with urban big data. *IEEE Transactions on Sustainable Computing* **2(4)**: 333–344. 10.1109/TSUSC.2017.2728805

Lwin KK, Sekimoto Y and Takeuchi W (2018) Estimation of hourly link population and flow directions from mobile CDR. *ISPRS International Journal of Geo-Information* **7(11)**: Article 11. 10.3390/ijgi7110449

Ma M, Preum SM, Ahmed MY *et al.* (2019) Data sets, modeling, and decision making in smart cities: a survey. *ACM Transactions on Cyber-Physical Systems* **4(2)**: 14:1–14:28. 10.1145/3355283

Mamta and Nagpal CK (2018) Urban computing: key challenges and issues of traffic management system. *International Journal of Computer Applications* **179(26)**: 18–21.

Maskrey S, Vilcan T, O'Donnell E and Lamond J (2020) Using Learning and Action Alliances to build capacity for local flood risk management. *Environmental Science & Policy* **107**: 198–205.

Mattingly M (1994) Meaning of urban management. *Cities* **11(3)**: 201–205.

Mavrokapnidis D, Mohammadi N and Taylor J (2021) Community dynamics in smart city digital twins: a computer vision-based approach for monitoring and forecasting collective urban hazard exposure. In *Proceedings of the 54th Hawaii International Conference on System Sciences*, pp. 1810–1818. http://hdl.handle.net/10125/70832 (accessed 1/09/2024).

Mayaud JR, Anderson S, Tran M and Radić V (2019) Insights from self-organizing maps for predicting accessibility demand for healthcare infrastructure. *Urban Science* **3(1)**: Article 1. 10.3390/urbansci3010033

McGill R (1998) Urban management in developing countries. *Cities* **15(6)**: 463–471.

McGill R (2020) Urban resilience–An urban management perspective. *Journal of Urban Management* **9(3)**: 372–381.

McHugh A and Thakuriah V (2018) Developing smart statistics for urban mobility: challenges and opportunities. https://dgins2018.statisticsevents.ro/wp-content/uploads/2018/10/23-Developing-smart-statistics-for-urban-mobility.pdf (accessed 1/09/2024).

McMeekin N, Wu O, Germeni E and Briggs A (2020) How methodological frameworks are being developed: evidence from a scoping review. *BMC Medical Research Methodology* **20(1)**: 173. 10.1186/s12874-020-01061-4

Michalec AO, Hayes E and Longhurst J (2019) Building smart cities, the just way. A critical review of 'smart' and 'just' initiatives in Bristol, UK. *Sustainable Cities and Society* **47(2019)**: 101510.

Miles D (2021) Managing the challenges of a post-pandemic world: cities 4.0. *Econsult Solutions, Inc*. https://econsultsolutions.com/rise-of-cities-4-0/ (accessed 1/09/2024).

Min K, Yoon M and Furuya K (2019) A comparison of a smart city's trends in urban planning before and after 2016 through keyword network analysis. *Sustainability* **11(11)**: Article 11. 10.3390/su11113155

Mingers J (1997) Towards critical pluralism. In *Multimethodology: The Theory and Practice of Combining Management Science Methodologies*. John Wiley & Sons, Chichester, UK, pp. 407–440.

Mingers J (2006) *Realising Systems Thinking: Knowledge and Action in Management Science*. Springer US, New York.

Mohammadi N and Taylor J (2020) Knowledge discovery in smart city digital twins. In *Proceedings of the 53rd Hawaii International Conference on System Sciences*, pp. 1656–1664.

Muir J (1911) *My First Summer in the Sierra*. Houghton Mifflin, Boston, MA.

Nallaperuma D, Nawaratne R, Bandaragoda T *et al*. (2019) Online incremental machine learning platform for big data-driven smart traffic management. *IEEE Transactions on Intelligent Transportation Systems* **20(12)**: 4679–4690. 10.1109/TITS.2019.2924883

National Digital Twin Programme (2021). Digital twin toolkit. https://digitaltwinhub.co.uk/files/file/62-digital-twin-toolkit/ (accessed 1/09/2024).

National Infrastructure Commission (2017) *Data for the Public Good*. National Infrastructure Commission Report.

National Infrastructure Commission (2020) *System Mapping for UK Infrastructure Systems Decision Making*. https://nic.org.uk/app/uploads/Systems-mapping-for-UK-infrastructure-systems-decision-making.pdf

Negri E, Fumagalli L and Macchi M (2017). A review of the roles of digital twin in CPS-based production systems. *Procedia Manufacturing* **11**: 939–948. 10.1016/j.promfg.2017.07.198

Nochta T, Parlikad A, Schooling J, Badstuber N and Wahby N (2019) *The Local Governance of Digital Technology – Implications for the City-Scale Digital Twin* [Report]. CDBB. 10.17863/CAM.43321

Nochta T, Wan L, Schooling JM and Parlikad AK (2021) A socio-technical perspective on urban analytics: the case of city-scale digital twins. *Journal of Urban Technology* **28(1–2)**: 263–287. 10.1080/10630732.2020.1798177

Nourian P, Ohori KA and Martinez-Ortiz C (2018) Essential means for urban computing: specification of web-based computing platforms for urban planning, a Hitchhiker's guide. *Urban Planning*, **3(1)**: 47–57.

Nübel K, Bühler MM and Jelinek T (2021) Federated digital platforms: value chain integration for sustainable infrastructure planning and delivery. *Sustainability* **13(16)**: Article 16. 10.3390/su13168996

Olaisen JL (1991) Pluralism or positivistic trivialism: important trends in contemporary philosophy of science. In *Information Systems Research: Contemporary Approaches and Emergent Traditions*. Elsevier Science Publishers, pp. 235–265.

Oléron-Evans TP and Salhab M (2021) Optimal land use allocation for the Heathrow opportunity area using multi-objective linear programming. *Land Use Policy* **105**: 105353. 10.1016/j.landusepol.2021.105353

Orellana D and Guerrero ML (2019) Exploring the influence of road network structure on the spatial behaviour of cyclists using crowdsourced data. *Environment and Planning B: Urban Analytics and City Science* **46(7)**: 1314–1330. 10.1177/2399808319863810

Panagoulia E (2017) Open data and human-based outsourcing neighborhood rating: a case study for San Francisco Bay Area gentrification rate. In *Planning Support Science for Smarter*

Urban Futures (Geertman S, Allan A, Pettit C and Stillwell J (eds.)). Springer International Publishing, Cham, pp. 317–335. 10.1007/978-3-319-57819-4_18

Panagoulia E (2020) Human-centered approaches in urban analytics and placemaking. In *Sustainability in Urban Planning and Design* (Almusaed A, Almssad A and Hong LT (eds.)). BoD – Books on Demand, pp. 149–176.

Pang J, Huang Y, Xie Z, Li J and Cai Z (2021) Collaborative city digital twin for the COVID-19 pandemic: a federated learning solution. *Tsinghua Science and Technology* **26(5)**: 759–771. 10.26599/TST.2021.9010026

Parrott A and Warshaw L (2017) Industry 4.0 and the digital twin. Manufacturing meets its match. *Deloitte Insights*. https://www2.deloitte.com/content/www/us/en/insights/focus/industry-4-0/digital-twin-technology-smart-factory.html (accessed 1/09/2024).

Pattee HH (1973) *Hierarchy Theory; The Challenge of Complex Systems*, 1st edn. G. Braziller, New York, USA.

Pavlovskaya M (2006) Theorizing with GIS: a tool for critical geographies? *Environment and Planning A: Economy and Space* **38(11)**: 2003–2020. 10.1068/a37326

Penn A and Al Sayed K (2017) Spatial information models as the backbone of smart infrastructure. *Environment and Planning B: Urban Analytics and City Science* **44(2)**: 197–203. 10.1177/2399808317693478

Pettit C, Bakelmun A, Lieske SN *et al.* (2018) Planning support systems for smart cities. *City, Culture and Society* **12**: 13–24. 10.1016/j.ccs.2017.10.002

Piascik B, Vickers J, Lowry D *et al.* (2012) Materials, structures, mechanical systems, and manufacturing roadmap. *NASA TA*, pp. 12–2.

Pluchinotta I, Pagano A, Vilcan T *et al.* (2021) A participatory system dynamics model to investigate sustainable urban water management in Ebbsfleet Garden City. *Sustainable Cities and Society* **67**: 102709.

Plunz RA, Zhou Y, Carrasco Vintimilla MI, Mckeown K, Yu T, Uguccioni L and Sutto MP (2019) Twitter sentiment in New York City parks as measure of well-being. *Landscape and Urban Planning* **189**: 235–246. 10.1016/j.landurbplan.2019.04.024

Popper K (1957) *The Poverty of Historicism*. Routledge and Kegan Paul, Abingdon, UK.

Pregnolato M, Gunner S, Voyagaki E *et al.* (2022) Towards Civil Engineering 4.0: concept, workflow and application of digital twins for existing infrastructure. *Automation in Construction* **141**: 104421. 10.1016/j.autcon.2022.104421

Rao PV, Azeez PMA, Peri SS *et al.* (2020) IoT based waste management for smart cities. In *2020 International Conference on Computer Communication and Informatics (ICCCI)*, pp. 1–5. 10.1109/ICCCI48352.2020.9104069

Resch B, Sudmanns M, Sagl G *et al.* (2015) Crowdsourcing physiological conditions and subjective emotions by coupling technical and human mobile sensors. *GI_Forum* **1**: 514–524.

Ríos J, Hernández JC, Oliva M and Mas F (2015) Product avatar as digital counterpart of a physical individual product: literature review and implications in an aircraft. In *Transdisciplinary Lifecycle Analysis of Systems* (Curran R, Verhagen WJC, Stjepandić J, Wognum N, and Borsato M (eds.)), pp. 657–666.

Rittel HW and Webber MM (1973) Dilemmas in a general theory of planning. *Policy Sciences* **4(2)**: 155–169.

Rogage K, Clear A, Alwan Z, Lawrence T and Kelly G (2019) Assessing building performance in residential buildings using BIM and sensor data. *International Journal of Building Pathology and Adaptation* **38(1)**: 176–191. 10.1108/IJBPA-01-2019-0012

Rosen R, Fischer J and Boschert S (2019). Next generation digital twin: an ecosystem for mechatronic systems? *IFAC-PapersOnLine* **52(15)**, 265–270. 10.1016/j.ifacol.2019.11.685

Saunders M, Lewis P and Thornhill A (2009) *Research Methods for Business Students*. Pearson Education.

Schleich B, Anwer N, Mathieu L and Wartzack S (2017) Shaping the digital twin for design and production engineering. *CIRP Annals* **66(1)**: 141–144. 10.1016/j.cirp.2017.04.040

Schooling J, Enzer M and Broo DG (2021) Flourishing systems: re-envisioning infrastructure as a platform for human flourishing. *Proceedings of the Institution of Civil Engineers-Smart Infrastructure and Construction* **173(1)**: 166–174.

Seuring S and Gold S (2012) Conducting content-analysis based literature reviews in supply chain management. *Supply Chain Management: An International Journal* **17(5)**: 544–555. 10.1108/13598541211258609

Shelton T, Zook M and Wiig A (2015) The 'actually existing smart city.' *Cambridge Journal of Regions, Economy and Society* **8(1)**: 13–25.

Shirowzhan S, Trinder J and Osmond P (2020) New metrics for spatial and temporal 3D urban form sustainability assessment using time series lidar point clouds and advanced GIS techniques. In *Sustainability in Urban Planning and Design* (A. Almusaed A. Almssad and L.T.-Hong (eds)). BoD – Books on Demand, pp. 53–68.

Shuster LA (2021) Social justice, equity, and infrastructure. *Civil Engineering Magazine Archive* **91(6)**: 28–37.

Sideris N, Bardis G, Voulodimos A, Miaoulis G and Ghazanfarpour D (2019) Using Random Forests on real-world city data for urban planning in a visual semantic decision support system. *Sensors* **19(10)**: Article 10. 10.3390/s19102266

Simon HA (1960) *The New Science of Management Decision* (pp. xii, 50). Harper & Brothers, New York. 10.1037/13978-000

Simon HA (1977) The organization of complex systems. In *Models of Discovery: And Other Topics in the Methods of Science* (Simon HA (ed)). Springer Netherlands, Dordrecht, pp. 245–261. 10.1007/978-94-010-9521-1_14

Simon HA (1996) *The Sciences of the Artificial*, 3rd edn, MIT Press, Cambridge, MA.

Simone C, Iandolo F, Fulco I and Loia F (2021) Rome was not built in a day. Resilience and the eternal city: insights for urban management. *Cities* **110**: 103070.

Smith V, Devane D, Begley CM and Clarke M (2011) Methodology in conducting a systematic review of systematic reviews of healthcare interventions. *BMC Medical Research Methodology* **11(1)**: 15. 10.1186/1471-2288-11-15

Smith A and Martín PP (2021) Going beyond the smart city? Implementing technopolitical platforms for urban democracy in Madrid and Barcelona. *Journal of Urban Technology* **28(1–2)**: 311–330.

Söderberg R, Wärmefjord K, Carlson J S and Lindkvist L. (2017) Toward a Digital Twin for real-time geometry assurance in individualized production. *CIRP Annals* ***66*(1)**: 137–140. 10.1016/j.cirp.2017.04.038

Stefik M (1981) Planning with constraints (MOLGEN: Part 1). *Artificial Intelligence* **16(2)**: 111–139. 10.1016/0004-3702(81)90007-2

Stren R (1993) 'Urban management' in development assistance: an elusive concept. *Cities* **10(2)**: 125–138.

Sunim H (2018) *The Things You Can See Only When You Slow Down: How to Be Calm in a Busy World* (C.-Y. Kim, Trans.; 1st edn). Penguin Life.

Tang L, Gao J, Ren C *et al.* (2019) Detecting and evaluating urban clusters with spatiotemporal big data. *Sensors* **19(3)**: Article 3. 10.3390/s19030461

Tao F, Cheng J, Qi Q *et al.* (2018) Digital twin-driven product design, manufacturing and service with big data. *The International Journal of Advanced Manufacturing Technology* **94(9)**: 3563–3576. 10.1007/s00170-017-0233-1

Tao F, Qi Q, Wang L and Nee AYC (2019) Digital twins and cyber–physical systems toward smart manufacturing and Industry 4.0: correlation and comparison. *Engineering* **5(4)**: 653–661. 10.1016/j.eng.2019.01.014

Thakuriah PV, Tilahun NY and Zellner M (eds.) (2017) Big data and urban informatics: innovations and challenges to urban planning and knowledge discovery. In *Seeing Cities Through Big Data*. Springer, Cham, pp.11–45.

The Institution of Engineering and Technology (2019) Digital twins for the built environment. An introduction to the opportunities, benefits, challenges and risks. https://www.theiet.org/media/8762/digital-twins-for-the-built-environment.pdf (accessed 1/09/ 2024).

Tomko M and Winter S (2019) Beyond digital twins–A commentary. *Environment and Planning B: Urban Analytics and City Science* **46(2)**: 395–399.

Townsend AM (2013) *Smart Cities: Big Data, Civic Hackers, and the Quest for a New Utopia*. WW Norton & Company.

Un-Habitat (2012) *Enhancing Urban Safety and Security: Global Report on Human Settlements 2007*. Routledge.

Urban Innovation Labs (2019) *Integration and Optimisation of Services Embedded in the Built Environment* [Report]. CDBB. 10.17863/CAM.40459

Vachálek J, Bartalský L, Rovný O *et al.* (2017) The digital twin of an industrial production line within the Industry 4.0 concept. In *2017 21st International Conference on Process Control (PC)*, pp. 258–262. 10.1109/PC.2017.7976223

Varela FJ (1981) Autonomy and autopoiesis. In *Self-Organizing Systems: An Interdisciplinary Approach*. (Roth G and Schwegler H (eds.)) Campus Verlag, pp.14–23.

Verrest H and Pfeffer K (2019) Elaborating the urbanism in smart urbanism: distilling relevant dimensions for a comprehensive analysis of Smart City approaches. *Information, Communication & Society* **22(9)**: 1328–1342.

Wahba S (2021) *Integrating Infrastructure in U.S. Domestic & Foreign Policy: Lessons from China*. Wilson Center. https://www.wilsoncenter.org/article/integrating-infrastructure-us-domestic-and-foreign-policy-lessons-china (accessed 11/07/2021).

Walsham G (1995) Interpretive case studies in IS research: nature and method. *European Journal of Information Systems* **4(2)**: 74–81.

Wang F-Y (2010) The emergence of intelligent enterprises: from CPS to CPSS. *IEEE Intelligent Systems* **25(4)**: 85–88.

Wang Y, Zhang W, Zhang F *et al.* (2020) Analysis of subway passenger flow based on smart card data. In: *2020 6th International Conference on Big Data Computing and Communications (BIGCOM)*, pp.198–202. 10.1109/BigCom51056.2020.00034

Warfield JN (2002) *Understanding Complexity: Thought and Behavior*. AJAR.

Werna E (1995) The management of urban development, or the development of urban management? Problems and premises of an elusive concept. *Cities* **12(5)**: 353–359.

White G, Zink A, Codecá L and Clarke S (2021) A digital twin smart city for citizen feedback. *Cities* **110**: 103064. 10.1016/j.cities.2020.103064

Whyte J, Coca D, Fitzgerald J *et al.* (2019) *Analysing Systems Interdependencies Using a Digital Twin* [Report]. CDBB. 10.17863/CAM.43314

Witteborg A (2021) Digital twins for wastewater infrastructure. *Royal HaskoningDHV Digital*. https://global.royalhaskoningdhv.com/digital/blogs-and-news/publications/digital-twins-for-wastewater-infrastructure (accessed 1/09/2024).

World Bank (1991) *Urban Policy and Economic Development: An Agenda for the 1990s*. The World Bank. https://elibrary.worldbank.org/doi/abs/10.1596/0-8213-1816-0

Xia T, Song X, Zhang H *et al.* (2019) Measuring spatio-temporal accessibility to emergency medical services through big GPS data. *Health & Place* **56**: 53–62.

Xie Y, Gupta J, Li Y and Shekhar S (2018) Transforming smart cities with spatial computing. In *2018 IEEE International Smart Cities Conference (ISC2)*, pp. 1–9. 10.1109/ISC2.2018.8656800

Yabe T and Ukkusuri SV (2019) Integrating information from heterogeneous networks on social media to predict post-disaster returning behavior. *Journal of Computational Science* **32**: 12–20. 10.1016/j.jocs.2019.02.002

Ye C, Butler L, Calka B *et al.* (2019) *A Digital Twin of Bridges for Structural Health Monitoring*. DEStech Publications, Inc, Lancaster. 10.17863/CAM.63903

Yigitcanlar T, Kamruzzaman M, Buys L *et al.* (2018) Understanding 'smart cities': intertwining development drivers with desired outcomes in a multidimensional framework. *Cities* **81**: 145–160.

Zheng Y, Capra L, Wolfson O and Yang H (2014) Urban computing: concepts, methodologies, and applications. *ACM Transactions on Intelligent Systems and Technology (TIST)* **5(3)**: 1–55.

Zhu Z (2011) After paradigm: why mixing-methodology theorising fails and how to make it work again. *Journal of the Operational Research Society* **62(4)**: 784–798.

Ramy Elsehrawy and Bimal Kumar
ISBN 978-1-83662-235-2
https://doi.org/10.1108/978-1-83662-234-520251006
Copyright © 2025 by Ramy Elsehrawy and Bimal Kumar. Published under exclusive licence by Emerald Publishing Limited

Chapter 4
Designed amalgam of twinning for urban management (DATUM): the way forward

4.1. Introduction

To reiterate the key points discussed in the previous chapters: designing an effective artefact requires three essential elements

- a thorough understanding of the problem situation and the current state
- a clear vision of a desirable target state
- a well-defined path that bridges the present to the envisioned future.

The first requirement was addressed in Chapter 3 through a comprehensive systematic literature review. The focus of this chapter is on fulfilling the remaining two requirements. To envision a desirable target state for the discipline of UM and to provide well-grounded recommendations for its progression, Section 4.5 analyses the evolution of UM over time using a three-layered perspective, which is introduced in Sections 4.2 to 4.4.

Subsequently, in Section 4.6, a conceptual design of the theoretical artefact, the designed amalgam of twinning for urban management (DATUM), is proposed as a means of achieving the desirable target state. It is argued that transforming the discipline from its current state, as outlined in Chapter 3, to the envisioned state described in this chapter, can be accomplished through the implementation of the DATUM and its proposed design. Finally, the chapter presents the evaluation criteria that will be used to assess the effectiveness of the DATUM.

4.2. Three-layered perspective

This research presents a three-layered framework for understanding the evolution of UM and provides recommendations for how the discipline should advance. This layered approach is illustrated in Figure 4.1 and explained below, moving from the outermost layer (third) to the innermost layer (first).

(a) 3rd Layer: The outermost layer focuses on understanding the evolution of UM research and practice through a 'Kuhnian' lens. As discussed in Section 3.3 and supported by other authors (Ding and Lai, 2012), UM is viewed as a scientific discipline. Consequently, any claims regarding the emergence of a new paradigm for UM, such as *DT for UM*, must be grounded in a theoretical understanding of how scientific disciplines evolve and how new paradigms emerge or perish. This layer draws from Thomas Kuhn's (1970) conceptualisation of scientific revolutions, providing a theoretical basis for recommendations on how the UM discipline can grow and how new paradigms can mature and develop. Section 4.4 provides an overview of Kuhn's theory on the structure of scientific revolutions.

Figure 4.1 Three-layered perspective of the evolution of UM (author's own)

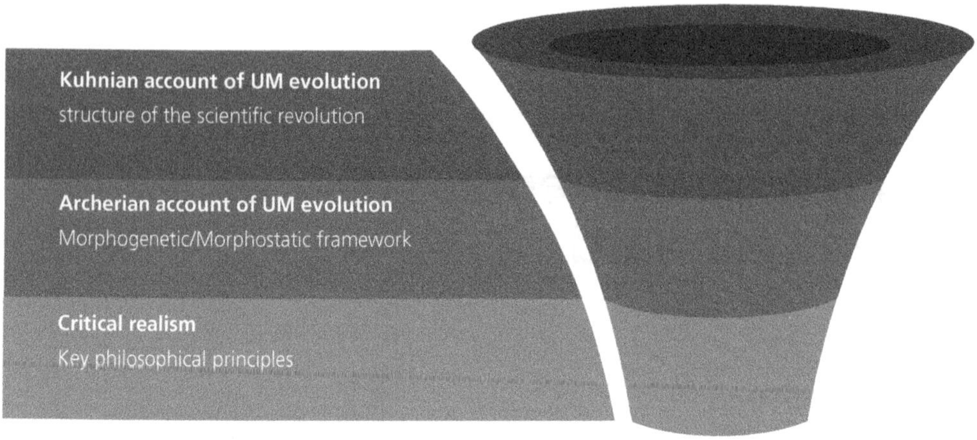

(b) 2nd Layer: UM is considered an 'applied science' (Ding and Lai, 2012), and since science is not merely a collection of knowledge but a social activity (Sayer, 1992), it is appropriate to incorporate social sciences into the analysis of UM's evolution. The second layer of this framework draws on the morphogenetic/morphostatic [M/M] framework developed by Margaret Archer (1995). The M/M framework analyses the relationship and interaction between humans and the reality they encounter, particularly the interplay between 'socio-cultural interactions' (the actions of UM discipline members) and the 'cultural system' (the collection of theories, beliefs and ideologies that constitute UM) over time. Section 4.3 provides a brief explanation of the M/M framework.

(c) 1st Layer: When integrating social theories, it is important to address the philosophical concerns that arise, particularly those highlighted by philosophers such as Giddens (1984). The involvement of social sciences inevitably brings philosophical issues to the forefront, as described by Latour *et al.* (2011: p. 41).

> *'I'm like a dog following its prey, and then the prey arrive in the middle of a band of wolves which are called professional philosophers ... My intention was not to fall in with the wolves and to have to answer all of these guys while trying to catch my prey.'*

This deepest layer of this framework is rooted in the philosophy of CR, which underpins Archer's M/M approach. This layer forms the philosophical foundation of the three-layered lens. Archer's development of CR, as presented in her 1995 work, is crucial to this foundational layer.

4.3. Layer 2: The morphogenetic/morphostatic framework

4.3.1 Time factor

The concept of analytical dualism, as explained in Section 2.2.6, forms the foundation of the M/M framework. Analytical dualism is central to Archer's critique of the three approaches that conflate structure and agency, as discussed in Section 2.2.6. The common flaw in these approaches, and the primary reason Archer critiques them, is that they obscure the examination of the dynamic relationship between social structure and agency. By conflating the two, there is no space to analyse how agency interacts with the social context in which it is embedded. However, by treating structure and agency as ontologically distinct, the M/M framework enables the exploration of their interplay. To facilitate this analysis, the M/M framework introduces the dimension of time, recognising its critical role in the emergence of events from the interaction of generative mechanisms. Archer (1995: p. 14) emphasises that emergence 'takes time since it derives from interaction and its consequences which necessarily occur in time.' Acknowledging that agents' actions and interactions unfold over time (Næss, 2004), and subsequently give rise to emergent powers that influence the structure, necessitates the incorporation of time into the analysis. Only by considering time can we adequately examine the interplay between agents and their contextual conditions as it evolves over time.

4.3.2 The morphogenetic/morphostatic cycle

Archer illustrates the M/M framework using the diagram in Figure 4.2. The M/M framework is a non-conflationary approach that operationalises the concept of analytical dualism. This approach involves a cycle comprising three analytical phases. The first phase, 'structural conditioning,' occurs at time (T1). Since humans can neither reproduce nor transform something that does not already exist, the social structure must pre-exist, forming the context they encounter. This structure conditions and influences subsequent human actions in a non-deterministic manner. The second phase unfolds during the time interval (T2–T3), during which agents, conditioned by the circumstances they face, act and interact. In this period, agents are simultaneously influenced by their agential powers and the ideologies they hold. The final phase occurs at time (T4), where human activities and social interactions either reproduce the initial conditions (morphostasis) or alter them, leading to a transformation of the social structure (morphogenesis). In both scenarios, the outcome becomes the context for the next cycle.

4.3.3 The morphogenetic/morphostatic framework applied to culture

As detailed in Section 2.2.6, Archer extends the M/M framework beyond the interplay of social structure and agency to also encompass the relationship between the CS and S-C. Archer's approach begins by challenging two key assumptions that contribute to what she describes as the myth of cultural integration. The first assumption she disputes is the idea that a culture is inherently homogeneous, consisting of a set of internally coherent and integrated ideas. The second assumption is that the members of a particular culture uniformly share and embrace these coherent ideational projects and practices. Archer instead introduces two key acknowledgments that challenge these assumptions.

Figure 4.2 The morphogenetic/morphostatic framework. Source: Archer (1995)

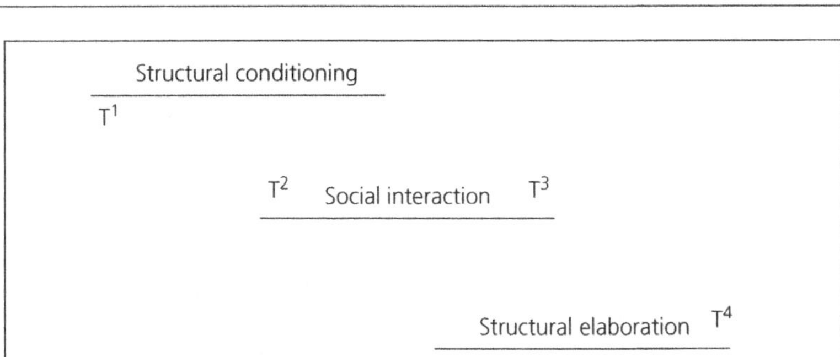

- Cultural heterogeneity: A culture does not necessarily consist of a set of integrated ideas; rather, it can plausibly include contradictory and conflicting ideational projects.
- Diverse endorsement of ideas: Members of a culture might not equally share or endorse the ideas or beliefs that the culture comprises.

These acknowledgments have significant implications for the application of the M/M framework to cultural analysis.

In Archer's framework, the CS is understood as a structure that exists ontologically, independent of any individual knower. It encompasses a body of theories and ideas, composed of various propositions that are related in ways that may be either complementary or contradictory. The nature of these relationships within the CS gives rise to specific causal powers at the cultural systemic level. These powers influence S-C interactions which occur independently at a different ontological stratum of reality.

In essence, the way in which ideas are related within the CS determines the 'situational logic' in which individuals who adopt these ideas find themselves. This situational logic conditions their actions and interactions within the socio-cultural domain, highlighting the complex and non-linear relationship between cultural ideas and social practices. In Archer's (1996: p. 145) words:

> *'the maintenance of ideas [at the C.S. level] which stand in manifest logical contradiction or complementarity to others, places their holders [at the S-C level] in different ideational positions. The logical properties of their theories or beliefs create entirely different situational logics for them'.*

4.4. Layer 3: The structure of scientific revolution

UM, recognised as an 'applied science' (Ding and Lai, 2012: p. 1), can be understood through the lens of Thomas Kuhn's theory of scientific revolutions. In his influential book *The Structure of Scientific Revolutions* (Kuhn, 1970), Kuhn provides a compelling framework for analysing scientific development. His ideas have profoundly influenced research across various disciplines and offer a useful perspective for examining the evolution of UM.

Kuhn describes scientific progress as a cycle of successive paradigms, which is illustrated in Figure 4.3. To fully grasp the significance of this cycle in relation to UM, it is important first to understand what Kuhn means by 'paradigm,' a central concept in both his theory and its application here. Kuhn acknowledges the challenges in providing a precise definition of 'paradigm.' In a postscript to his 1970 work, he elaborates on this concept, defining a paradigm as a 'constellation of beliefs, values, techniques, and so on shared by the members of a given community' (Kuhn, 1970: p. 175). This concept is essential to Kuhn's view of scientific revolutions. Kuhn refers to a paradigm as a 'disciplinary matrix,' where 'matrix' denotes the interconnected and organised nature of the elements within it. The term 'disciplinary' highlights that these elements – beliefs, values, techniques – are collectively held by a specific group or community.

4.4.1 Stage 1: Normal paradigm

In the first stage of scientific development, the 'normal paradigm' represents a well-established phase in which the discipline operates under a dominant framework (Kuhn, 1970: p. 65). This paradigm not only defines the scientific problems to be addressed but also establishes the rules for solving them, guiding researchers towards in-depth exploration of specific areas. While this focus may narrow the scope of problems investigated, it also ensures the achievement of precise and detailed insights, which, as Kuhn notes, would be impossible through random or unfocused inquiry.

Figure 4.3 Structure of scientific revolution. Adapted from Kuhn (1970)

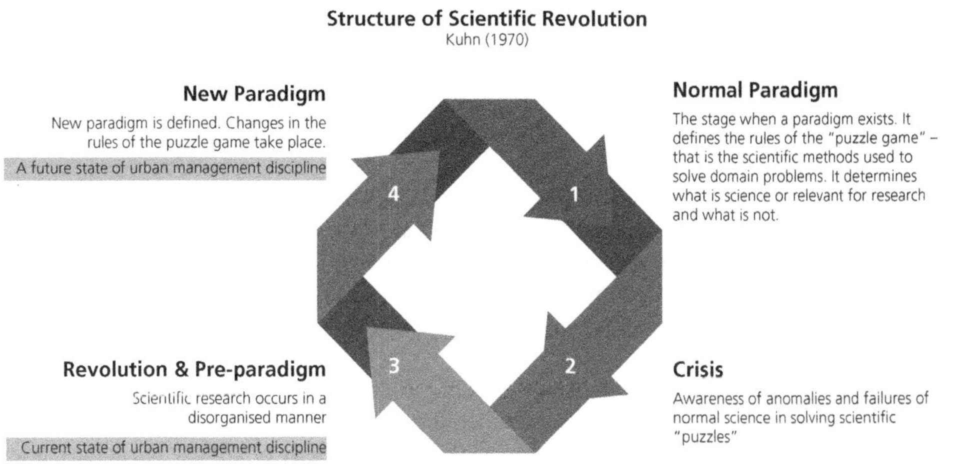

4.4.2 Stage 2: Crisis

However, this concentrated approach inevitably leads to limitations. While certain relationships between variables are clarified, others remain neglected or obscured. Phenomena that do not fit into the conceptual 'box' the *normal paradigm* supplies 'are often not seen at all' (Kuhn, 1970: p. 24). Over time, as anomalies – instances where the real world fails to conform to the paradigm's expectations – become more apparent, the inadequacies of the paradigm are exposed. These unresolved anomalies create a growing sense of dissatisfaction within the scientific community. They start to question the structure of the normal paradigm's conceptual 'box' instead of trying to assimilate the persisting anomalies into it. Eventually, 'deep debates over legitimate methods, problems and standards of solution' (Kuhn, 1970: p. 48) take place, followed by an 'expression of explicit discontent' (Kuhn, 1970: p. 91), and a loss of confidence in the normal paradigm, ultimately leading to a period of 'crisis'.

4.4.3 Stage 3: Revolution and pre-paradigm

As the crisis deepens, so do debates about the validity of the prevailing paradigm. The discipline begins to question its foundational assumptions, re-evaluating the theories that underpin its framework. When a paradigm is no longer able to accommodate the anomalies it encounters, it may be replaced by a new conceptual structure. This period of revolution allows for a significant overhaul of the existing worldview, leading to the resolution of the most pressing issues that the old paradigm could not address. However, in the interim, before a new paradigm emerges, the discipline enters a *pre-paradigm* phase characterised by fragmentation and disorganisation. This is followed by a 'consequent conflict between competing schools of scientific thought' (Kuhn, 1970: p. 96), indicating what Kuhn recognises as an 'immature science'. Under these conditions, little opportunity for collective progress is offered.

4.4.4 Stage 4: New paradigm

The final stage occurs when one of the competing ideas gains traction, leading to the formation of a new paradigm. This breakthrough fosters widespread consensus, and the discipline moves forward with renewed focus and coherence. The establishment of a new normal paradigm is crucial for continued scientific advancement. According to Kuhn, significant progress can only be made when the majority of the discipline aligns itself with a single, accepted framework, allowing the field to mature and evolve effectively (Kuhn, 1970).

4.5. The evolution of urban management

Viewing UM through a three-layered lens (Figure 4.1) allows for a synthesis of Archer's and Kuhn's theories. This integration has the following implications.

First, from the perspective of Archer's M/M framework, a discipline like UM consists of two core elements. On the one hand, there is the CS that encompasses the entire body of ideas, theories and knowledge related to the discipline, existing independently of whether individuals are aware of or interpret it. On the other hand, UM involves S-C interactions that reflect the activities, ideational projects and motivations of its practitioners and researchers.

By adopting the M/M approach, it becomes possible to disentangle UM's cultural system from its S-C interactions, thereby allowing for a detailed analysis of the relationship between the two. This approach facilitates an examination of how UM's evolving body of knowledge at the CS level interacts with the activities, motivations, and contributions of the discipline's members at the S-C level, offering insights into the evolution of UM as a discipline.

A key strength of the M/M framework is its ability to operate across different temporal scales. As Archer points out, 'analytical dualism can be used by any researcher to gain theoretical purchase on much smaller problems where the major difficulty of seeing the wood from the trees becomes much more tractable if they can be sorted out into the components of temporal cycles of morphogenesis – however short the time-span involved may be' (Archer, 1996: p. 228). This flexibility allows for an exploration of how the interplay between these components operates during specific stages of Kuhn's scientific revolution – such as the normal paradigm – as well as throughout the entire revolutionary cycle, from the established paradigm to the emergence of a new one.

In the analysis that follows, the CS/S-C interplay is examined on two distinct temporal scales, revealing two different M/M cycles. The first, a short-term cycle, leads to the reproduction of the existing UM paradigm, reflecting a morphostatic process. The second, a longer-term cycle, culminates in a paradigm shift, signifying a morphogenetic process that ultimately gives rise to a new paradigm.

4.5.1 M/M short-span cycle

4.5.1.1 Cultural conditioning (time: T1)

The cycle under examination here, focusing on the short-term perspective (Figure 4.4), begins with an established CS that consists of a highly coherent and consistent body of ideas, theories, beliefs and doctrines. This body of knowledge forms an interconnected web of what Archer (2005) refers to as 'concomitant compatibilities', where ideas or theories are both necessarily related and logically aligned, reinforcing and supporting one another. The result is a durable and deeply ingrained body of knowledge that underpins the normal paradigm of the discipline.

While a comprehensive historical review of the CS associated with UM's *normal paradigm* is beyond the scope of this discussion[1], it is essential to highlight two prominent themes that characterised UM

[1]McGill (1998) provides a brief yet panoramic view of the *normal paradigm* of UM.

Figure 4.4 Short-span morphostatic cycle during normal paradigm stage (author's own)

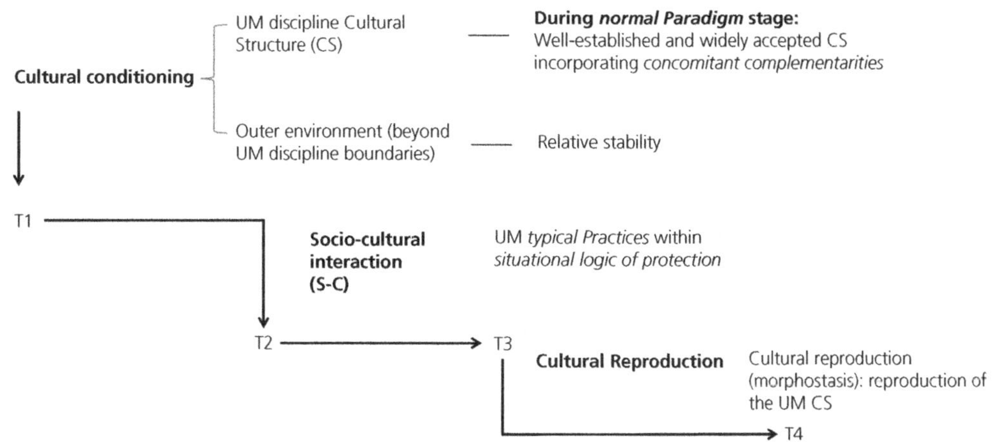

debates during this period. First, there was significant focus on the roles of government institutions and their ability to empower officials to make decisions regarding the allocation of urban resources (Leonard, 1982; Williams, 1978). Second, the era saw a strong emphasis on a mono-dimensional infrastructure approach, marked by large-scale engineering projects. This trend was later attributed to the siloed sectoral structure of governmental organisations (Baker, 1989).

4.5.1.2 Socio-cultural interactions (time: T2–T3)

The CS of a discipline exerts both enabling and constraining powers. During the period between T2 and T3, these influences, attributed to the CS, impact the discipline's members in a conditional manner, rather than deterministically. These powers shape the worldviews that members adopt, the theories they seek to defend and the practices they choose to implement. In the context of UM's well-established normal paradigm, described at T1 of the short-span cycle, the CS creates what Archer (2005) calls a 'protective insulation', where individuals find themselves in a 'situational logic of protection.' In this state, ideas and projects that reinforce the dominant paradigm are welcomed, while 'typical practices' – those regarded as standard, natural, and widely accepted across the discipline – are encouraged[2]. This environment also explains why normal paradigms often 'suppress fundamental novelties,' as these are inherently disruptive to the prevailing system's core principles (Kuhn, 1970, p.5).

From a critical realist perspective, the situational logic of protection can be viewed as a closed system. According to CR, a system exhibits stable behaviour when it meets both intrinsic and extrinsic closure conditions (Bhaskar, 1975) (more on conditions of closure in Section 6.2.3.1.). The intrinsic condition relates to maintaining relationships between the system's internal components, while the extrinsic condition is fulfilled when the system's interactions with its external environment

[2] *Typical practices* may involve adjustments to performance that might be relatively unique at a finer scale or a shorter spans (Porpora, 2015). In context of UM, minor modifications to CS may help endow it with flexibility, enhance its practical adequacy and thus, extend its life. Modifications in the definition of the concept of UM (e.g. McGill, 1998) is an example of such adjustments. Nonetheless, it is seldom the case that any of these alterations would cause any significant disruption to the general form of the existing CS, the *normal paradigm* it constitutes, or the typical practices concomitant to it.

remain unchanged. When both conditions are satisfied, the situational logic of protection prevails. This state enables the system to 'brook no rivals from outside and repress rivalry inside' (Archer, 2005, p.28) at the socio-cultural level.

4.5.1.3 Cultural reproduction (T4)

Conditioned S-C interactions often lead to the reproduction of the CS, as typical practices contribute to the 'systematisation' of the discipline, further stabilising the pre-existing normal paradigm and reinforcing its internal structure (Archer, 2005). This process perpetuates the same systemic conditioning, resulting in a morphostatic cycle. Although this book does not delve deeply into the history of UM, the persistent focus on key themes in the UM literature – particularly the emphasis on institutional roles in allocating urban resources and the mono-dimensional infrastructure approach – throughout the 1970s and 1980s provides some evidence of how the CS remained relatively closed and stable over time. These dominant themes suggest that the closure of the UM CS was largely maintained during this period.

4.5.2 M/M long-span cycle

4.5.2.1 Cultural conditioning (time: T1)

The ideas that persist through successive short-span cycles establish the initial cultural conditions for a longer-term cycle (Figure 4.5). However, over time, global structural changes within the broader context of the discipline begin to emerge. These changes can reveal inadequacies in the systematised CS of the normal paradigm, as it becomes less effective in addressing the most pressing issues relevant to the discipline. As Kuhn (1970: p. 37) notes, a paradigm 'can ... even insulate the community from those socially important problems that are not reducible to the puzzle form, because they cannot be stated in terms of the conceptual ... tools the paradigm supplies.' When the normal paradigm consistently fails to address these problems, a crisis slowly becomes apparent, exacerbating the issue as anomalies persist unresolved.

In the context of UM, three global trends – 'the diffusion of sustainability, the spread of urbanisation, and the rise of ICT' (Bibri, 2018a: p. 53) – have played a significant role in shaping the discipline. Urbanisation and the rapid growth of urban environments have become increasingly pressing and

Figure 4.5 Long-span morphogenesis cycle during crisis and revolution stage (author's own)

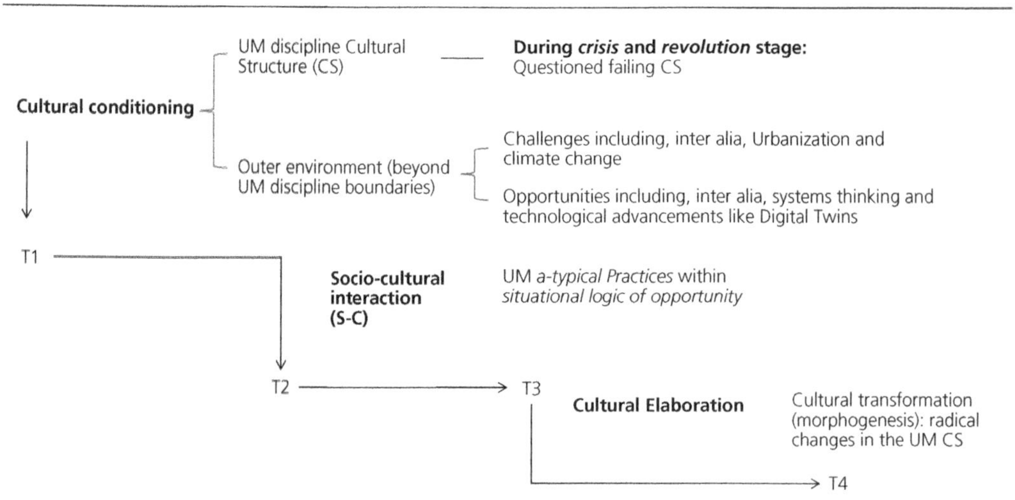

complex phenomena. UM researchers have acknowledged the intricate nature of the urban environment, recognising the essential need to 'consider the problem in its entirety' and integrate 'the socio-cultural, economic and environmental elements, which all go towards the construction of that complex set of relations we call city' (Camagni, 1998: p. 18). Researchers have focused on various aspects of urban complexity, with Ding and Lai (2012) categorising UM problems into physical and non-physical domains, emphasising the technical dimension of complexity. Physical issues typically concern technical, or 'hard', elements such as land use, infrastructure and the built environment, while non-physical problems pertain to socio-economic or 'soft' elements (see Section 3.2.5).

Chakrabarty (2001: p. 331) highlighted that urban professionals could significantly improve the urban environment by addressing 'the conflicting interests of multiple stakeholders and achieving equity', focusing on political complexity. He also emphasised the importance of collaboration and integration among urban organisations, marking a shift towards organisational complexity. Engin *et al.* (2020) brought structural complexity to the forefront, arguing that modifications to existing urban structures could lead to both expected and unexpected, as well as desirable and undesirable, outcomes. The notion of unpredictable emergent outcomes has gained increasing importance, particularly for global issues like climate change. Recent UM studies have increasingly focused on the natural environment as a major concern, alongside the ongoing challenge for urban managers of balancing economic development and well-being (e.g. A collaboration of leading figures in the built environment, 2021; Bibri, 2018; Sharma, 1989).

Another significant global shift relevant to UM is the rapid advancement of technology, particularly in ICT, which has introduced new concepts like DT. The idea of a DT can be traced back to Gelernter's (1991) concept of 'mirror worlds,' which envisioned a city-scale visualised model fed by streams of urban data, enabling urban managers to explore city dynamics across various scales. Initially developed in industries like manufacturing and aerospace, the DT concept has gradually gained traction in fields such as the built environment and UM (Al-Sehrawy and Kumar, 2021).

4.5.2.2 Socio-cultural interactions (time: T2–T3)

As discipline members grow increasingly frustrated by the inability of their *typical practices* to address the escalating crisis (Werna, 1995), they become more inclined to explore 'contingent complementarities' (Archer, 2005). These are ideas that, while not necessarily related, can complement one another when agents choose to leverage their potential. The emergence of such complementarities creates a 'situational logic of opportunity' (Archer, 2005), presenting possible solutions to the ongoing crisis. Identifying and seizing these opportunities requires agents to engage in atypical practices, involving innovative research and applications that demand higher levels of reflexivity.

As Kuhn suggests, 'external conditions may help to transform a mere anomaly into a source of acute crisis' (Kuhn, 1970: p. 10). This has certainly been the case for the UM discipline, where external factors such as urbanisation and increasing urban complexity have amplified the shortcomings of the UM normal paradigm. As urban environments became more complex and urban managers became increasingly aware of this complexity, their realisation of the normal paradigm's inadequacies also grew. At the S-C level, discipline members began to question the paradigm itself, along with the modes of thinking that underpin it. Anomalies once overlooked or dismissed could no longer be ignored – for instance, the mono-dimensional approach characteristic of the normal paradigm proved less effective than initially thought. The United Nations Development Programme (UNDP) highlighted this, stating that 'one of the most important lessons learnt from the distant and recent past is the failure of … isolated projects' (UNDP, 1989: p. 60).

Some authors noted that the sectoral organisational structure, while effective for simpler challenges, struggled to cope with 'a very broad and highly integrated nature' (Baker, 1989: p. 33). The siloed nature of these organisational structures reinforced a reductionist view, which treated naturally complex urban phenomena as discrete parts to be managed independently. This approach often led to undesirable outcomes. Chakrabarty (2001: p. 332) illustrated how the 'adoption of a conventional approach … [where] each discipline working in the urban sector tends to look at urban problems only from its own angle … may create inequity instead of achieving the welfare objective.' As the crisis deepened, it became clear that UM needed to adopt a broader perspective. McGill (1998: p. 464) stressed that 'urban management must take a wider view of things' and questioned, with regards to the 'sectoral thinking versus the inter-sectoral nature of the city', whether there is 'a way to ensure an institutional complexity to match the urban complexity it is dealing with'.

Recognising the value of complementarity, many in the UM field advocated for the application of systems thinking. The mono-dimensional approach and siloed thinking were criticised in favour of a more holistic, systemic approach that acknowledged 'the multisectoral nature of urban activities' (Harris, 1992: p. xxi) and the complex interdependencies within urban environments. Researchers began to assert that a systemic, integrated approach to UM was crucial for addressing these interconnections and their consequences (Chakrabarty, 2001; Stren *et al.*, 1992). Mattingly (1994) made a distinction between UM as project management and UM as development management, the latter being more focused on managing the increasing demands of cities in a systemic way by transforming or rebuilding urban policies, structures and networks. Camagni (1998: p. 18) calls for an evolutionary approach underpinned by systems thinking.

> *'The various systems making up the city (economic, social, physical—built and cultural heritage—and environmental) must be considered together and in their dynamic interactions (externalities, feedback, increasing returns, synergies). We cannot just put different aspects together and expect them to add up to a proper sum. We must take up an evolutionary approach characterized by full consideration of the complexity involved, with its components of non-linearity, cumulativity and irreversibility.'*

Engin *et al.* (2020) drew parallels between UM and programme management to demonstrate how systems thinking is central to both disciplines. Bibri and Krogstie (2017: p. 185) argued for 'newfangled ways of urban thinking grounded in a holistic approach', while others called for managing the built environment – encompassing 'everything we've built' – as 'a system of systems' (A collaboration of leading figures in the built environment, 2021).

It has become evident that advocating systems thinking is essential for anyone who wishes to address the complexity of the urban environment in a meaningful way. Clinging to the normal paradigm, with its isolationist approach and tendency to break complex problems into separate, simplified parts, has proven ineffective and ultimately futile.

> *'The fact that it faces us with the task of analyzing forbiddingly complex environmental interactions gives us no more of an excuse to isolate organizations conceptually than the proverbial drunk had when searching for his [sic] lost watch under the street lamp because there was plenty of light when he knew he had lost it in the dark alley'* (Emery, 1969: p. 14).

For nearly three decades, much has been said about the need to adopt systems thinking in UM, yet little has been implemented in practice. While Allen *et al.* (2007: p. 403) argue that 'complex systems thinking offers us a new, integrative paradigm', UM authors, such as Bibri and Krogstie (2017: p. 6), maintain that urban 'intractable problems require evidently an unprecedented paradigm change to disentangle and overcome.' McGill's question, 'is there a way to ensure an institutional complexity to match the urban complexity it is dealing with?' remains largely unanswered, with years of continued dissatisfaction proving that systems thinking, although well-argued, is much easier said than done.

Nevertheless, changes in global trends may present additional opportunities to operationalise systems thinking and provide more practical solutions. Kuhn (1970: p. 10) emphasises that 'conditions outside the sciences may influence the range of alternatives available' to those seeking to end a crisis by proposing revolutionary reforms. Technology has historically been a catalyst for paradigm shifts, as seen in the Einsteinian revolution, driven by advances in X-rays, radioactivity and other technological developments (Gillies, 2016). Mutch (2017: p. 499) similarly argues that 'the same or similar beliefs can result in significantly different practices depending on factors such as advances in technology.' Therefore, even as UM practitioners embrace systems thinking and recognise urban environments as complex systems, technological advancements can significantly influence the outcomes of their actions. Engin *et al.* (2020: p. 141) suggest that technology can help 'address many of the challenges found within fast-paced, complex urban environments.' Ding and Lai (2012: p. 1) echo this sentiment, highlighting the need for sophisticated tools and modelling to manage urban systems of extreme complexity.

The rise of ICT and the concept of DT has increased the situational logic of opportunity for UM members. Many now see DT as a potential enabler for implementing systems thinking in practice. The Data for Public Good report (National Infrastructure Commission, 2017) initiated efforts to develop a national DT of the UK's infrastructure system. Recent initiatives like the Gemini papers advocate for the creation of a web of 'connected digital twins,' integrating systems thinking and DT technology to support UM. Gemini Council and Lamb's (2022: p. ii) paper argues that 'due to its complexity, it is difficult to understand the entire picture of the built environment and the social and environmental layers with which it intersects' and to this end, 'connected digital twins are a potent tool to help do that.' The same paper continues.

> *'An ecosystem of connected digital twins breaks down the complexities of understanding the system as a whole. This system-based outlook will help us all understand the knock-on effects and the trade-offs that a decision could cause. By connecting physical assets, processes and systems with shared digital connections, we will gain insights that will enable improvements, optimisations, and better interventions across scales ... By sharing data across organizational or sector boundaries we can identify interdependencies. This leads to improved decision making that just wouldn't happen immediately if each silo of data were considered independently.'*

The Climate Resilience Demonstrator [CReDo] project by the National Digital Twin programme [NDTp] exemplifies how DT technologies can foster a holistic approach to managing a city's infrastructure systems, improving climate resilience (Akroyd, 2022). Similarly, a report from *SmartCitiesWorld* advocates for liberating data from silos through cooperative data systems to enhance understanding of urban dynamics. The report gives examples, such as how a port can integrate air quality and operations data to address pollution, highlighting the value of interconnected data systems in uncovering hidden relationships (SmartCitiesWorld, 2022: p. 9).

4.5.2.3 Cultural elaboration (time: T4)

Atypical practices eventually give rise to what Archer (2005: p. 32) describes as 'novel areas of intensive specialisation'. When added to the discipline's CS and body of knowledge, these areas of specialisation may signal the dawn of a revolution, positioning the discipline at the cusp of a

Figure 4.6 Suggested morphogenesis cycle during current pre-paradigm stage (author's own)

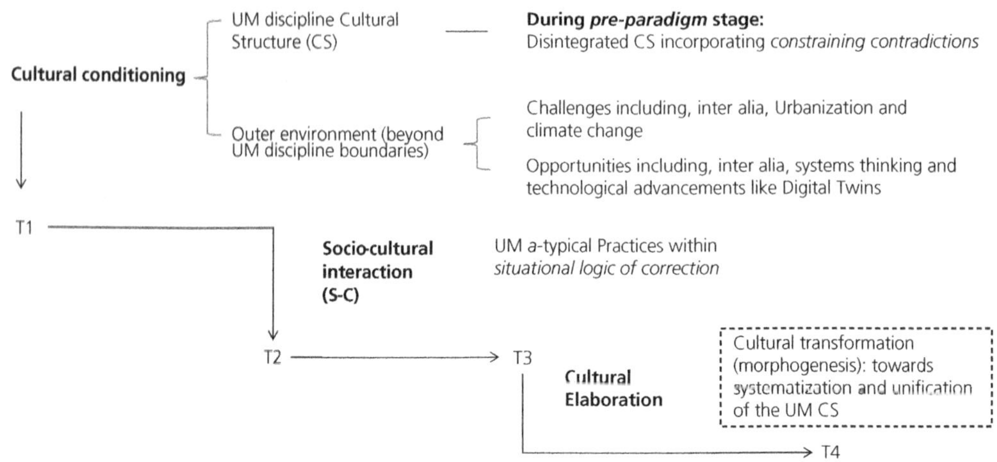

new paradigm. At the S-C level, this typically leads to an exponential increase in the number of discipline members drawn to contribute to the emerging structural variety. This results in further diversification, creating a 'positive feedback loop' (Archer, 2005: p. 32). The resultant richness and growing ideational diversity of this loop often generates a range of diverse, and even contradictory, ideologies at the systemic CS level. This can lead to research and practice becoming somewhat disorganised, a hallmark of the pre-paradigm phase in the Kuhnian cycle (Figure 4.6).

In the context of *DT for UM*, this nascent paradigm is currently experiencing a high degree of ideational diversity, characterised by a variety of methods and approaches. According to Ferré-Bigorra *et al.* (2022: p. 1), 'rapid growth has led to a fragmented situation.' The comprehensive review of the relevant literature in Chapter 3 resulted in the development of a taxonomy, outlining the various DT uses, tools, and techniques emerging across the UM field. Furthermore, Section 3.8.4 conceptualises four distinct DT approaches, referred to as *tech-driven, cognitive, disruptive* and *humanistic*, each representing different forms of DT practices.

However, these four approaches are rooted in four incommensurable paradigms: functionalism, radical structuralism, interpretivism and radical humanism, respectively. This philosophical tension, although evident, is argued to be a necessary one, shaped by the practical demands of the real world. UM practitioners face a highly complex urban environment, where problems cannot be adequately addressed without integrating and combining different DT approaches in a pluralistic manner. Given the multi-faceted nature of urban challenges, it is simply not feasible to advocate for one DT approach while dismissing the others[3].

This necessary yet contradictory relationship is an example of what Archer (2005) refers to as a 'constraining contradiction'. The unique tension that arises from the practical necessity to endorse contradictory ideological positions is what we have termed the 'dilemma of pluralism' (Section 3.8.5).

[3]Ashby (1961) makes a similar argument to establish the law of requisite variety.

4.5.3 Way forward

From a Kuhnian perspective, the pre-paradigm phase is typically followed by the establishment of a new paradigm. However, as inferred from the earlier discussion, we are currently in the midst of a positive feedback loop characterised by ideational heterogeneity. Every day, more diverse research and applications emerge regarding the use of DT in UM. Moreover, the relationship between the various DT approaches adopted by UM practitioners has become an essential one, meaning that the inherent contradictions between these approaches cannot simply be resolved by advocating one particular theory or school of thought and dismissing others.

The 'constraining contradictions' faced by members of the UM discipline, hence the 'dilemma of pluralism', place them in a 'situational logic of correction' (Archer, 2005). According to Archer's principle of analytical dualism, agents at the S-C level hold powers that are both related to and independent of the CS. This allows them to work towards resolving these contradictions, much like they previously discovered and exploited 'contingent complementarities' during periods of crisis and revolution. To overcome these constraints, UM agents must engage in ideological corrections, working towards the reconciliation and unification of conflicting worldviews.

As a result, the new paradigm phase will commence, marking the beginning of a longer-span morphogenetic cycle. As Archer (2005: pp. 26–27) puts it.

> *'What the "constraining contradiction" does in practice is to confront those committed to (A), who also have no option but to live with (B) as well, with a particular situational logic. According to this logic, given their continuing dedication to (A) (its abandonment is always possible because conditioning is never determinism), then they are constrained to deal with (B) in a specific manner. Since (A) and (B) are logically inconsistent, then no genuine resolution is possible between them, but if (B) remains unaltered, it threatens the credibility or tenability of (A). Consequently, the situational logic directs that continued adherence to (A) entails making a correction of its relationship with (B) mandatory. Corrective action involves addressing the contradiction and seeking to repair it by reinterpretation of the ideas involved. The generic result will be some form of syncretism that brings about union between the antithetical but indispensable sets of ideas.'*

The successful unification of ideas, theories and approaches at the CS level would initiate a negative feedback mechanism, counteracting the cultural diversity produced by the positive feedback loop discussed earlier. This process would occur through cultural densification, gradually consolidating the diverse ideas into a more cohesive whole. Eventually, the discipline would re-establish a 'situational logic of protection' as agents, at the S-C level, reproduce necessary or concomitant complementarities. Through multiple short-span cycles of cultural densification, this would result in a firmly systematised CS that reinforces the dominant paradigm.

4.6. Designed amalgam of twinning for urban management [DATUM]

4.6.1 A 'conceptual box' for the new paradigm

The previous argument demonstrates that the discipline of UM is currently experiencing the emergence of a CS marked by cultural diversity. The utilisation of opportunities discovered by agents, such as systems thinking and DT, through *atypical practices*, has led to the proliferation of various ideologies, philosophical perspectives and practical approaches in the implementation of DT for UM. As a result, it has been suggested that for UM to progress meaningfully in the era of DTs, systematisation of the discipline is required. This entails structuring a CS for the new paradigm,

Figure 4.7 The structure of the designed amalgam of twinning for urban management (DATUM) (author's own)

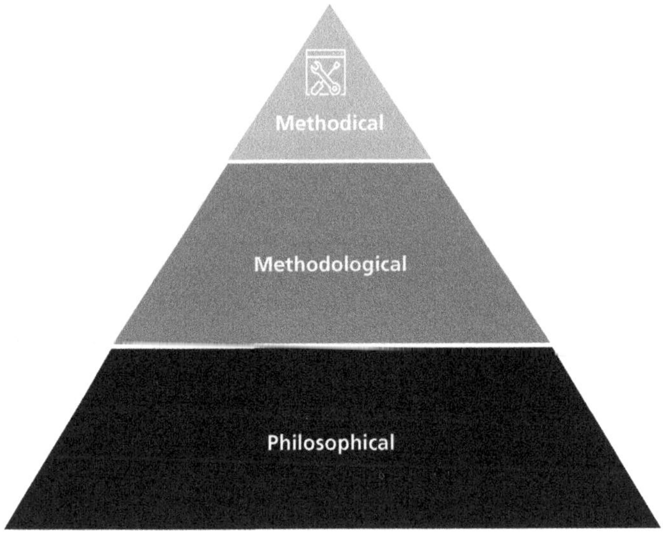

DT for UM, with a coherent set of related theories capable of unifying the discipline and guiding its members' efforts.

The main aim of this book is to develop a conceptual structure for a CS that crystallises the new paradigm, *DT for UM*. This structure, referred to as DATUM, is depicted in Figure 4.7. DATUM focuses on defining a universal or generic approach, proposed in this book as a foundation for good practice. Its objective is to provide a model or exemplar that unifies research and practice in *DT for UM*. Although fully developing DATUM is a substantial task beyond the scope of this book, the basic structure of DATUM is introduced (Figure 4.7), with key components to be further elaborated in the following chapter.

4.6.2 DATUM components

Kuhn highlighted that textbooks representing well-established paradigms typically present the body of accepted 'scientific concepts and theories.' Additionally, they may include 'at a level lower or more concrete than that of laws and theories … a multitude of commitments to preferred types of instrumentation and to the ways in which accepted instruments may legitimately be employed' (Kuhn, 1970: p. 40). In this sense – aligning with the critical realist perspective of a stratified reality – one can conceptualise the key components of a paradigm as a set of stratified elements at various levels of abstraction. At the most abstract level are philosophical assumptions and worldviews. At a more practical level, guidelines are offered to help contextualise abstract theories in practical terms. At the most concrete level, propositions concerning the pragmatic use of accepted tools and techniques are outlined. Similarly, researchers and practitioners across various fields, including systems thinking and project management, which often overlap with UM, have found it useful to perceive paradigms as stratified, with layered elements at different levels of abstraction (Checkland, 1999; Fitzgerald and Howcroft, 1998; Jackson, 2019; Mingers and Brocklesby, 1997; Ragsdell, 2000; Remington and Pollack, 2016).

It is essential to stress that the stratified view of DATUM, as depicted in Figure 4.7, should not be interpreted as hierarchical. This does not imply that philosophy is more valuable or unquestionable than methods. If that were the case, it would be difficult to acknowledge how 'encountering a problem in practice may signal a philosophical inadequacy' (Midgley, 2000: p. 21). However, 'philosophy, methodology and tools [are] mutually supportive' (Midgley, 2000: p. 21). Stratification, therefore, accurately reflects reality by showing how different forms of practice implicitly endorse philosophical assumptions. These assumptions, in turn, guide practitioners to adopt specific methodologies or approaches, characterised by the use of particular methods to analyse urban phenomena and engage in real-world interventions[4]. As Remington and Pollack (2016: p. 78) noted, 'methodology, and subsequently theory, become embodied in practice, through informing both the selection of tools and how they are applied in the project'. The three key components of DATUM are explained below.

4.6.2.1 Philosophy

The philosophical element forms one of the foundational pillars of this structure (Figure 4.7). It provides the intellectual framework and conceptual context that guide more practical matters at the upper levels. Furthermore, it offers a basis for practitioners to reflect on or critique their actions, and against which the consistency of practices can be assessed. In essence, the philosophical component of DATUM provides a worldview for interpreting and examining reality, 'spanning ontology, epistemology and methodology' (Healy and Perry, 2000: p. 121), and defines 'the nature of possible research and intervention' (Mingers, 1997: p. 429) – for example, it will be argued in the following chapter that CR would provide a suitable philosophical foundation for the implementation of DT in UM.

4.6.2.2 Methods

By contrast, the methodical element, at the uppermost stratum, is purely concerned with practicalities. It encompasses models, tools and techniques, such as sensors or the internet of things (IoT) for the collection of urban big data (Rathore *et al.*, 2018), or, more fundamentally, 'standards for their use' (Remington and Pollack, 2016: p. 77), including standard terminology and naming conventions. An example of the latter is the digital twin uses and classification system (DTUCS) developed in Chapter 5, which aims to offer a common language among practitioners and to create standard methods across the discipline. These methods can be employed to refer to various DT uses, specify different DT specifications or features, and model DT use case scenarios. Methods are often implicitly linked to particular philosophical positions – for instance, quantitative analytical methods associated with the use of DT for UM applications, like machine learning, align with the doctrine of functionalism and the positivist stance (Kitchin, 2014). As such, although methods rarely refer directly to philosophical propositions, they enable their practical application and testing.

4.6.2.3 Methodology

The methodological element of DATUM bridges the gap between philosophy and methods. Whereas 'a technique tells you 'how' and a philosophy tells you 'what'', a methodology contains elements of both 'how' and 'what' (Checkland, 1981: p. 162). It does so by first drawing on underlying philosophical assumptions and highly abstract theoretical principles to offer procedural guidelines or more practical approaches. It then informs the selection and combination of particular methods to achieve desirable outcomes (Mingers, 2006; Remington and Pollack, 2016). The four DT approaches

[4]It is for this logical sequence, from philosophy to methodology to methods, that it seemed more appropriate to put the philosophical element at the base of DATUM (Figure 4.7) despite being at the highest abstract level, and despite how earlier illustrations proposed the same layering but in reverse, with philosophy on top and methods at the base (e.g. Remington & Pollack, 2016).

(i.e. tech-driven, disruptive, cognitive and humanistic), based on the four distinct philosophical paradigms (i.e. functionalism, radical structuralism, interpretivism and radical humanism), are an example of how philosophical stances can be transformed into more actionable forms of practice (Section 3.8.4). Additionally, other publications provide generic practical guidelines (e.g. Gemini Council and Lamb, 2022) and methodological principles, such as the Gemini Principles (Bolton *et al.*, 2018), to support the development and use of *DT for UM* in a broader context (Section 3.8.1).

4.6.3 The value of DATUM

The value of having a conceptual framework or for the CS of the new paradigm and the benefits of breaking it into constituent parts (Figure 4.7) are threefold. The mixed Kuhnian-Archerian perspective allows us to justify the call for a systematising and unifying CS for the new paradigm. However, this recommendation remains somewhat abstract. This high level of abstractness and generalisability may weaken its applicability and limit the opportunity to realise what it proposes – the actual unifying CS. Therefore, the first benefit of DATUM is that it offers more practical guidance by providing a conceptual framework to which future research can refer or contribute, thus bringing order to future research. Secondly, when examining research within a well-established normal paradigm, Kuhn (1970: p. 5) describes it as 'a strenuous and devoted attempt to force nature into the conceptual boxes supplied by professional education'. He then questions 'whether research could proceed without such boxes'. Hence, DATUM, in Kuhn's terms, provides the shape and internal structure of the conceptual box that defines the paradigm in focus (*DT for UM*). Finally, if all past relevant studies are mapped onto DATUM, discipline members can visualise the research gaps within the emerging paradigm. They would be able to identify which elements of DATUM – philosophy, methodology, or methods – are overstudied and which are understudied. Thus, DATUM can direct future research towards addressing gaps that hinder the growth of the paradigm.

4.6.4 DATUM general requirements

For DATUM to successfully systematise *DT for UM*, it needs to satisfy two essential conditions. The first condition is intrinsic to the structure of DATUM. The three key elements of DATUM (i.e. philosophy, methodology, methods) need to be coherent and consistent. When the philosophical propositions complement the methodological principles they underpin, and the latter support the implementation of methods, DATUM as a whole is reinforced. These intrinsic and necessary complementarities generate a situational logic of protection – for example, a philosophy like functionalism, supporting a positivist position, would complement the use of a *tech-driven* DT approach (Section 3.8.4.1), which would in turn encourage the use of quantitative DT methods and numerical analytical tools in practice. In such a situation, alterations or disruptions to DATUM's internal configuration or any of its elements are normally discouraged.

DATUM would also need to satisfy a condition that is extrinsic to its structure and more related to the outer environment in which it operates to sustain its role as a unifying CS. Drawing upon Simon's (1996: p. 3) perspective, an artefact has an inner environment and operates within an outer environment. As such, he argues that 'if the inner environment is appropriate to the outer environment, or vice versa, the artefact will serve its intended purpose.' Accordingly, even if DATUM's key components, constituting its inner environment, are in perfect harmony, satisfying the intrinsic condition explained above, they need to collectively form a whole that is relevant to the outer environment and adequate to address the problems it raises. Using DATUM, when it shows signs of failure or inability to deal with the wicked nature of the urban environment and the extremely complex problems it imposes, would be nothing but a desperate attempt to fit the urban reality into the wrong 'conceptual box' – for example, if we acknowledge that urban problems are extremely complex, DATUM should embrace systems thinking and a holistic approach. Similarly, if urban problems are viewed as multi-faceted phenomena, DATUM should enable pluralism and provide practitioners with the requisite variety needed to tackle these different facets, and so on. Unless this connection between DATUM's inner and outer environments is realised, DATUM would hardly be of any significant help to discipline members. In such cases, the attempts to systematise the discipline and firmly establish a widely accepted CS for the new paradigm would eventually fail.

4.6.5 DATUM evaluation criteria

As recommended above (Section 4.5.3), the ultimate aim of the proposed artefact (DATUM) is to unify the new paradigm (*DT for UM*). To this end, the unifying ability of DATUM will be evaluated according to the evaluation criteria detailed in Figure 4.8 below.

4.7. Summary

A considerable amount of research questioning the effectiveness of UM or critiquing the fruitfulness of its traditional practices was the starting point of this book. However, any recommendations on how to enhance the performance of the UM discipline must be provided based on a deeper understanding of its current state. To this end, a novel three-layered theoretical lens was used to examine how the discipline of UM has evolved and to better understand its current state. The first layer captures the underlying philosophical tenets, rooted in critical realism, that necessarily underpin Archer's M/M framework. The latter, lying at the second layer of the lens, provides means for appreciating the social nature of scientific progress. It creates practical means that enable examining the relationship between the corpus of ideas and theories pertaining to the discipline of UM and the practices of discipline members and the interactions between them. The third layer includes Kuhn's account of the structure of scientific revolution, which offers a way of conceptualising the stages a discipline goes through in its evolution.

Figure 4.8 The criteria for evaluating DATUM (author's own)

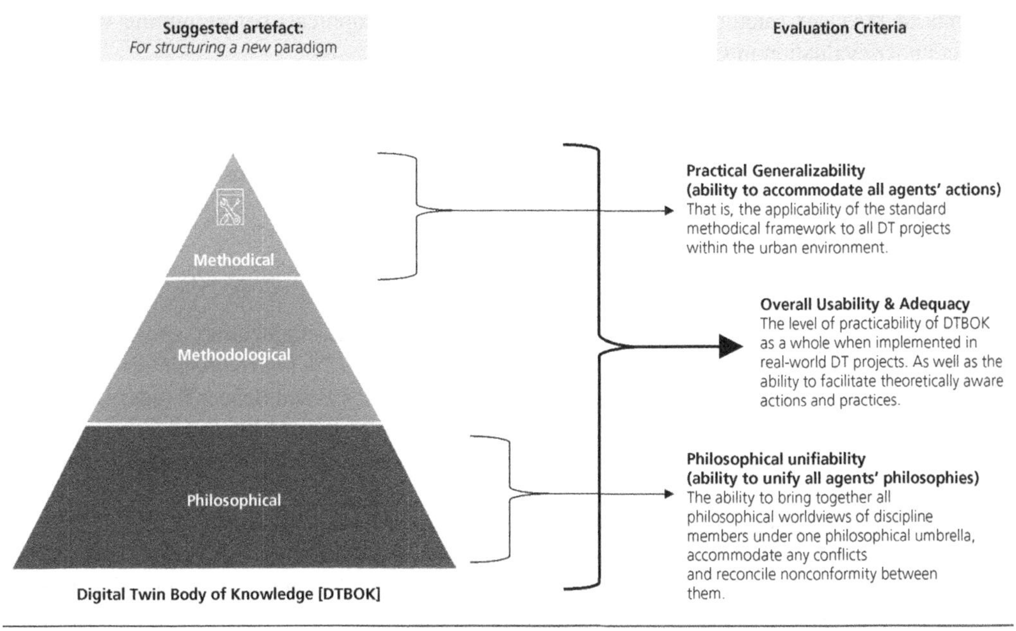

Supplementing Kuhn's account by adopting an M/M approach draws our attention not only to the interplay between the discipline's CS and the actions of its members over time within a particular stage of the Kuhnian cycle, but also to how this interplay leads, in the longer term, to the transition from one Kuhnian stage to another. Moreover, it ensures that explanations pertaining to how UM evolves are not divorced from the broader context within which the discipline exists. This wider environment involves worlds of increasing complexity, rapid proliferation of ICT, pressing climate change and other phenomena.

This, in turn, helped in understanding how *typical practices*, performed by discipline members during times of the normal paradigm, including, among other things, siloed thinking and mono-dimensional interventions, failed to address the urban problems relevant to UM. The rise of innovative ideologies and concepts like systems thinking and DT has created a situational logic of opportunity within which UM discipline members have been placed. *Atypical practices*, being creative and reflexive, utilised the way these new ideas complement UM to pave the way for a new paradigm. Subsequently, diversity within the UM CS has been created, provoking even more diversity, and giving rise to distinct and, to a greater extent, contradicting ideational projects, which is a key feature of the Kuhnian pre-paradigm stage. It was argued that the complex and multi-faceted real-world problems that urban managers encounter forced them to adopt these contradicting ideas simultaneously. Accordingly, UM discipline members are conditioned within a CS that incorporates 'constraining contradictions', where endorsing only one while renouncing the rest is not an option. Based on the M/M approach, it is argued that such a CS places people in a situational logic of correction, where they would need to intervene by 'modifying current logical relationships and introducing new ones' (Archer, 1996: p. 227), in a way that would bring about a more unified version. The unification of the CS would then facilitate the systematisation of the discipline and accelerate its growth, as is the case during periods of transition to a new paradigm.

The DATUM, proposed in this chapter, is suggested as a theoretical artefact or, in Kuhn's words, a 'conceptual box', for the new paradigm, *DT for UM*. It can offer a common ground upon which UM discipline members can be united. The following chapters embark on developing the three elements of DATUM (methodical, methodological and philosophical) before putting the whole artefact under evaluation in Chapter 8.

REFERENCES

A collaboration of leading figures in the built environment (2021) *Our Vision for the Built Environment*. CDBB. https://www.cdbb.cam.ac.uk/news/industry-unites-behind-vision-built-environment (accessed 11/07/2021).

Akroyd J (2022) *CReDo Methodology Papers: Implementation*. CDBB.

Allen PM, Strathern M and Baldwin JS (2007) Complexity and the limits to learning. *Journal of Evolutionary Economics* **17(4)**: 401–431.

Al-Sehrawy R and Kumar B (2021) Digital twins in architecture, engineering, construction and operations. A brief review and analysis. In *Proceedings of the 18th International Conference on Computing in Civil and Building Engineering* (Toledo Santos E and Scheer S (eds)). Springer International Publishing, Cham, pp. 924–939. 10.1007/978-3-030-51295-8_64

Archer MS (1995) *Realist Social Theory: The Morphogenetic Approach*. Cambridge University Press, Cambridge, UK.

Archer MS (1996) *Culture and Agency: The Place of Culture in Social Theory*. Cambridge University Press, Cambridge. https://books.google.co.uk/books?id=ljpbPeHdJL0C (accessed 11/07/2021).

Archer MS (2005) Structure, culture and agency. In *The Blackwell Companion to the Sociology of Culture*. Blackwell Publishing, Oxford, UK, pp. 17–34.

Ashby WR (1961) *An Introduction to Cybernetics*. Chapman & Hall Ltd, London, UK.

Baker R (1989) Institutional innovation, development and environmental management: an 'administrative trap' revisited. Part I. *Public Administration and Development* **9(1)**: 29–47.

Bhaskar R (1975) *A Realist Theory of Science*. Leeds Books Ltd, Leeds, UK.

Bibri SE (2018) *Smart Sustainable Cities of the Future: The Untapped Potential of Big Data Analytics and Context-Aware Computing for Advancing Sustainability*. Springer, Cham.

Bibri SE and Krogstie J (2017) Smart sustainable cities of the future: an extensive interdisciplinary literature review. *Sustainable Cities and Society* **31**: 183–212.

Bolton A, Butler L, Dabson I *et al.* (2018) *Gemini Principles* [Report]. CDBB. 10.17863/CAM.32260

Camagni R (1998) Sustainable urban development: definition and reasons for a research programme. *International Journal of Environment and Pollution* **10(1)**: 6–27. 10.1504/IJEP.1998.002228

Chakrabarty BK (2001) Urban management: concepts, principles, techniques and education. *Cities* **18(5)**: 331–345.

Checkland P (1981) *Systems Thinking, Systems Practice*. John Wiley & Sons, Chichester, UK.

Checkland P (1999) *Systems Thinking, Systems Practice: Includes a 30-Year Retrospective*. John Wiley & Sons Ltd, Chichester, UK.

Ding C and Lai S-K (2012) Challenges in urban management. *Journal of Urban Management* **1(2)**: 1–2.

Emery FE (1969) *Systems Thinking: Selected Readings*. Penguin Books, London, UK.

Engin Z van Dijk J, Lan T *et al.* (2020) Data-driven urban management: mapping the landscape. *Journal of Urban Management* **9(2)**: 140–150.

Ferré-Bigorra J, Casals M and Gangolells M (2022) The adoption of urban digital twins. *Cities* **131**: 103905. 10.1016/j.cities.2022.103905

Fitzgerald B and Howcroft D (1998) Towards dissolution of the IS Research Debate: from polarization to polarity. *Journal of Information Technology* **13(4)**: 313–326. 10.1177/026839629801300409

Gelernter DH (1991) *Mirror Worlds, Or, The Day Software Puts the Universe in a Shoebox: How it Will Happen and What it Will Mean*. Oxford University Press, Oxford. https://books.google.co.uk/books?id=PmCoPwAACAAJ (accessed 11/07/2021).

Gemini Council and Lamb K (2022) *Gemini Papers: Why Connected Digital Twins*. Centre for Digital Built Britain. 10.17863/CAM.82237

Giddens A (1984) *The Constitution of Society: Outline of the Theory of Structuration*. Polity Press, Cambridge, UK.

Gillies D (2016) Technological origins of the Einsteinian revolution. *Philosophy & Technology* **29(2)**: 97–126.

Harris N (1992) *Cities in the 1990s: The Challenge for Developing Countries*. UCL Press, London.

Healy M and Perry C (2000) Comprehensive criteria to judge validity and reliability of qualitative research within the realism paradigm. *Qualitative Market Research: An International Journal* **3(3)**: 118–126. 10.1108/13522750010333861

Jackson MC (2019) *Critical Systems Thinking and the Management of Complexity*. Wiley, Chichester, UK. https://books.google.co.uk/books?id=1jmNDwAAQBAJ (accessed 11/07/2021).

Kitchin R (2014) Big Data, new epistemologies and paradigm shifts. *Big Data & Society*, April-June, pp. 1–12. 10.1177/2053951714528481

Kuhn TS (1970) *The Structure of Scientific Revolutions*, 2nd edn. University of Chicago Press, Chicago, IL, USA.

Latour B, Harman G and Erdélyi P (2011) *The Prince and the Wolf: Latour and Harman at the LSE*. John Hunt Publishing, Alresford, UK.

Leonard S (1982) Urban managerialism: A period of transition? *Progress in Human Geography* **6(2)**: 190–215.
Mattingly M (1994) Meaning of urban management. *Cities* **11(3)**: 201–205.
McGill R (1998) Urban management in developing countries. *Cities* **15(6)**: 463–471.
Midgley G (2000) *Systemic Intervention: Philosophy, Methodology, and Practice*. Springer US, New York. 10.1007/978-1-4615-4201-8_6
Mingers J (1997) Towards critical pluralism. In *Multimethodology: The Theory and Practice of Combining Management Science Methodologies* (Mingers J and Gill A (eds.)). John Wiley & Sons, Chichester, UK, pp. 407–440.
Mingers J (2006) *Realising Systems Thinking: Knowledge and Action in Management Science*. Springer US, New York. https://books.google.co.uk/books?id=C-UcDQEACAAJ (accessed 11/07/2021).
Mingers J and Brocklesby J (1997) Multimethodology: Towards a framework for mixing methodologies. *Omega* **25(5)**: 489–509. 10.1016/S0305-0483(97)00018-2
Mutch A (2017) Practices and morphogenesis. *Journal of Critical Realism* **16(5)**: 499–513. 10.1080/14767430.2017.1361289
Næss P (2004) Prediction, regressions and critical realism. *Journal of Critical Realism* **3(1)**: 133–164. 10.1558/jcr.v3i1.133
National Infrastructure Commission (2017) *Data for the Public Good*. National Infrastructure Commission Report.
Porpora DV (2015) *Reconstructing Sociology: The Critical Realist Approach*. Cambridge University Press, Cambridge, UK. https://books.google.co.uk/books?id=oPhfCgAAQBAJ (accessed 11/07/2021).
Ragsdell G (2000) 'Engineering a paradigm shift? An holistic approach to organisational change management', *Journal of Organizational Change Management* **13(2)**: 104–120. 10.1108/09534810010321436
Rathore MM, Paul A, Hong W-H *et al.* (2018). Exploiting IoT and big data analytics: defining Smart Digital City using real-time urban data. *Sustainable Cities and Society* **40**: 600–610. 10.1016/j.scs.2017.12.022
Remington K and Pollack J (2016) *Tools for complex projects*. Routledge, Milton, UK.
Sayer RA (1992) *Method in Social Science: A Realist Approach*. Psychology Press. https://books.google.co.uk/books?id=kr8QvOpM2RwC
Sharma SK (1989) Municipal management. *Urban Affairs Quarterly–India* **21(4)**: 47–53.
Simon HA (1996) *The Sciences of the Artificial* (third edition). MIT Press, Cambridge, MA, USA.
SmartCitiesWorld (2022) *Insight Report: Boosting operations in smart places with cooperative data ecosystems*. SmartCitiesWorld. https://www.smartcitiesworld.net/whitepapers/whitepapers/insight-report-boosting-operations-in-smart-places-with-cooperative-data-ecosystems
Stren R, Bhatt V, Bourne L *et al.* (1992) *An Urban Problematique: The challenge of urbanization for development assistance*. Centre for Urban and Community Studies, University of Toronto, Toronto.
UNDP (1989) *Urban transition in developing countries: Policy issues and implications for technical co-operation in the 1990s. Programme Advisory Note*. United Nations Development Programme, NY, USA.
Williams P (1978) Urban managerialism: A concept of relevance? *Area* **10(3)**: 236–240.
Werna E (1995) The management of urban development, or the development of urban management? Problems and premises of an elusive concept. *Cities* **12(5)**: 353–359.

Ramy Elsehrawy and Bimal Kumar
ISBN 978-1-83662-235-2
https://doi.org/10.1108/978-1-83662-234-520251007
Copyright © 2025 by Ramy Elsehrawy and Bimal Kumar. Published under exclusive licence by Emerald Publishing Limited

Chapter 5
Methods: digital twin uses classification system (DTUCS)

5.1. Introduction

As explained in Chapters 3 and 4, the CS of the UM discipline is currently witnessing increasing diversity and ideational heterogeneity. Consequently, it was argued that the systematisation of the new paradigm, *DT for UM*, is the recommended way forward for it to grow and mature. This involves structuring a CS that encompasses a comprehensive set of complementary and necessarily related theories to unify this paradigm and the endeavours of its members. Therefore, in Section 4.6, a conceptual design for this CS, namely the designed amalgam of twinning for urban management (DATUM), is proposed. It is made up of three constituent elements: philosophical, methodological and methodical (Figure 4.8).

This chapter addresses the methodical element of proposed CS for the paradigm *DT for UM*, which has been proposed as a means to unify and advance the increasingly diverse and ideationally heterogeneous field of UM. While earlier chapters introduced the conceptual design of this CS, termed DATUM, this chapter focuses on developing DATUM's methodical element. Specifically, it outlines a set of standard methods essential for the systematisation of *DT for UM*. These methods are designed to aid in (a) specifying and classifying DT, (b) defining the various functions and applications of DTs, and (c) modelling and documenting DT use case scenarios. By developing this methodical element, the chapter contributes to creating a structured, cohesive framework that partially constitutes DATUM, along with the philosophical and methodological elements developed in the following two chapters.

5.2. Digital twin uses classification system

As demonstrated in Chapter 3, DT case studies, pilot projects and proof-of-concepts are proliferating. Such endeavours support the idea of using DT to gradually mature UM. They provide practical evidence of the value and potential of DT and help secure buy-in from key stakeholders and decision makers within urban and built environments (National Digital Twin Programme, 2021). However, as discussed in Sections 3.6 and 3.7, these contributions lack consistency. The absence of a widely accepted standard common language across the field is impeding its growth. Essentially, any technical language within a discipline is made up of (a) technical words or terms with specific meanings, and (b) a set of rules (akin to a grammatical system) that allows for putting these technical terms together, forming meaningful sentences or models.

Accordingly, it is argued here that the systematisation of *DT for UM* and the creation of 'some form of syncretism that brings about union' (Archer, 2005: p. 27) at the methodical level begins with constructing these two fundamental elements of a proposed standard common language. To this end, the digital twin uses and classification system (DTUCS) is presented (Figure 5.1). It is made up of the following three prongs.

Figure 5.1 DTUCS

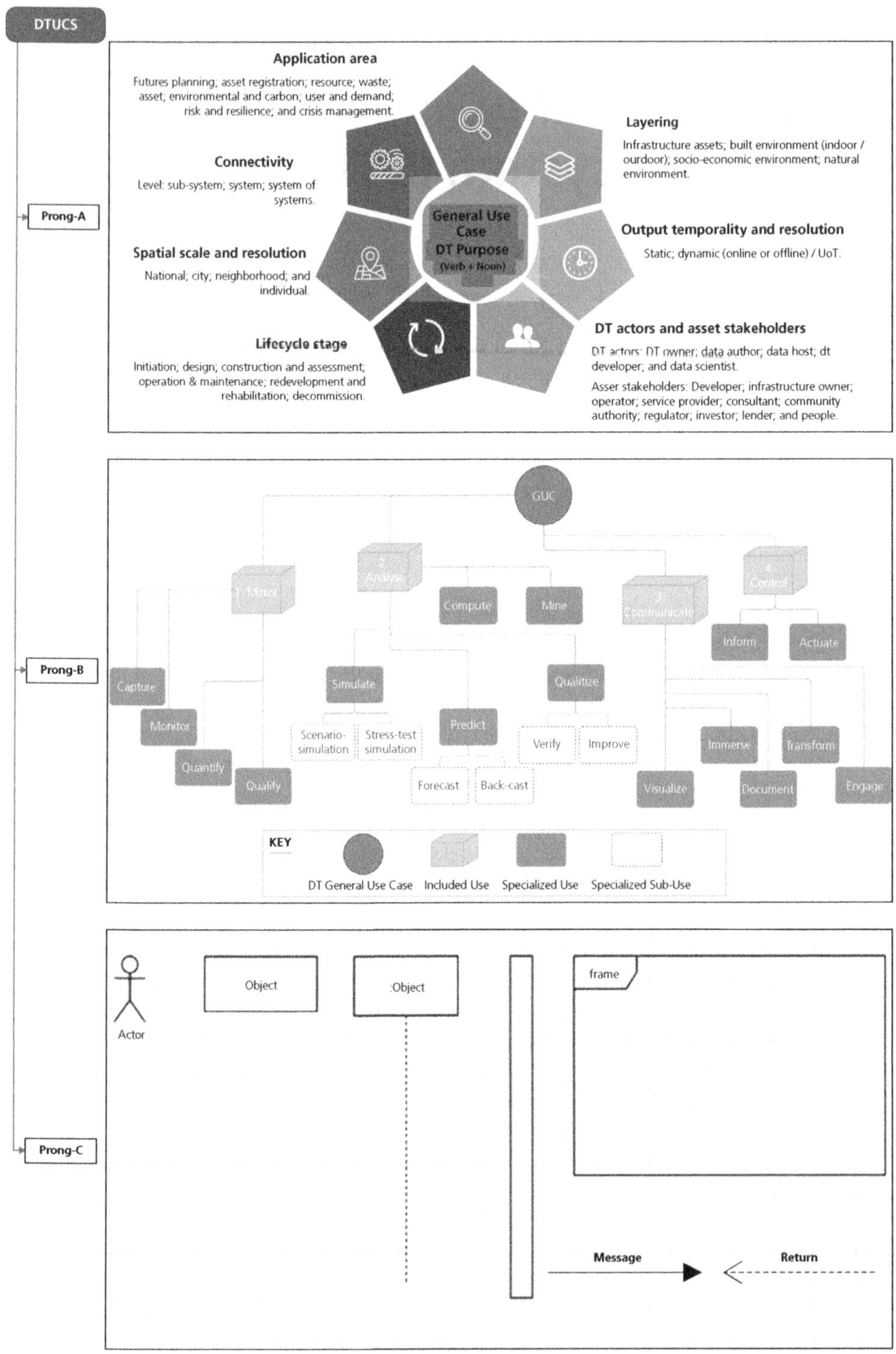

5.2.1 Prong-A

Prong-A encompasses the DT use case multi-dimensional classification framework, developed in Section 3.6 in response to the systematic literature review question [Q1]. This framework consists of technical terms that describe various features or characteristics of a DT constructed for a specific use case.

5.2.2 Prong-B

Prong-B, on the other hand, is the DT uses taxonomy, synthesised in Section 3.7 based on the reviewed body of relevant studies, in response to the systematic literature review question [Q2]. While both prongs, A and B, propose sets of technical terms with specific meanings, they fundamentally differ, representing two distinct levels of complexity. To elaborate, it is useful to recall Checkland's (1999: p. 78) statement, quoted in Section 3.2.3.

> *"The shape of an apple', although the result of processes which operate at the level of the cells, organelles, and organic molecules which comprise apple trees, and although, we hope, eventually explicable in terms of those processes, has no meaning at the lower levels of description. The processes at those levels result in an outcome which signals the existence of a new stable level of complexity – that the whole apple itself – which has emergent properties, one of them being the apple's shape.'*

This hierarchical perspective clarifies the distinction between the technical vocabularies offered by Prongs A and B. While Prong-A is equivalent to 'the shape of an apple' – that is, the constructed DT in our context – Prong-B is focused on formulating terms relevant to the processes at the lower 'level of the cells', corresponding to DT uses. These DT uses defined by Prong-B, when combined for a particular use case, give rise to the DT itself, which possesses unique features described by Prong-A.

5.2.3 Prong-C

Prong-C is concerned with employing unified modelling language (UML) to model any DT UCS. A UCS is the sequence of DT-actor interactions executed throughout a DT project to achieve a general use case (GUC) (Section 3.6.1). Over the years, UML has evolved into a standard expressive language for specifying and visualising IT systems and use cases, thereby facilitating communication and reducing misunderstandings among stakeholders (for a detailed explanation of UML, see Booch, 2005). In the context of DT, Prong-C leverages this powerful language to model and document DT-actor interactions occurring within a UCS in a standard format.

At a high level, UML diagrams are built on four key elements: 'systems', 'actors', 'use cases' and 'relationships'. A 'system' is the artefact we develop, which, in our context, corresponds to the DT itself, including its various subsystems used to *mirror*, *analyse*, *communicate* and *control* (Section 3.7.3). An 'actor' is someone or something that interacts with the 'system'. For a DT, these could be any of the DT actors (see Section 3.6.3.8 for more details on DT actors). In UML, a 'use case' represents a function the system performs to achieve the actor's goal, which corresponds to the notion of DT uses described by Prong-B. By definition, an 'actor' uses the system for a specific goal. A 'relationship' represents the interaction between the 'actor' and the 'system', or between different subsystems within the overarching system.

Thus, while Prong-B offers standard terminology for publishing UCSs, Prong-C is where these technical terms are assembled using UML to structure and model any UCS. In other words, while Prong-B supplies the 'story-teller' with the building blocks (i.e. DT uses) for describing a UCS, Prong-C provides the rules and modelling methods that facilitate putting these blocks together.

Figure 5.2 The role of DTUCS in supplementing the information management framework, presented by the NDTP (Hetherington and West, 2020), with the ability to communicate modelled and standardised DT use case scenarios

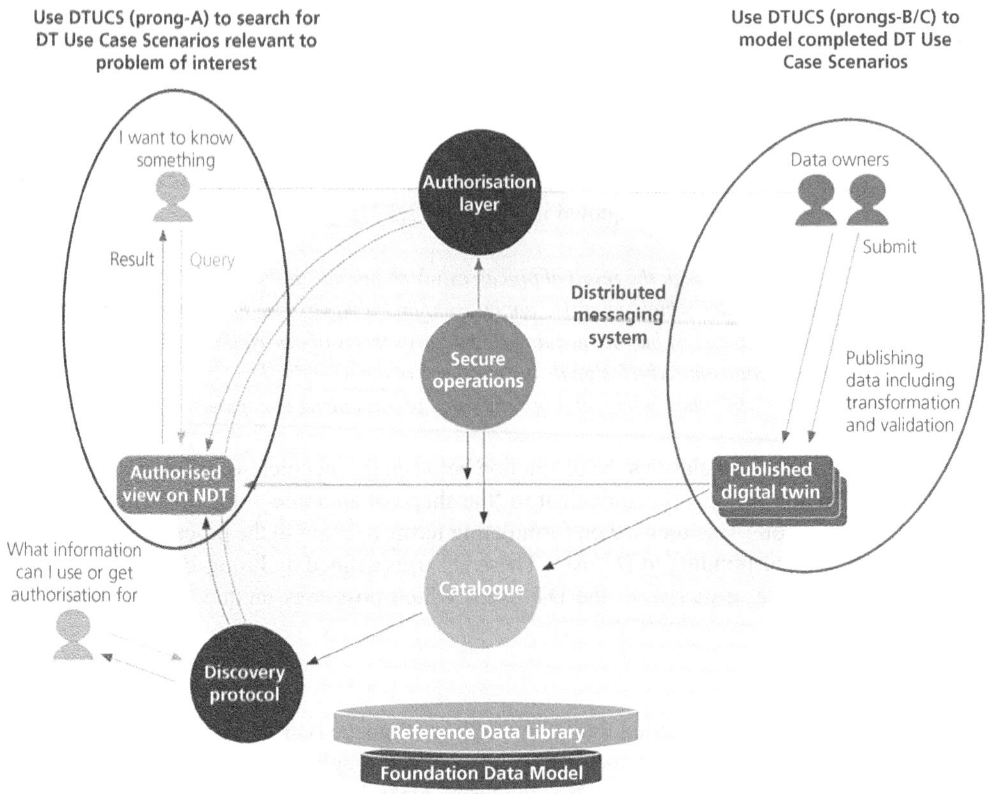

The final product is a modelled story illustrating how a particular GUC is pursued and achieved. Real-world applications of Prong-C are presented in Chapter 8 as part of evaluating DATUM.

An interesting outcome is how Prong-A can then be used by people to discuss, refer to, or search for 'stories' based on their features or characteristics. Figure 5.2 demonstrates how practitioners within the discipline can use the proposed standard common language (i.e. DTUCS) to disseminate information and knowledge. The suggested method is superimposed on the integration architecture (IA) of the information management framework (IMF) proposed by the National Digital Twin Programme (NDTP) (Hetherington and West, 2020).

5.3. The Gemini framework: reference architecture

The Gemini framework, developed by the lead author of this book and his team at Mott MacDonald in collaboration with the Digital Twin Hub[1], serves as a collaborative tool for DT Hub members to connect, compare and innovate in DT projects. At its core lies the reference architecture

[1]The Digital Twin Hub (DT Hub) (https://digitaltwinhub.co.uk/) holds the copyright for the Gemini framework.

Figure 5.3 The reference architecture. Source: https://digitaltwinhub.co.uk/gemini-framework/

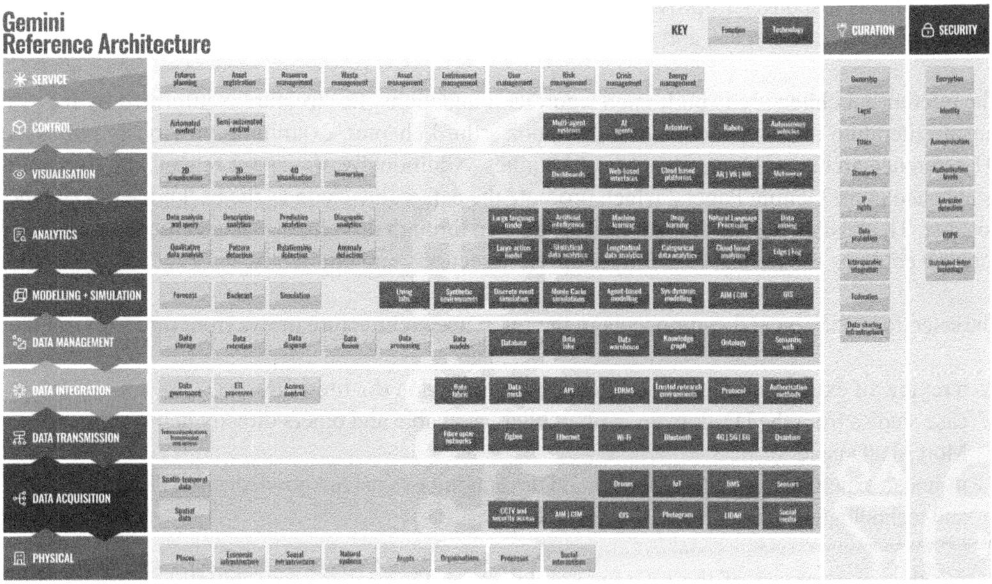

(Figure 5.3), a practical tool designed to offer a unified approach to mapping the data-value chain that underpins DT projects. The idea of unifying how the DT community demonstrates it DT journeys aligns seamlessly with the overarching aim of this book – to systematise and unify this field.

The reference architecture offers a clear visualisation of how individual DT projects fit within the broader DT ecosystem. By highlighting connections, gaps, and opportunities, it enables teams to identify areas for improvement and innovation. Its design is modular, tracking functions, technologies, data layers and their interconnections. Typically, data in a DT moves through several defined stages. It is collected, transmitted, managed, modelled, simulated, analysed and visualised. This sequential journey mirrors the structure of DTUCS Prong-B (Section 5.2.2.), which classifies this journey into hierarchical yet independent functions or *DT uses*.

- The *mirror* DT use in DTUCS Prong-B correspond to the *data acquisition, transition, integration, and management* layers of the reference architecture.
- The *analyse* DT use maps directly to the *modelling, simulation, and analytics* layers.
- *Communication* DT use aligns with the *visualisation* layer, while *control* DT use reflects the *control and service* layers.

Each layer of the reference architecture is composed of common functions and technologies relevant to specific stages in the data-value chain, supported by overarching security and curation principles, which apply across all layers. This hierarchical yet modular design allows the reference architecture to function flexibly. Layers can operate independently while maintaining seamless data flow between them, simplifying data tracking and visualisation.

This modular approach also extends the utility of the reference architecture. It can map fully operational DTs that encompass the entire data-value journey, as well as smaller-scale projects that may evolve into more complex systems. Practitioners and organisations can select relevant functions and technologies for their projects through a user-friendly web form on the DT Hub platform. The system collects responses to provide insights into the broader DT landscape, identifying trends, frequently used technologies, and gaps or areas requiring development.

The reference architecture is built upon rigorous foundations. Its structure is informed by the systematic literature review conducted in this book's third chapter, examining previously published DT reference architectures in the retrieved studies. Additionally, its layout reflects the principles established in the Gemini Papers, which promote the data-value chain as the backbone of DT functionality. The layered design progresses step-by-step through this chain, while vertical security and curation columns ensure that every layer operates securely and efficiently.

The selection of layers and components in the reference architecture draws from three key sources

(a) a review of existing reference architectures identified in the literature
(b) case studies from the DT domain, both publicly available and others directly delivered by Mott MacDonald
(c) in-house expertise from Mott MacDonald, leveraging extensive knowledge of DT capabilities and technologies.

During the development of the reference architecture, the project team initially proposed introducing a layer called 'people' above the topmost layer, 'service.' This layer aimed to categorise the various types of users interacting with digital twin services, ensuring that every DT project prioritised user- and human-centric design. However, recognising the fundamental importance of this aspect, the DT Hub opted to expand the concept of the 'people' layer into a dedicated piece of future work. This approach ensures that the most critical element of any DT project – the people it serves – is not merely represented as an additional layer, but rather as a central focus deserving its own detailed framework. By addressing this separately, the DT Hub underscores its commitment to creating truly user-driven digital twin ecosystems.

REFERENCES

Archer MS (2005) Structure, culture and agency. In *The Blackwell Companion to the Sociology of Culture* (Jacobs MD and Hanrahan NW (eds.)). Blackwell Publishing, Chichester, UK, pp. 17–34.

Booch G (2005) *The Unified Modeling Language User Guide*. Pearson Education India, India.

Checkland P (1999) *Systems Thinking, Systems Practice: Includes a 30-Year Retrospective*. John Wiley and Sons Ltd, Oxford, UK.

Hetherington J and West M (2020) *The Pathway Towards an Information Management Framework—A 'Commons' for Digital Built Britain* [Report]. CDBB. 10.17863/CAM.52659 (accessed 18/09/2024).

National Digital Twin Programme (2021) *Digital Twin Toolkit*. https://digitaltwinhub.co.uk/files/file/62-digital-twin-toolkit/ (accessed 18/09/2024).

Ramy Elsehrawy and Bimal Kumar
ISBN 978-1-83662-235-2
https://doi.org/10.1108/978-1-83662-234-520251008
Copyright © 2025 by Ramy Elsehrawy and Bimal Kumar. Published under exclusive licence by Emerald Publishing Limited

Chapter 6
Philosophical foundations: critical realism

6.1. Introduction

In Section 3.8, the literature was reviewed and analysed at a more abstract and deeper level, focusing on the philosophical assumptions underpinning the retrieved studies. It is rare for these philosophical assumptions to be explicitly stated or examined in DT research. However, the consequences of overlooking these assumptions in any research are significant. First, conducting research without being clear about its theoretical or philosophical foundations makes it difficult to ensure rigour and produce reliable findings. Second, it becomes challenging for other researchers to evaluate a study that is unclear about its philosophical underpinnings, as it lacks crucial evidence for such assessment.

These implications of a-theoretical research are widely acknowledged in other fields such as systems thinking (Jackson, 2019), Operational Research (Lane, 1999), and Information Systems research (Mingers, 2006). A-theoretical research limits reflexivity, provides no basis for well-grounded critique and, therefore, hinders the development of the emerging *DT for UM* paradigm. For this paradigm to mature, its members must reflect on the outcomes of their own and each other's practices, analyse them, explain why certain DT approaches are more effective, and contribute to public knowledge. Reflecting on or criticising practices that are ad-hoc and disconnected from a robust theoretical foundation is hardly constructive (Flood and Jackson, 1991). In short, 'practice which is not reflective about the ideas upon which it is based will abandon the chance to learn its way to better ways of taking action' (Checkland and Scholes, 1990: p. xiv). As Paton (2001: p. 100) notes, theory can help us 'move beyond simply using methods which merely work in the short term to understanding why and how they do so, and this enhances our ability both to communicate between practitioners and to evolve better methods.'

6.2. Way forward – three theoretical propositions

To address the 'dilemma of pluralism' discussed in Section 3.8.5, it is necessary to establish a theoretical foundation that allows for pluralistic DT practice without sacrificing theoretical coherence. This Section introduces three alternative philosophies: post-modernism, ontological flexibility and critical realism. These philosophies were chosen because they represent distinct perspectives on how pluralistic practices should be theoretically grounded.

Post-modernism advocates for a deliberate indifference towards theory, suggesting that practitioners should simply follow what feels right. Ontological flexibility supports discordant pluralism, encouraging different DT approaches to challenge one another. In contrast, critical realism promotes complementary pluralism, proposing that all DT ideal approaches should be combined within a single paradigm that can accommodate them all.

6.2.1 Post-modernism

An argument might suggest that pluralism aligns with post-modernism. Post-modernism embraces relativism and promotes 'difference' in a heterogeneous world that requires a high degree of variety and dynamism. At first glance, relying on post-modernism seems to offer the missing theoretical foundation for the flexible, ad-hoc mixing of DT methods. However, this book contends not only that the weaknesses of this proposition outweigh its strengths but, also, that it is inherently problematic.

A well-known critique of post-modernism comes from the English philosopher Roger Scruton, who stated, 'a writer who says that there are no truths, or that all truth is 'merely relative,' is asking you not to believe him (sic). So don't' (Scruton, 2012). This critique highlights post-modernism's inability to justify a particular action or the use of a specific DT approach in practical interventions. While it encourages pluralism and the use of all available means, post-modernism goes beyond healthy scepticism to extreme relativism, remaining neutral or indifferent to all proposed approaches. It suggests that all intervention plans and DT tools and techniques are equally and relatively valid, which 'seems to lack an imperative to action' (Ormerod, 1996). Furthermore, 'for theory to be valuable, it must enable action' (Remington and Pollack, 2016).

Consequently, post-modernism offers little promise of contributing to epistemology. As the philosopher Noam Chomsky stated, 'postmodernism is meaningless because it adds nothing to analytical or empirical knowledge' (Chomsky, 1995).

6.2.2 Ontological flexibility

Ontological flexibility (OF) advocates for a philosophically or theoretically informed pragmatism (Zhu, 2011) rather than the naïve or postmodernist approach of 'do what feels good' critiqued earlier. In OF, practitioners are not bound by a particular paradigm or worldview. It 'respects all kinds of ontology but accepts obligation to no one. It puts into use diverse ontologies in the face of changing circumstances. It examines and refines them in the light of practical consequences' (Zhu, 2012: pp. 3–4). Thus, the justification for adopting a particular DT approach lies in its practical adequacy and ability to effect desired changes. OF is explicitly pragmatic and supports pluralistic practice but is not a-theoretical. 'It has to know what theories it is using to understand and act upon the world, in order that it can decide which of them enable objectives to be achieved and which don't' (Jackson, 2019: p. 587).

Zhu views OF as a more fruitful alternative to paradigm-based theorisation, positioning it as action-oriented, whereas paradigm-based approaches hinder the free use of mixed methodologies. He advocates for 'moving beyond paradigm-based theorising. After paradigm, there are many opportunities' (Zhu, 2011: p. 784). Jackson (2019: p. 587) draws on OF to support his multi-paradigm, multi-methodology and multi-method critical systems practice. He highlights that 'it is humans who impose structure on the complexity of *the world*. Because we do not have direct access to the *external world*, we cannot judge our theories by how well they correspond to it. Instead, we must justify our beliefs and actions by their practical effectiveness. Further, because *reality* is not static, and our beliefs are important in constructing *the world*, we need to look for and employ concepts that are effective in helping us to achieve our goals and in bringing long-term benefits' (italicised emphasis added by author).

While OF offers significant advantages, such as unbounded flexibility that allows practitioners to embrace pluralism without concern for theoretical inconsistencies, it also presents challenges. First, DT practitioners would need to examine or select their theoretical stance each time they employ a specific DT method or practice, which is a complex and impractical process. Practitioners who are more focused on applications than on philosophical debates may find this convoluted, potentially widening the gap between practical application and theoretical reflection, contrary to the aims of this research.

Second, OF appears to operate at a meta-level beyond paradigms, allocating suitable methods to different aspects of a problem as necessary. The difficulty, as Tsoukas (1993: p. 315) notes, is that 'reality-shaping paradigms … are not a la carte menus; you don't just pick whatever suits you at any time.' Similarly, Luhmann (2013: p. 101) emphasised that 'the observer does not exist somewhere high above reality. He (sic) does not hover above things and look down from above to observe what is going on. Nor is he a subject … outside the world of objects. Instead, he is in the middle of it all.'

Therefore, OF can be viewed in one of two ways. First, it could be seen as a paradigm in its own right, which introduces a paradox since it claims to move beyond paradigms while simultaneously introducing a new one. The second, more rational, argument rests on two key OF assumptions. First, advocates of OF, such as Zhu and Jackson, acknowledge on multiple occasions (see emphases added above) an external pluralist reality that exists independently of and is inaccessible to humans. Second, OF regards all paradigms as falsifiable, with a paradigm's practical adequacy determining its suitability in a specific problem context. In this view, ontologies are seen as fallible theories about the inaccessible reality, available for practitioners to use as they see fit.

Given these assumptions, it becomes difficult to see why OF should not eventually be reduced to the doctrine of CR, which is the third proposition discussed below and argued to be the most appropriate theoretical foundation for a pluralistic *DT for UM*.

6.2.3 Critical realism

Another philosophical paradigm proposed by this book to underpin DATUM is CR. CR, along with its core principles, was discussed in Chapter 2. Two additional key principles of CR, which are integral to the DATUM and its philosophical dimension, are: (1) the view of the world as an open system, and (2) the ethical dimension of life. These principles are considered fundamental to the development of DATUM. Table 6.1 provides a summary of all CR principles, highlighting their relevance to DT practices, particularly in relation to the four DT *ideal* approaches conceptualised in Chapter 3. This summary supports the argument for promoting CR as the philosophical foundation for the DATUM.

6.2.3.1 World is an open system

CR views the world as 'open'. 'Closed' systems, typically found in natural science lab experiments, can be generated under specific conditions where two criteria of closure are met.

Intrinsic condition (i.e. the system is internally stable and unchanging): For an object (X) to be intrinsically closed, as shown in Figure 6.1, its internal structure (S) – the source of its causal powers (p) and liabilities (l) – must remain qualitatively unchanged. This means there must be no alterations to its components or their relationships. If such changes occur, the system's causal powers and tendencies will also shift.

Extrinsic condition (i.e. the system is isolated from, or maintains constant relations with, the surrounding environment): For extrinsic closure, the system must maintain constant interactions with other objects in its environment. As shown in Figure 6.1, the contextual conditions (c1, c2, c3, etc.) – which are contingently related to object (X) and within which its causal mechanisms operate – must remain constant to demonstrate regular changes or events (e1, e2, e3, etc.).

Table 6.1 Summary of critical realism key principles and their relevance to different DT practices (continued on next page)

CR principles	Description	Relevance to DT practices
Independent reality	At the heart of CR lies the explicit divorce between an ontological reality and our knowledge of it. This entails an acknowledgment of an independent reality 'out there' regardless of whether we perceive it or even know about it or not. As Trigg (1980) argues, what reality is and how we have conceived it are different questions since many things are beyond our conceptual and linguistic capacities. CR, thus, gives primacy to ontology and avoids committing what Bhaskar calls the 'epistemic fallacy' – that is, collapsing ontology into epistemology, where we let 'the question 'what can we know?' determine our notions of what exists' (Bhaskar, 1975: p. 36).	This justifies *tech-driven* endeavours to capture the reality 'out there' using sensing technologies and with high fidelity. Alternatively, spatiotemporal urban big data analytics can be used to reveal patterns of inequalities inherent to existing urban structures, as in the *disruptive* approach.
Epistemic relativism	Human knowledge is finite, contextual, value-laden, socially constructed and always fallible. Knowledge is dependent on and created by the knower, and, thus, changes from knower to another. Our limited capacity to perceive all levels of reality (see 'stratified reality' below) and the uniqueness of each individual's perspective make it impossible for us to hold perfect true knowledge of everything about the real world.	This encourages stakeholders' engagement and the utilisation of DT outcomes to help different people gain different insights as recommended by the *cognitive* approach. Moreover, this principle fosters the idea of citizen participation like the *humanistic* approach does.
Structure, stratification, hierarchy and emergence	For CR, the real world is stratified. It is nested into the three domains of *real*, *actual* and *empirical* (Figure 2.1). The *real* includes the generative mechanisms attributed to objects of reality by virtue of their structures. These mechanisms are what 'can cause something in the world to happen' (Danermark *et al.*, 2005: p. 55). The domain of actual involves the emerging events these generative mechanisms produce. An event is the occurrence resulting from the activation	For example, in the context of urban environment, socio-economic factors may act as mechanisms in the domain of the *real* which, when interacting together, might give rise to events in the domain of the *actual* like transportation trips, traffic jams and others which could then be measured or quantified in the domain of the *empirical* (Næss, 2015). This CR principle, first,

Table 6.1 Continued

CR principles	Description	Relevance to DT practices
	of one or more generative mechanisms. However, events are ontologically distinct from the mechanisms generating them (Bhaskar, 1975) – for instance, the enactment of a mechanism might not give rise to any events. Simply because the effect of a mechanism can be countervailed by counteracting effects generated by one or more other mechanisms (Gambetta, 1998). Conversely, the effects of an activated mechanism can be further exacerbated as a result of other mechanisms in play that produces reinforcing effects. The domain of *empirical* comprises the experiences and the subsets of events that have indeed been perceived.	encourages capturing big urban data using advanced technologies, as in *tech-driven* and *disruptive* approaches, to augment observations of the *empirical* domain. Second, it justifies the notion of gaining insights, as promoted by the *cognitive* approach, where upon observing DT outcomes urban managers can postulate the existence of causal mechanisms at the deeper strata of reality that are not directly accessible or observable yet playing a role in the emergence of the observed events.
Open and quasi-closed systems	CR views the world as open. In natural sciences closed systems can be generated under special circumstances satisfying the two conditions of closure: intrinsic condition (i.e. system is internally stable and unchanging) and extrinsic condition (i.e. system is either isolated from or maintaining constant relations with the surrounding environment). These conditions of closure, however, are seldom satisfied by social or socio-technical systems constituting the urban environment. Nonetheless, urban systems may manifest partial or short-lived states of closure, giving rise to quasi-closed or pseudo-closed systems. Such systems may occasionally and tentatively emerge, showing consistent or regular patterns of behaviour for a period of time which are known as demi-regularities or demi-regs (Lawson, 1997).	Thus, allowing for the use of big urban data analytics to detect these regularities in order to predict future trends as promoted by the *tech-driven* approach. Alternatively, spatiotemporal urban big data analytics can be used to reveal patterns of inequalities inherent to existing urban structures, as in the *disruptive* approach.

Table 6.1 Continued

CR principles	Description	Relevance to DT practices
Analytical dualism	According to CR, there is a relentless interplay between agency and social structure. Social structure is activity-dependent, nonetheless, once produced it retains emergent properties with relatively enduring causal powers exerting, in turn, influences on agents, shaping the conditions amid which human activities occur. In short, 'structure and agency are separate strata, that is, they possess completely different properties and powers, but the one is essential for how the other will be moulded.' (Danermark *et al.*, 2005: p. 181).	The implication of this principle on DT practices is twofold. Acknowledging the agential power, free-will, and reflexivity of people supports citizens engagement and participation like *humanistic* practitioners. Simultaneously, recognising the influences, constraints and enablements urban structures can exert on people calls for *disrupting* them to bring about radical change to their lives.
Ethical dimension of life	There is also an emancipatory dimension to CR that chimes with both radical change paradigms (i.e. radical structuralism and radical humanism). Falling back on epistemic relativism, CR accepts the fact that different individual perspectives, social context, and cultural traditions exist. However, beneath all lie 'moral truths' which are grounded in human nature, upon which 'universalized freedom for all' can be realized (Mingers, 2014).	A *disruptive* approach appears to be most effective in undertaking interventions explained in (a).
	(a) On the one hand, to CR, it is crucial to unveil the generative mechanisms or structures giving rise to false interpretations or coercive status quo hindering our realization of a moral-truths-based society. Once identified, such structures can be altered or deconstructed through deliberate interventions (Wilson and Greenhill, 2004). A *disruptive* approach appears to be most effective in undertaking such interventions.	Arguments in (b) are most related to the *humanistic* DT approach and its concerns about emancipation of the individual.

Table 6.1 Continued

CR principles	Description	Relevance to DT practices
	(b) On the other hand, Sayer (2011) views humans as naturally needy, evaluative, ethical beings who experience flourishing and suffering. Accordingly, we should be more attentive to people's first-person evaluative relation to the world. Drawing on the capabilities approach, Sayer suggests that when thinking about how people may flourish, we need not focus only on what they have but also on what they can do or be. In relation to datafication, Kennedy and Hill (2018: p. 25) highlight that 'capabilities might include being able to have control over one's own data, to choose to opt out of – or, better still, into – data gathering, and to make sense of data mining processes because they are made transparent to non-expert citizens, or accountable to expert others'.	
Judgmental rationalism	Although our account of what is 'out there' and our theories of how things are, are always fallible (see 'epistemic relativism' above), one can still judge a hypothesis about reality and how true or fallible it is compared to its rivals based its rationality and explanatory power.	This principle endows DT practitioners with the ability to rationally select, depending on the context and problem under study, the right methods or approaches deemed to be more appropriate for developing and implementing a DT.

In natural science, closed systems are deliberately created to satisfy both intrinsic and extrinsic conditions of closure, generating regularities that, for empiricists, equate to scientific laws. For critical realists, however, these regularities, produced in controlled environments, present an opportunity to understand the underlying generative causal mechanisms of a phenomenon. By distinguishing between the three layers of reality (as discussed in Section 2.2.5), CR appreciates the effort required to produce empirical invariances (Bhaskar, 1975). Essentially, from a CR perspective, the success of natural sciences is largely due to their ability to meet conditions of closure, which reveals the nature of objects and their causal powers (Sayer, 1992).

Figure 6.1 'Intrinsic' and 'extrinsic' conditions of closure. Adapted from Sayer, R.A. (1992)

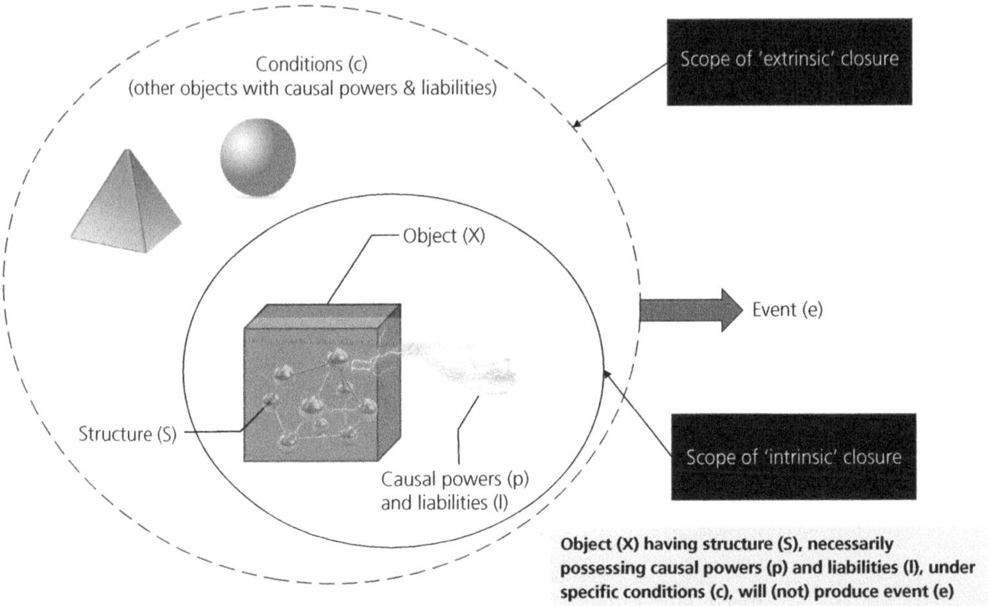

However, it is rarely the case that social or socio-technical systems, such as urban environments, satisfy either of the conditions of closure, let alone both. The considerable challenge in establishing stable behavioural patterns or regularities in these systems does not indicate immature science or unskilled scientists but, rather, reflects the fact that these systems are 'open'. Such systems typically violate closure conditions, making them extremely difficult to close, and as a result, they fail to demonstrate enduring conjunctions. Despite this, social and socio-technical systems may occasionally experience partial or short-lived states of closure, leading to what is referred to as *quasi-closed* systems (Sayer, 1992) or *pseudo-closed* systems (Lawson, 1997).

When quasi-closed systems sporadically and tentatively emerge, they may exhibit consistent or regular behavioural patterns for a limited period, known as demi-regularities or *demi-regs* (Lawson, 1997). A demi-reg is a non-enduring regularity, manifesting within restricted spatial and/or temporal boundaries, suggesting some level of stability among the underlying generative mechanisms and the contextual conditions in which they operate. In some instances, quasi-closed systems are artificially and deliberately created to sustain emerging patterns of regularity, allowing for better controllability in a dynamic world. However, if they do occur, these systems are superficial and temporary, unlike the fully closed systems produced in natural science experiments or lab-like environments.

Systems involving human activity are inherently open (as discussed in Section 3.2.5), though they may show signs of quasi-closure in various situations or contexts. Given the relatively greater stability, controllability or predictability that quasi-closed systems and the demi-regs they exhibit can provide, some critical realists have emphasised the significance of such systems in generating knowledge when dealing with largely 'open' social or socio-technical systems, as is the case in urban planning and urban research (Næss, 2004, 2015).

6.2.3.2 Ethical dimension of life

As Bhaskar asserts, 'some views of human nature will characteristically form the basis of the ethical ingredient' (2009: p. XXV). Building on this, Sayer (2011) presents his view of human nature as one where humans are needy, evaluative, and ethical beings who experience both flourishing and suffering. He argues that life has an ethical dimension, encompassing values, emotions, beliefs, desires and concerns, which can be expressed through the experiences of human flourishing and suffering. Furthermore, Sayer (2011) emphasises that while flourishing or suffering is real (i.e. objective), the ethics surrounding these experiences are shaped by culturally constructed, fallible views of how people flourish or suffer.

For CR, flourishing and well-being are understood as objective, yet pluralistic. This means that different cultural forms of flourishing and suffering may exist. However, adhering to the principle of judgmental rationalism (Section 2.2.3), CR allows for the assertion that some forms of flourishing are better than others. Sayer (2011) further operationalises his approach by drawing on the 'capabilities approach', pioneered by Martha Nussbaum and Amartya Sen (1993). This approach offers practical tools for comparing different forms of flourishing and suffering. Rather than focusing on how happy an individual is, the capabilities approach emphasises what people need to be capable of doing and being, should they choose to. This connection suggests that CR, through the capabilities approach, can offer a structured way to evaluate the ethical implications of digital and data-driven environments.

In the context of digitisation and the contemporary age of data, Kennedy (2018: p. 25) provides valuable insights on how the capabilities approach can guide the definition and pursuit of ethical data-related practices.

> *'Ideas about capabilities and flourishing might help us to understand what is important for people to live their lives well with data. In relation to datafication, capabilities might include being able to have control over one's own data, to choose to opt out of – or, better still, into – data gathering, and to make sense of data mining processes because they are made transparent to non-expert citizens, or accountable to expert others. The problem is that, whilst these issues are widely discussed amongst data activists, ordinary people's perspectives on whether they might result in living better with data are missing from these debates. This is why we need to produce the kinds of first-person evaluations of 'living well with data' that Sayer advocates.'*

Figure 6.2 illustrates how CR principles support the implementation of observed DT practices from the literature, which are in turn aligned with the relevant DT approaches. It is important to emphasise that combinations of various CR principles enable the application of a wider range of DT practices. However, capturing all potential combinations in a single figure is challenging due to the complexity involved – for instance, the principles of 'epistemic relativism' and 'stratified reality' together demonstrate how different individuals may derive diverse insights into the deeper layers of reality, beyond what is immediately observable.

Figure 6.2 The principles of CR promoting pluralism and encapsulating all DT ideal approaches (author's own)

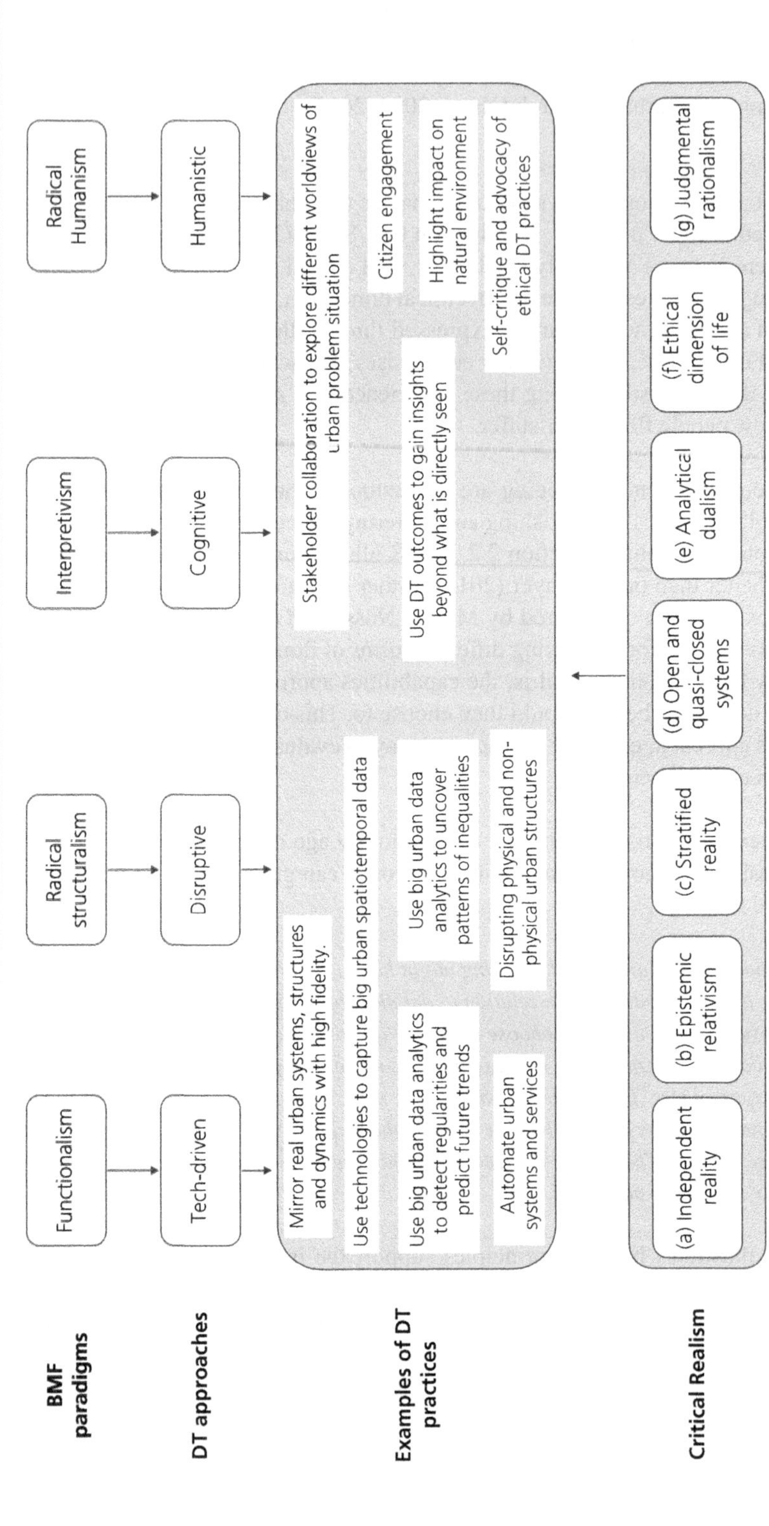

REFERENCES

Bhaskar R (1975) *A Realist Theory of Science*. Leeds Books Ltd, Leeds, UK.

Bhaskar R (2009) *Scientific Realism and Human Emancipation*. Routledge, New York, USA. 10.4324/9780203879849 (accessed 18/09/2024).

Checkland P and Scholes J (1990) *Soft Systems Methodology in Action*. Wiley, Oxford, UK.

Chomsky N (1995) *Chomsky on Postmodernism*. Z-Magazine's Online Bulletin Board.

Danermark B, Ekstrom M and Jakobsen L (2005) *Explaining Society: An Introduction to Critical Realism in the Social Sciences*. Routledge, New York, USA.

Flood RL and Jackson MC (1991) *Creative Problem Solving: Total Systems Intervention*. Wiley, Oxford, UK.

Gambetta D (1998) Concatenations of mechanisms. In *Social Mechanisms: An Analytical Approach to Social Theory* (Hedstrom P and Swedberg R (eds.)). Cambridge University Press, Cambridge, UK, pp. 102–124.

Jackson MC (2019) *Critical Systems Thinking and the Management of Complexity*. Wiley, Oxford, UK. https://books.google.co.uk/books?id=1jmNDwAAQBAJ (accessed 18/09/2024).

Kennedy H (2018) Living with data: aligning data studies and data activism through a focus on everyday experiences of datafication. *Krisis: Journal for Contemporary Philosophy* **2018(1)**: 18–30.

Kennedy H and Hill RL (2018) The feeling of numbers: Emotions in everyday engagements with data and their visualisation. *Sociology* **52(4)**: 830–848.

Lane DC (1999) Social theory and system dynamics practice. *European Journal of Operational Research* **113(3)**: 501–527.

Lawson T (1997) *Economics and Reality*. Routledge, Milton, UK. 10.4324/9780203195390 (accessed 18/09/2024).

Luhmann N (2013) *Introduction to Systems Theory*. Polity Press, Cambridge, UK.

Mingers J (2006) *Realising Systems Thinking: Knowledge and Action in Management Science*. Springer US, New York. https://books.google.co.uk/books?id=C-UcDQEACAAJ (accessed 18/09/2024).

Mingers J (2014) *Systems Thinking, Critical Realism and Philosophy: A Confluence of Ideas*. Routledge, New York, USA.

Næss P (2004) Prediction, regressions and critical realism. *Journal of Critical Realism* **3(1)**: 133–164. https://doi.org/10.1558/jcr.v3i1.133 (accessed 18/09/2024).

Næss P (2015) Critical realism, urban planning and urban research. *European Planning Studies* **23(6)**: 1228–1244. 10.1080/09654313.2014.994091 (accessed 18/09/2024).

Nussbaum M and Sen A (1993) *The Quality of Life*. Clarendon Press, London, UK. https://books.google.co.uk/books?id=QurkDwAAQBAJ (accessed 18/09/2024).

Ormerod RJ (1996) Combining management consultancy and research. *Omega* **24(1)**: 1–12.

Paton G (2001) A systemic action learning cycle as the key element of an ongoing spiral of analyses. *Systemic Practice and Action Research* **14(1)**: 95–111.

Remington K and Pollack J (2016) *Tools for Complex Projects*. Routledge, New York, USA.

Sayer RA (1992) *Method in Social Science: A Realist Approach*. Psychology Press, Hove, UK. https://books.google.co.uk/books?id=kr8QvOpM2RwC (accessed 18/09/2024).

Sayer RA (2011) *Why Things Matter to People: Social Science, Values and Ethical Life*. Cambridge University Press, Cambridge, UK.

Scruton R (2012) *Modern Philosophy: An Introduction and Survey*. A&C Black, London, UK.

Trigg R (1980) *Reality at Risk: A Defence of Realism in Philosophy and the Sciences*. Harvester Press. https://books.google.co.uk/books?id=8Y7pqn1MQBUC (accessed 18/09/2024).

Tsoukas H (1993) The road to emancipation is through organizational development: a critical evaluation of total systems intervention. *Systems Practice* **6(1)**: 53–70.

Wilson M and Greenhill A (2004) Theory and action for emancipation: elements of a critical realist approach. In *Information Systems Research*. Springer, New York, pp. 667–674.
Zhu Z (2011) After paradigm: why mixing-methodology theorising fails and how to make it work again. *Journal of the Operational Research Society* **62(4)**: 784–798.
Zhu Z (2012) *Essai: Process Thinking Without Process Ontology: Unpublished Paper.*

Ramy Elsehrawy and Bimal Kumar
ISBN 978-1-83662-235-2
https://doi.org/10.1108/978-1-83662-234-520251009
Copyright © 2025 by Ramy Elsehrawy and Bimal Kumar. Published under exclusive licence by Emerald Publishing Limited

Chapter 7
Methodology: data-driven multimethod methodology [DM2]

7.1. Introduction

At the methodical level, the DTUCS is structured around three core components. Prong-A encompasses the multi-dimensional classification framework for DT use cases, while Prong-B features the DT uses taxonomy. Prong-C employs the UML to model DT UCS. Alongside this, the philosophical foundation of DATUM is established in Chapter 6, drawing on the principles of CR. CR is recognised as an inherently pluralistic philosophy, offering a solution to the dilemma of pluralism discussed in Chapter 3.

However, there is an evident disconnect between the methodical and philosophical elements. The tangible, practical methods outlined in DTUCS lack a direct connection to the underpinning philosophy, while the philosophy of CR remains too abstract to be directly applicable for practitioners without further development. To bridge these two dimensions, the methodological component of DATUM, known as the data-driven multi-method (DM2) methodology, is introduced in this chapter.

First, existing CR-informed methodologies are reviewed, providing the foundation for the proposed approach. The subsequent section then builds on this literature to outline the key steps and guiding principles upon which the methodological aspect of DATUM is constructed. The methodology is tailored to suit the specific processes of DT and is designed to integrate its capabilities effectively.

7.2. Existing CR-informed methodologies

The methodologies informed by CR, drawn from the literature, are summarised in Table 7.1. While all these methodologies use CR to derive their guiding principles, they differ in the types of research they aim to support. The first two align with Gibbons *et al.*'s (1994) distinction between 'Mode 1' and 'Mode 2' research. Mode 1, or 'to-know' research, is purely explanatory and seeks to answer the question of 'why' the world functions as it does. Mode 2, or 'to-do' research, focuses on practical knowledge, seeking to solve specific problems or realise desirable outcomes. Mode 2 is typically employed in application-driven contexts, such as design and problem-solving. The third type of research is what we refer to as 'Mode 3', or 'critical research', which concentrates on critiquing the oppressive conditions that inhibit human emancipation and freedom.

These three modes of research are complementary, rather than competitive. Mode 2 does not replace Mode 1, but rather builds on it (Dresch *et al.*, 2015). Similarly, Mode 3 does not supplant Modes 1 and 2; instead, it introduces a new dimension, broadening the way our understanding of the world (gained through Mode 1) and our practical interventions (executed through Mode 2) can contribute to human flourishing and freedom. As Wilson and Greenhill (2004: p. 671) explain, 'revealing the way things are (achieved through Mode 1) is a necessary step to demonstrate the

Table 7.1 Existing CR-informed methodologies

Research mode	(Bhaskar, 1975)	(Danermark *et al.*, 2005)	(Wilson and Greenhill, 2004)	(Mingers, 2006)	(Sayer, 2011)	(Wynn and Williams, 2012)	(Fletcher, 2017)
Mode 1	Resolution	Description		Appreciation		Explication of events; Explication of Structure and Context	Abduction
	Redescription	Redescription (Abduction)					
	Retrodiction	Retroduction		Analysis		Retroduction	Retroduction
	Elimination	Comparison; Evaluation; Concretization		Assessment		Empirical corroboration	Identification of demi-regularities
	Identification						
						Triangulation and Multi-methods	
Mode 2	Action			Action			
Mode 3	'Ethical tetrapolity'	Critique	'Critical element': Construction of alternatives; Questioning status quo; Deconstruct dominant ideology; Equality and Inequality; Emancipation		Capabilities approach		

place of human acts (guided by Mode 2) in the 'reproduction of social structures and relations that stand in the way of emancipation" (Ackroyd and Fleetwood, 2000: p. 23).

The authors cited in Table 7.1 provide comprehensive accounts of the methodologies they have developed. It is beyond the scope of this book to fully explore their contributions. Therefore, what follows is a brief summary of the most significant and well-developed aspects of each author's methodological contribution. This will enable us to draw on the strengths of each CR-informed methodology to support the formulation of DM2.

7.2.1 Mode 1 research

When Bhaskar (1975) introduced CR, he outlined two explanatory models, DREI(C) and RRREI(C), to distinguish between the approaches suitable for natural and social sciences, respectively. The DREI(C) model is primarily applicable to the 'closed systems' often found in controlled laboratory environments (see Section 6.2.3.1 for more on open and closed systems from a CR perspective). This model follows a series of steps.

Description – Identify and describe an observed pattern or consistent behaviour.

Retroduction – Propose plausible causal mechanisms that could explain the observed phenomena. Unlike traditional deductive or inductive reasoning, retroduction involves postulating hypothetical mechanisms based on CR's ontological assumptions of structure, hierarchy, and stratified reality (see Sections 2.2.4 and 2.2.5).

Eliminate – Use rational judgment to rule out less likely explanations.

Identify/Correct – Pinpoint or adjust the generative mechanism at work, corroborating the explanation with further experiments or advanced technology to uncover unobserved mechanisms.

The RRREI(C) model, on the other hand, is more suited to the 'open systems' typical of the social world and socio-technical environments. Its steps include the following.

Resolution – Break down the complex event under investigation, defining the boundaries and key components of the situation.

Redescription – Uniquely, this model assumes that existing theories and mechanisms can be applied to redescribe the situation identified during resolution.

Retrodiction[1] – Utilise available theories to suggest plausible mechanisms that could explain the occurrence of the event.

Elimination – Eliminate alternative explanations and choose the one deemed most rational and powerful.

Identification/correction – Corroborate or refine the chosen explanation using triangulation and empirical evidence.

Bhaskar's models (DREI(C) and RRREI(C)) laid the groundwork for subsequent methodologies. These later developments, while focusing on different aspects, remain consistent with Bhaskar's

[1]Retroduction and retrodiction both refer to the same logic form. Retrodiction refers to falling back on previously identified mechanisms to explain events emerging in new contextual settings or conditions. Retroduction, however, refers to postulation of new mechanisms that can justify the emergence of observed events in case no previously identified mechanisms were found to be adequate to do so. For simplicity, the term retroduction is used henceforth to refer to both.

original frameworks. Danermark *et al.* (2005) expanded on these models by adding 'concretisation', emphasising the use of empirical evidence to record how specific mechanisms produce observed events (Danermark *et al.*, 2005: p. 194).

Similarly, Wynn and Williams (2012) developed a set of methodological principles for conducting and evaluating CR-based case studies. Like Danermark and his team, they emphasised 'empirical corroboration', 'ensuring that the proposed mechanisms adequately represent reality, and have both sufficient causal depth and better explanatory power than alternative explanations for the focal phenomenon' (Wynn and Williams, 2012: p. 801). While pluralism and the use of multiple methods are widely advocated by critical realists (Mingers, 2006), Wynn and Williams took this further by introducing the principle of 'triangulation and multi-methods', supporting 'causal analysis based on a variety of data types and sources, analytical methods and theoretical perspectives' (Wynn and Williams, 2012: p. 803).

One of their significant contributions was bridging theory and practice by contextualising these principles within information systems research, making explicit links to CR's ontological and epistemological foundations. Still within Mode 1 research, Fletcher (2017) offers a practical example of a CR-informed framework for qualitative research, placing particular emphasis on identifying 'demi-regularities', or 'demi-regs' (Section 6.2.3.1), through quantitative methods to guide retroduction.

7.2.2 Mode 2 research

Mingers (2006) brings CR directly into Mode 2 research, focusing on developing a methodology aimed at interventions that lead to tangible improvements in reality. Bhaskar (1993: p. 243) briefly touched on applying CR for problem-solving with his DEA model, which stands for 'Diagnose the problem, Explain it, and take appropriate Action to resolve it.' However, this model is considered less developed compared to the DREI(C) and RRREI(C) frameworks (Mingers, 2006).

In response, Mingers (2006) proposed a more comprehensive CR-informed methodology, consisting of four key stages.

> *Appreciation* – This initial phase involves stakeholders and practitioners describing the situation at hand, identifying key concerns, and defining the system's boundaries as understood by the participants. A variety of data collection methods are employed during this stage.
>
> *Analysis* – Here, the gathered information is examined to explain why the situation exists in its current form. This involves retroduction, where hypothetical mechanisms are suggested that, if they exist in the deeper reality, would account for the observed phenomena.
>
> *Assessment* – Once a hypothetical explanation is established, it is tested through corroboration or refutation using empirical evidence derived from multiple quantitative and qualitative methods. Simultaneously, the explanation is compared with alternative hypotheses. This step aligns with the processes of elimination and identification/correction in Bhaskar's model.
>
> *Action* – The final stage focuses on using the knowledge obtained from the previous steps to guide practical actions that bring about the desired changes. This represents the transition from Mode 1 research, which seeks to understand the world, to Mode 2 research, which focuses on actively shaping it.

This methodology exemplifies the shift from theory to practice, providing a structured approach to applying CR in real-world contexts.

7.2.3 Mode 3 research

Beyond Modes 1 and 2, CR encompasses an additional emancipatory and ethical dimension. Bhaskar, in his works on dialectical critical realism (1993, 1994), lays out a framework for morality rooted in four key stages, which Mingers (2014) succinctly summarises.

> *Value-laden nature of social science:* Unlike the claim that social science is value-free, Bhaskar argues it is inherently value-laden and evaluative. This resonates with Kant's view in his *Critique of Practical Reason*, where he posits that humans, as phenomena in the realm of appearances, are subject to causality, but as noumena (things in themselves), they transcend our knowledge and perception. This transcendence justifies the principle of morality (see Section 3.2.5 for more on the unique and teleological role of human agency).
>
> *Deriving 'ought' from 'is':* Bhaskar further asserts that it is possible to deduce ethical imperatives from reality. In other words, certain structures or conditions can be deemed immoral if they perpetuate false beliefs or impose unnecessary constraints that limit human freedom.
>
> *Commitment to action:* Once moral judgments are made, there is an obligation to act and dismantle oppressive structures and harmful causal mechanisms. Critiquing an immoral situation naturally leads to the necessity of changing it.
>
> *Universalisation of actions:* Rather than addressing isolated issues, Bhaskar advocates for actions aimed at eliminating all forms of similar societal constraints, thus pursuing broader emancipation.

While Danermark and others (2005), along with Wilson and Greenhill (2004), based their Mode 3 research methodologies on Bhaskar's tetrapolity, Mingers (2014) highlights certain limitations in its practical application. One major critique is the assumption that people universally share the same set of values for identifying and addressing immoral situations. Mingers argues that this overlooks the inherent cultural diversity and pluralism in how flourishing or suffering is understood across different societies.

To address these concerns, Sayer (2011) turns to the notions of independent reality and epistemic relativism, proposing a pluralistic yet real conception of flourishing and suffering. In support of his argument's practical application, Sayer draws on the capabilities approach (see Section 6.2.3.2), shifting the focus from defining happiness to considering what individuals can do or become. This acknowledges that people's aspirations and notions of well-being vary according to their cultural contexts and beliefs.

7.3. Data-driven multi-method methodology [DM2]

The proposed data-driven multi-method methodology (DM2) is a CR-informed approach designed specifically for digital twinning. Drawing on the fundamental principles of CR (as outlined in Table 7.1) and the CR-informed methodologies reviewed earlier (see Section 7.2), DM2 offers a practical framework for professionals working with digital twins. As illustrated in Figure 7.1, DM2 is built upon

- a set of principles derived from CR, which serve as foundational guidelines for its implementation
- a group of key steps that practitioners follow, which ensure that the methodology is aligned with the core aspects of CR
- the notion of trialectic interplay, a key feature of DM2, which is explained in more detail below. This concept emphasizes the dynamic interaction between different layers or aspects of reality, aligning with CR's stratified view of the world.

Figure 7.1. Data-driven multi-method methodology [DM2] (author's own)

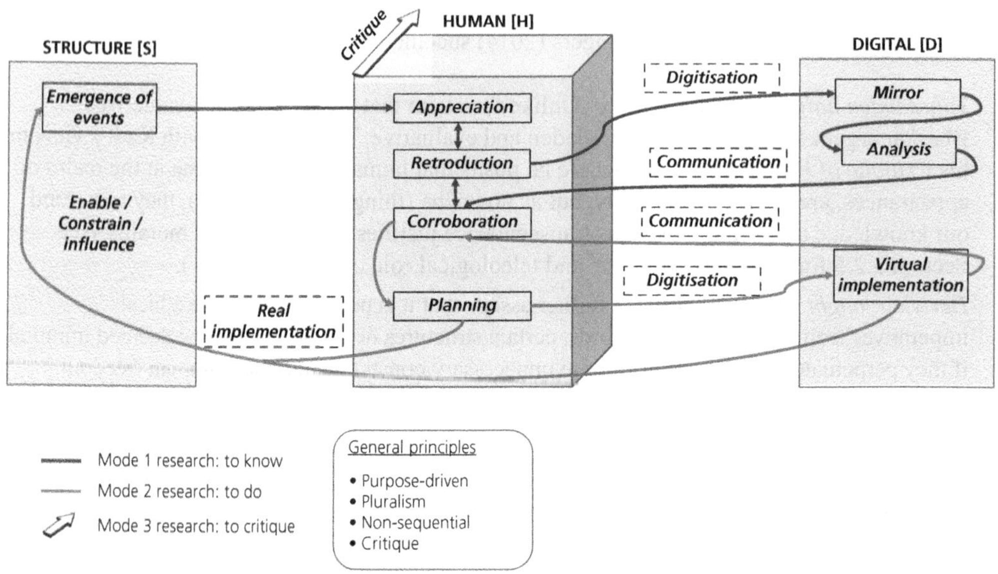

The DM2 methodology aims to integrate CR principles with the practical demands of digital twinning, offering a flexible yet rigorous approach that combines theory and practice for the benefit of DT projects.

7.3.1 The trialectic interplay: Structure – Human – Digital

In Section 3.7.4, four key capabilities of DT were discussed. These unique capabilities, we now argue, carry significant ontological and epistemological implications, which serve as the foundation for the contextualisation and development of the DM2 methodology.

Building on the concept of analytical dualism (Section 2.2.6), social, cultural, or socio-technical reality is traditionally seen as a dialectic between structure and human agency. However, due to the unique capabilities of DTs, it is more accurate to conceptualise urban reality as a 'trialectic', comprising *structure, human agency* and *digital*. The distinctive attributes of DTs have profound ontological and epistemological ramifications, supporting this notion of a trialectic interplay.

From an ontological perspective, the control capability of DTs suggests that DT possesses a form of 'agency', enabling it to 'react, predict, and act' (Tomko and Winter, 2019: pp. 395–397). This introduces the idea that DTs have agential power, allowing them to pursue predefined goals and act to achieve them. As Danermark *et al.* (2005: p. 179) explain, 'an agent has an intention. To fulfil a wish, that is to obtain a goal, the agent uses a means to that end. This is obviously the most important difference between social structures and agents: social structures cannot set up goals and they cannot act; only humans can – agents are the only effective causes of society.' Therefore, given their control capability, DTs are clearly more than mere structures, yet not equivalent to human agents. They represent a third ontological entity, actively contributing to the evolution of urban reality.

On the epistemological front, certain DT capabilities further support the argument that DTs should be considered independent entities, distinct from both *structure* and *human agency*. For instance, their analytical capabilities demonstrate how DTs can generate new knowledge beyond the reach of *human agency*. Whether by uncovering hidden patterns and demi-regularities within urban dynamics or by calculating the outcomes of a highly intricate web of causal factors, the knowledge generated by a DT has the potential to transform human understanding of urban reality, enabling interventions that would otherwise be inconceivable.

It is important to clarify that conceptualising a trialectic interplay does not imply that the *digital* component of reality is equivalent to *human agency*. Rather, it underscores the uniqueness of the DT, which exists neither as part of the structure nor as a human agent. DTs possess exceptional characteristics that afford them a distinct role within the dynamics of urban environments, allowing them to significantly influence the evolution of these environments. This largely theoretical argument is, in fact, implicitly acknowledged by DT practitioners, as evidenced by their perception of the relationships between the physical, social and digital worlds (Figure 3.13).

7.3.2 Emergence of events [S][2]

At this stage, new events emerge, prompting different approaches to research. These could follow Mode 1 (seeking to understand the nature of these events and the reasons behind their occurrence), Mode 2 (focusing on managing, controlling, or altering the situation brought about by these events), and/or Mode 3 (aimed at liberating individuals impacted by the events). The concept of new events emerging is grounded in three key principles of CR. First, it assumes the existence of an ontological reality that is independent of our knowledge of it. Second, it views this reality as an open system, which accounts for the ongoing internal and external transformations within its structures. Third, it embraces a stratified understanding of reality, where the events we observe are only a subset of those generated by numerous unseen mechanisms.

7.3.3 Appreciation [H]

The step of appreciation stems from CR's principles of a stratified reality, epistemic relativism and an open-system perspective of the world. This step focuses on thoroughly understanding the system, events or problem situation at hand. It involves abstracting, breaking down and resolving the experienced events into relevant components or structures. In the complex, open-system environment of cities, entities from the physical, social or digital realms may influence the emergence of the phenomena or events being studied.

To fully appreciate the phenomenon, it is crucial to consider the bigger picture. Just like everyone else, the researcher's or DT practitioner's perspective is inherently fallible. Therefore, gaining a comprehensive view and establishing a holistic boundary around the problem requires extensive investigation and a pluralistic approach. This can be achieved by mapping out the involved stakeholders, engaging them in the process, and incorporating their unique perspectives to generate a richer understanding of the situation. Employing multiple methods to define the problem's boundaries and components is a central tenet of DM2, as highlighted by the principle of pluralism (see Section 7.3.13). The outcome of the appreciation process typically includes details of events, variables and relevant structures, providing a holistic account of the situation.

[2]The letter in square brackets (i.e.: [S], [H], or [D]) associated with every DM2 step indicates the primary component of reality in this particular step. Two letters, however, (e.g. [H-D]) indicates that this step is mainly transitional, leading to the shift of focus from one component (e.g. [H]) to another (e.g. [D]).

Taking an open-system view encourages exploring the connections between the system in question and its broader environment. As Næss (2004, p. 143) argues, 'unpacking' multiple layers of explanations should 'be a goal', hence, 'repeatedly asking 'why' questions may be fruitful'. However, Næss acknowledges that there are limits to how deeply one can delve into these chains of explanation, due to factors such as time constraints, resources and the principle of parsimony (Wynn and Williams, 2012). This may require a careful balance, avoiding the temptation to 'sweep in' too many viewpoints into the system's boundaries (Jackson, 2019: p. 349). As Simon (1977: p. 258) puts it, 'everything is connected, but some things are more connected than others.' Therefore, DT practitioners must carefully weigh the purpose of the DT and the relevance of various factors to that purpose. Ultimately, the appreciation of a problem is a constrained, fallible process that must be driven by purpose to yield valuable and actionable insights.

Appreciation is also a continuous, iterative process – for instance, during the step of retroduction (Section 7.3.4), which involves hypothesising the existence of causal mechanisms, the practi tioner may need to revisit the appreciation stage to include structures that were initially overlooked but which influence those mechanisms. Similarly, if the proposed explanation fails to be corroborated (Section 7.3.5), the events might need to be redescribed, prompting a fresh round of appreciation.

7.3.4 Retroduction [H]

This key step in the DM2 methodology, equivalent to retroduction as presented in other CR-informed methodologies (Section 7.2), is grounded in the CR assumptions of a stratified reality, structure, hierarchy, emergence and judgmental rationalism. The goal of the appreciation phase was to describe and clarify the events and issues at hand, identifying relevant factors or structures.

However, it is through retroduction that one moves beyond mere descriptions of observations and structures, to make reasoned assumptions about the generative mechanisms and interactions between these structures. These mechanisms, existing at deeper levels of reality, are considered responsible for the events experienced. Retroduction is a way of determining, albeit tentatively, what causal powers or tendencies these structures possess, and how they contribute to the occurrence of observed phenomena. In essence, retroduction seeks to answer the question, 'What is it about the structure that might produce the effects in question?' (Sayer, 1992: p. 95). This process is aided by drawing on existing and previously corroborated theories that might explain the given events. If no current theory offers a sufficient or rational explanation, a new set of generative mechanisms may be proposed which, 'if they were to exist and act in the postulated way, would account for the phenomenon in question' (Bhaskar, 1975: p. 12).

At this stage in DM2, after completing retroduction, three fundamental assumptions can be made.

(a) *Structures* – These are the relevant components and structures involved in the situation and its contextual conditions.
(b) *Generative mechanisms* – These are the causal powers or tendencies within the involved structures, which enable them to exert influence or cause specific effects in the situation.
(c) *Interconnections* – This includes a detailed account of how the mechanisms interact, considering their levels of power, frequency of occurrence and how they influence or counteract one another. It also provides a systemic understanding of how these interactions give rise to the observed phenomena.

Following the appreciation and retroduction steps, the explanation might take the form of

(a) 'X' and 'Y' are components/structures relevant to the problem
(b) they possess generative mechanisms/causal powers 'P1' and 'P2', respectively
(c) the interaction or counteraction of 'P1' and 'P2' has resulted in the emergence of the observed event 'E'.

Like appreciation, retroduction is an iterative, evolving process. It can be thought of as a series of 'thought trials' (Weick, 1989) aimed at identifying the generative mechanisms and the contextual conditions that activate them, resulting in the observed phenomena. Retroduction often starts implicitly during the early stages of appreciation. Having a preliminary hypothesis in mind can lead to focusing more on relevant stakeholders and structures, while disregarding less relevant ones. As data is gathered and more perspectives are shared, the description of events may evolve, new structures or entities previously overlooked may be introduced, and the system boundary may need to be adjusted. This, in turn, may necessitate revisiting retroduction to refine the explanation in light of these new insights.

Furthermore, at any later stage of DM2, such as after reviewing the results of the DT analysis or virtual implementation steps of DM2 (as described later), the DT may provide insights that either corroborate, challenge, or refine the proposed explanation, prompting another round of retroduction.

7.3.5 Corroboration [H]

According to Bhaskar (1975), corroboration is the process by which the reality of postulated mechanisms is subjected to empirical scrutiny. This step in DM2 allows one to increase confidence in the proposed explanation and its underlying assumptions. The process of corroborating any hypothesis is grounded in the ontological belief that there is an independent reality, which may correspond in ways that confirm or challenge the accuracy of our knowledge. Moreover, judgmental rationalism permits us to claim, albeit tentatively, that a particular explanation or hypothesis is more plausible and rational than others.

During the appreciation and retroduction phases, participants are typically invited to share their perspectives on the problem under investigation. The outcome is a comprehensive view of the situation, which includes assumptions about the involved structures, generative mechanisms and the relationships between them. To validate this postulated understanding, it is crucial to scrutinise its foundational hypothesis by asking several key questions.

- Are the identified structures, variables or factors relevant to the problem and its context, or have some important structures been overlooked?
- Are the generative mechanisms assumed to be inherent within these structures real, active and effective, or have their effects been overestimated or underestimated due to the presence of countervailing mechanisms?
- Are the interconnected mechanisms truly responsible for the observed events or would their implications have led to different outcomes if they existed as postulated? Essentially, does the assumed network of interactions between the generative mechanisms accurately reflect their real relationships?

These questions are not exhaustive and can be expanded. Wynn and Williams (2012), for example, present a series of 'causal test questions' to aid in evaluating and corroborating proposed causal explanations.

While these questions outline what should be corroborated, they do not specify how corroboration should be carried out. In this regard, two key activities are essential for effective corroboration. First, adopting a pluralistic approach is crucial and is detailed as part of the general principles of DM2 (Section 7.3.13). Second, the utilisation of DT's unique capabilities – specifically its mirroring, analysis and communication functions – is vital. Although corroboration can often rely solely on direct empirical observation, the unique epistemological contributions of DT (see Section 7.3.1), and the new knowledge it generates, can greatly enhance the corroboration process. As Price and Martin (2018: p. 95) observe, the development of 'new technologies' enables researchers to 'literally see it' (i.e. the empirical evidence of a hypothesised theory).

Wynn and Williams (2012: p. 801) further explain that 'the application of accepted analytical methods, compelling logic, creativity, and intuition to the empirical data generates confidence that the hypothesized theoretical mechanisms approximate the powers and tendencies derived from the real structures.' Therefore, it is important to emphasise that corroboration is not achieved immediately through a single step. Instead, it occurs progressively as confidence increases, supported by accumulating empirical evidence that backs the initial hypothesis. In the context of DM2, corroboration unfolds gradually, beginning with *analysis*, followed by the *communication* of results, and culminating in insights drawn from this process.

7.3.6 Digitisation [H-D]

The process of digitisation serves as a crucial transition, marking the shift from human agency to the digital realm. This involves translating the entire postulated explanation – comprising various structures and interconnected mechanisms – into a digital, machine-readable format. In essence, digitisation creates the digital representation of the problem as understood through the stages of appreciation and retroduction.

The first task for DT practitioners is to clearly define the general use case (GUC), which represents the main objective of the proposed DT. This could involve understanding (mode 1) or solving (mode 2) the problem at hand. Once the GUC is established, the next step involves identifying the necessary digital datasets and linking each dataset to the relevant structures or variables. This forms the digital version of the mechanisms and their interconnections. Additionally, a conceptual design for the DT must be developed. Methods such as DTUCS (Figure 5.1) are useful here. DTUCS Prong-A helps to specify the technical features and characteristics of the DT that will deliver the defined GUC, Prong-B determines the uses the DT must perform, and Prong-C offers methods for modelling the expected DT UCSs.

It is important to emphasise that the digitisation or computerisation of the situation is not synonymous with operationalisation or quantification. Digital datasets and analysis methods can be either quantitative or qualitative. A common approach when developing a DT might be to operationalise or quantify captured structures and mechanisms, allowing for robust statistical calculations or enabling advanced big data analytics. However, in some cases, certain structures or mechanisms may be better digitised in a qualitative format if this approach is more appropriate for representing particular aspects of the problem – for instance, air temperature might be best measured quantitatively, whereas gathering citizens' feedback on their comfort levels could be more effectively captured in a qualitative format. Sayer (1992: p. 177) offers some guidance on whether quantification would be a suitable or a practically adequate way of modelling a particular object, process or property.

> *'Practically adequate forms of quantifying using interval scales can only be developed for objects and processes which are qualitatively invariant, at least in their fundamentals. As such, they can be split up and combined without changing their nature. We can measure them at different times or places in*

different conditions and know that we are not measuring different things ... Context-dependent actions or properties such as attitudes might therefore be considered unsuitable for quantification ... Whether process can be adequately represented mathematically depends on the type of change involved, on whether it is purely quantitative, or reducible to the movement of qualitatively unchanging entities, or irreducibly qualitative. The latter possibility might be divided into cases where individuals still retain their identity (e.g. the process of ageing) and cases where they cease to be identifiable.'

7.3.7 Mirroring [H-D]

Mirroring marks the initial methodological step of DM2, where the digital aspect of reality plays a central role, utilising the mirroring capability of the DT (Section 3.7.4.1). Through mirroring, the real-world structure is reflected in the digital realm, although inevitably shaped by human worldviews.

This process can take on various features or specifications, which can be determined with the help of DTUCS Prong-A. These features may include temporality – whether the mirroring is static or dynamic, and in the case of the latter, whether it operates in real-time (online) or at intervals (offline). Temporal resolution is also a consideration – how often the data should be updated – as well as the spatial scale and resolution or granularity (Section 3.6.3). Furthermore, during the mirroring stage, several data-related questions typically arise. These include who is responsible for producing the data, how it is collected and in what format, who owns the data, where it should be stored or hosted, what capacity is required for storage, and how long the data should be retained before it is destroyed.

Much like the appreciation and digitisation steps, the mirroring process is heavily influenced by the purpose of the intervention. It is a purpose-driven stage that ensures the DT aligns with the specific objectives of the task at hand. As Wynn and Williams (2012: p. 799) point out:

'Motivations such as why a particular research project is being conducted, at a specific point in time, and within a particular locale can help determine things like the relevant time horizon (e.g., days, week, months, years, etc.) and also the appropriate boundaries for the inquiry (e.g., this department, this firm, this system/software, these types of situations in general).'

A significant benefit of the mirroring process is the generation of longitudinal data (Wynn and Williams, 2012), which captures how a process, property or phenomenon evolves over time. This data adds an important dimension to understanding the problem at hand, enhancing the capacity to evaluate and, ultimately, corroborate the proposed explanation and the mechanisms involved.

By accessing data from a dynamic environment that exhibits shifting behavioural patterns in varying contextual conditions, DT practitioners can better explain emergent events and gain deeper insights into why they unfold as they do. This can lead to the refinement, refutation or corroboration of assumptions regarding the generative mechanisms linked to different structures. As Wynn and Williams (2012: p. 802) emphasise, 'methodologically, we can use the temporal unfolding of events and longitudinal data to corroborate the proposed mechanisms by developing confidence that we have captured the essence of the mechanism and its efficacy relative to alternative explanations.' In essence, longitudinal data serves as a crucial tool for enriching understanding and strengthening the validity of the initial hypotheses within a dynamic and evolving context.

7.3.8 Analysis [D]

The purpose of this step is to generate empirical evidence through the DT's analytical capabilities (see Section 3.7.4.2) and by employing DT's quantitative analytical methods such as mining, learning, predicting and simulating (see Section 3.7.3.3). These methods aim to corroborate, refine or reshape the initial understanding or explanation of the problem.

The broader concept of learning, as applied through machine learning, involves training an algorithm on historical data to predict the likelihood of specific outcomes. Data mining, on the other hand, is primarily focused on discovering patterns and regularities to support decision-making. However, the ability to predict outcomes or identify regularities is dependent on a central assumption: that the underlying mechanism or interacting mechanisms remain stable over time. Similarly, simulations of future events using artificial intelligence techniques are based on the assumption that the mechanisms observed during the mirroring and analysis stages will persist, ensuring that predictions or simulations provide meaningful and actionable insights.

This critical assumption, however, raises an essential question: what kind of system are we dealing with when trying to learn, predict or simulate its behaviour in the context of urban challenges? From a CR standpoint, systems are generally categorised as open, closed, or quasi-closed (see Section 6.2.3.1). In open systems, predicting future events is almost impossible. Closed systems, however, allow for accurate predictions based on mathematical, 'law-like' representations of emergent behavioural patterns, without necessarily requiring causal explanations of the mechanisms behind these behaviours. Yet, most urban problems occur within socio-technical systems, where technical elements are deeply intertwined with social actors. As such, urban systems are rarely purely open or entirely closed but tend to exist somewhere in between – classified as quasi-closed systems.

Even within this quasi-closed classification, it is important to recognise that different urban systems exhibit varying degrees of openness and closure. Rather than simplifying urban challenges into binary categories, such as 'tame' as against 'wicked' or 'difficulties' as against 'messes' (see Section 3.2.5), or into the three system types (open, closed, quasi-closed), it is more constructive to view these systems as existing on a continuum. As discussed in Section 3.2.4, system boundaries may vary along this continuum, ranging from concrete, tangible physical boundaries to highly abstract, intangible social system boundaries (see Figure 3.2). This approach acknowledges the complexity and fluidity of urban systems and allows for a more nuanced understanding of how they function and evolve.

In the context of the relationship between land use planning and travel behaviour, Næss (2004: pp. 149–150) posits a similar argument.

> *'I think it would be more fruitful to talk about different degrees of closure and openness, that is, a continuum from the completely closed systems characterizing some experiments of the natural sciences, to the completely open systems characterizing some research contexts within the social sciences... In particular, for the trips made regularly as a part of daily life, such as journeys to work or education, trips to the grocery store, and trips to bring children to kindergarten or school, a relatively high degree of closure – or restraint – exists ... these trips are therefore to some extent predictable. Trips of a more optional character, including visits, trips to follow children to non-compulsory and irregular activities, and journeys to recreational areas or culture and entertainment facilities, are situated closer to the 'open systems' pole, but we are still not dealing with completely open conditions.'*

In response to the earlier question regarding the type of system at play when addressing urban challenges, urban socio-technical systems are generally understood to be quasi-closed, albeit with varying degrees of closure. It is this 'limited and conditioned' closure (Danermark *et al.*, 2005:

p. 186) that makes cities, societies and other urban systems somewhat controllable, rather than entirely chaotic. With this context in mind, we can now explore how the analysis phase adds value and informs other stages of the DM2 process.

By applying analytical techniques such as data mining, correlation analysis or multivariate analysis through a DT in the analysis stage, one can uncover demi-regs (demi-regularities) that emerge within the quasi-closed system under investigation. These demi-regs often manifest at different scales of time and space, making them difficult to detect without advanced analysis tools, followed by effective visualisations (see communication in Section 7.3.9). The search for demi-regs can be approached from two dimensions

- detection against non-detection of patterns or regularities
- explanatory against exploratory modes of investigation.

Both detection and non-detection of demi-regs offer valuable insights. Detecting expected demi-regs serves to corroborate the explanation of events by providing empirical support for the initial outcomes of retroduction. However, the discovery of unexpected demi-regs, also known as 'contrastive demi-regs' (Wynn and Williams, 2012: p. 794), may prompt a review of the appreciation and retroduction phases. These unexpected findings can indicate that an active mechanism has been overlooked or underestimated.

On the other hand, the failure to detect typical demi-regs can signal changes within the quasi-closed system, potentially pointing to the activation of new mechanisms or the deactivation of old ones. This could suggest a shift towards a different system with distinct elements or properties. Additionally, failure to detect demi-regs may arise from factors such as spatial or temporal scales used during the mirroring phase being too broad or too narrow. In such cases, the DT might be operating at a different level of analysis from where the demi-regs are observable. Miscommunication or inappropriate visualisation methods could also distort or conceal existing demi-regs.

The observation of these signs – whether detection or non-detection – by way of appropriate communication methods (see Section 7.3.9) allows for the development of useful insights. However, it is essential to note that some failures in recognising demi-regs can be attributed to human factors. Overconfidence in initial assumptions may lead to rigid thinking, preventing practitioners from considering alternative perspectives or gaining a fuller understanding of the situation.

Beyond the challenge of detecting demi-regs, the analysis phase can follow either an explanatory or exploratory mode. Explanatory research (Table 7.2) uses qualitative methods to explain strong correlations or regularities revealed through quantitative analysis. In this mode, a DT can help uncover patterns or relationships that were not previously observable in the empirical domain, 'suggesting the existence of a mechanism' (Næss, 2004: p. 151). Revisiting the appreciation and retroduction phases within DM2 can then lead to the formulation of generative mechanisms that explain why and how these regularities or correlations occurred, given the contextual conditions.

Conversely, if a detailed explanation of the observed events is proposed initially, DT analysis capabilities can use exploratory research (Table 7.2) to identify relationships that either corroborate or challenge the proposed explanation. A theoretical framework guides data collection, processing and analysis, ensuring that the insights gained are meaningful, rather than random or accidental. This approach helps prevent the detection of irrelevant or coincidental relationships and keeps the analysis focused on what is most likely to yield valuable conclusions. As Kitchin (2014: p. 6) argues:

Table 7.2 Main multi-method research designs and their relevance to DM2

Research design	Logic	Relevance to DM2
Complementary	'Use one method to add to another and measure overlapping but different Phenomena' (Varga, 2018: p. 550).	To address the 'limitedness' of our observations. This endows practitioners with adequacy and requisite variety needed to inquire the multiple facets or structures constituting complex reality.
Triangulation	'Interpretation, validation or contradiction via comparison or juxtaposition' (Varga, 2018: p. 550).	To address the 'uniqueness' of our observations and eliminate bias. This is through collecting data about the same phenomenon but using different methods. Also, to adequately address the multi-faceted reality 'composed of many types of structure' (Wynn and Williams, 2012: p. 803).
Exploratory	'Qualitative data and methods are followed by quantitative research to explain the relationships in qualitative data' (Varga, 2018: p. 551).	This is relevant to how relationships between the various mechanisms, hypothesized using qualitative investigation during *appreciation* and *retroduction*, are *corroborated* upon the *communication* of quantitative empirical evidence and *analysis* results (see Section 7.3.8).
Explanatory	'Quantitative data is explained by qualitative investigation' (Varga, 2018: p. 550).	This design is relevant to how 'demi-regularities' discovered using quantitative *analysis* can be used to guide the qualitative explanation formulated using *appreciation* and *retroduction*. Nonetheless, uncovering some demi-regs would require an initial hypothesis or a basic understanding or implicit explanatory account presumed in the background in the first place (see Section 7.3.8).
Multi-phase	'Interrelates other design for common research objective' (Varga, 2018: p. 550). Researchers, thus, 'employ other designs but recognize that the purpose of the study is significant to its design' (Varga, 2018: p. 551).	This is relevant to how the purpose of the DT and the overall aim of intervention can motivate interrelating and combining different research designs in a unique way that is fit for purpose. This is in clear harmony with the DM2's general principle, 'purpose-driven', discussed above.
Transformative	'Address a social issue to bring about change' (550).	This is relevant to the ethical dimension of life, *critique*, and mode 3 of research incorporated into DM2.

> *'The data are not subject to every ontological framing possible, or every form of data-mining technique in the hope that they reveal some hidden truth. Rather, theoretically informed decisions are made as to how best to tackle a data set such that it will reveal information which will be of potential interest and is worthy of further research.'*

Practitioners must remain mindful during the search for demi-regs that patterns and correlations between two variables do not necessarily indicate direct causal relationships. The notion of stratified reality suggests that observed events emerge at a level distinct from the one at which the underlying mechanisms operate. Mechanisms in the domain of the real may interact, either reinforcing or counteracting each other's causal effects (Section 2.2.5). Therefore, demi-regs do not always imply a straightforward, single-cause relationship between the variables observed. Instead, these patterns provide an opportunity for further theoretical exploration, qualitative research and additional rounds of retroduction to develop a rational explanation for their emergence.

The continuous flow of data captured by a DT in the mirroring phase supports advanced analytical techniques such as ML, which can create substantial value. Every mechanism has a unique effect, characterised by its 'strength and frequency of occurrence' (Næss, 2004: p. 146). As more events unfold and a DT gathers increasing amounts of data, ML can more confidently determine the relative strength and frequency of various mechanisms, provided the quasi-closed system and its contextual conditions remain stable. Once results are effectively communicated, a clearer understanding emerges of how certain mechanisms might exert stronger influence or be activated more frequently than others – for example, a city's DT might reveal that, although disruptions to the power grid occur infrequently, they have the greatest impact on telecommunication services; a finding that can be qualitatively explained once communicated to practitioners.

Since its inception, DT has been recognised for its predictive capabilities, as seen in early adopting industries like manufacturing and aerospace, typically operating within closed systems. This has inspired applications in the built environment, such as predictive maintenance. However, predicting concrete events with certainty in open systems is an exceedingly challenging task due to the 'intrinsic' and 'extrinsic' variations that such systems experience over time (Section 6.2.3.1). These variations undermine attempts to foresee future developments, rendering model-fitting techniques inadequate in explaining how and why systems with a degree of openness behave in a certain way. No matter how well an ML algorithm fits a model to large datasets, 'a successful fit does not necessarily demonstrate a successful causal explanation but rather the contrivance of a calculating device, albeit one which will not predict the future development of open systems' (Sayer, 1992: p. 211) (see also Sayer, 1992: p. 184).

However, as mentioned earlier, in UM, we are usually dealing with socio-technical systems that possess some degree of closure – quasi-closed systems – rather than completely open ones. Næss (2004: p. 157) explains that this allows for 'qualitative and crude predictions of aggregate-level effects resulting from certain urban developmental strategies.' Although urban systems exhibit some openness, this does not preclude the possibility of identifying regularities within a specific temporal and spatial context. These regularities may persist for extended periods, as mechanisms such as rules, physical infrastructure, and people's reactions to changes evolve over longer timescales, enabling the formulation of rough predictions with greater certainty. Thus, the ability to make cautious, albeit fallible, predictions can significantly aid in understanding how the dynamics of the urban environment may evolve over time.

Employing DT methods, such as simulations during the analysis phase, can help uncover new insights depending on the intervention's purpose. Given a set of mechanisms – each with its unique influence, strength, and frequency of occurrence – it can be valuable to understand which mechanisms, if reinforced or counteracted, might compromise the system's overall resilience. In practical terms, a DT can stress-test a quasi-closed system, revealing its underlying vulnerabilities and deepening our understanding of how resilient the system truly is.

7.3.9 Communication [D-H]

Communication plays a crucial role in DM2 as it marks the transition from the digital realm to human agency. This step involves conveying mirrored data or the results of analysis in a way that allows humans to interpret and act upon them. Techniques such as visualisation, or other methods for presenting data (as outlined in Section 3.7.3.4 on communication DT uses), are considered fundamentally qualitative, even though they are key to understanding quantitative data (Pavlovskaya, 2006: p. 2012). Essentially, qualitative methods of communication complement the quantitative analytical techniques discussed in the analysis phase, enabling the creation of knowledge and unlocking of value.

Pavlovskaya (2006: p. 2012) highlights the power of visualisation, suggesting that it often reduces the manipulation of data to merely querying and displaying the results, as it makes information 'immediately accessible to our minds.' She further explains that, regardless of whether the data are qualitative or quantitative, 'we must 'see' the data to assess their quality, suitability, or completeness.' This demonstrates that the primary value of communication lies in the ability to 'see' the information. The concept of 'seeing' can be understood in two ways. First, it refers to literally gaining visual access to empirical evidence or analytical results from a specific 'level of reality' (Price and Martin, 2018: p. 95). Second, it refers to a deeper understanding – gaining insights into previously incomprehensible information and 'seeing beyond what is currently seen' (National Infrastructure Commission, 2017: p. 63). For this deeper level of insight to occur, the observer must be open to fully interpreting the communicated results.

Regarding this second meaning of 'seeing' (i.e. as in gaining insights), two important points must be considered, one related to digital communication and the other to human perception. First, the communication step must involve selecting the appropriate DT platform or designing an interface that conveys the intended message with the right level of detail for the purpose of the intervention. Second, in line with epistemic relativism (Section 2.2.2), it is recognised that each individual has a unique perspective (Section 3.7.4.1). Pavlovskaya (2006: p. 2013) supports this by noting that the 'power of the visual impact … is itself dependent upon emotions and other irrational sentiments.' As a result, the way individuals perceive and understand communicated information can vary significantly. Therefore, it is highly recommended that stakeholders collectively observe and discuss DT communications to reach a shared and rational understanding.

7.3.10 Planning [H]

Drawing from the insights and knowledge gained about reality, urban managers can begin planning the necessary interventions and actions for implementation in the real world. At this stage, Mode 2 research becomes more prominent and integral to the execution of the DM2 process. With stakeholders having developed a clearer understanding of the problem and envisioned an ideal future state, the key question this planning phase seeks to address is: which generative mechanisms need to be established and activated as an intervention into the real world to bring about the desired event or outcome?

7.3.11 Virtual implementation [D]

Alternative answers to the planning question may be considered. According to the stratified view of reality (Section 2.2.5), different configurations of generative mechanisms can lead to the same observable events, and vice versa. Thus, the virtual implementation phase leverages the analytical capabilities of the DT (Section 3.7.4.2) to aid in selecting the most effective intervention plan for the urban environment. Practically, this can be achieved by simulating various intervention plans within the DT before any actual implementation takes place. Such simulations can enhance confidence in the expected outcomes and systemic implications that may arise from executing the chosen plan, as well as the likelihood of achieving the desired situation. In essence, the central question the virtual implementation stage is trying to address is: what would the real world look like if this specific plan or intervention were executed?

However, one compelling argument could undermine the concept of virtual implementation if not adequately addressed. This pertains to the possibility that the urban system targeted for the intervention is an open system. In such a case, predicting the consequences of a proposed intervention becomes problematic. An open urban system is likely to experience intrinsic (related to its own structure) and extrinsic (related to its contextual environment) changes, resulting in a version of the system that may differ significantly from the one considered in the earlier DM2 steps. This could lead to the emergence of unexpected or undesirable outcomes. To counter this argument, it is essential to maintain the same assumption made during the analysis phase – that the system in question is not open, but rather quasi-closed (as discussed in Section 7.3.8, which addresses the type of system being examined). This assumption allows us to consider that the real urban system and its constituent structures and mechanisms will remain largely consistent with their digitised representation.

Another potential challenge to the virtual implementation stage is the possibility that even if the system is deemed quasi-closed and assumed to remain relatively unchanged during the intervention processes and DM2 steps, the understanding of the system and its components may still be flawed, which would compromise the results of the virtual implementation. In other words, the hypotheses concerning the generative mechanisms at work within the system, upon which the entire intervention planning is based, are necessarily fallible from a critical realist perspective, much like any other proposed explanation. This fallibility inevitably extends to the DT, as Simon (1996: p. 14) observes, 'a simulation is no better than the assumptions built into it', and 'a computer can do only what it is programmed to do.' Therefore, the outcomes of the virtual implementation cannot be any more reliable than the assumptions or explanatory framework underpinning the DT.

Admittedly, this is a valid and compelling argument. All perspectives or hypotheses are fallible, and this fallibility will be reflected in the developed DT. But if we accept this premise, how then can the virtual implementation process provide any value? The key to answering this question lies in understanding that the purpose of virtual implementation is not to rectify a potentially fallible initial hypothesis upon which the DT is constructed. Rather, the primary aim of virtual implementation is to reveal the expected systemic implications of the interventions carried out or the perturbations introduced into a complex system of systems within the digital realm. As Simon (1996: p. 15) asserts:

> *'... simulation can provide new knowledge ... even when we have correct premises, it may be very difficult to discover what they imply. All correct reasoning is a grand system of tautologies, but only God can make direct use of that fact. The rest of us must painstakingly and fallibly tease out the consequences of our assumptions. ... the idea is that we already know the correct basic assumptions ... but we need the computer to work out the implications of the interactions of vast numbers of variables starting from complicated initial conditions.'*

In conclusion, DT practitioners can leverage virtual implementation to acquire new insights, but they must keep two crucial considerations in mind to avoid overestimating the value of the outcomes.

First, it is essential to recognise that while a DT can aid in understanding the implications of a vast number of variables arising from complex initial conditions, these variables and conditions themselves reflect a fallible perspective and comprehension of reality. This inherent limitation adds value to the virtual implementation process, as it provides an opportunity to refine and enhance the DT along with the assumptions on which it is based by comparing predicted results with actual outcomes. Consequently, although the computational processes can be remarkably precise, the anticipated events or consequences of an intervention plan will inevitably carry the same level of fallibility as the initial premises.

Second, even if the assumptions were exceptionally accurate during the virtual implementation phase, the likelihood that the simulated intervention will yield the expected positive results when applied in the real world is contingent upon the degree to which the quasi-closed system under study maintains its transitory state of 'closure' throughout the execution of the intervention and until its consequences manifest. In other words, a failure to achieve the desired state of reality may represent not only a degree of openness of the involved system but also the fallibility of our understanding of that system and the efficacy of the intervention itself.

7.3.12 Real implementation [H/D]

This is the phase in which interventions in the real world are executed according to the established and agreed-upon plan of action. Naturally, this step is feasible by virtue of CR tenets, which posit the existence of an independent reality beyond our perceptions. This reality encompasses structures and interacting generative mechanisms that can lead to various outcomes if they are altered or modified. The level of human involvement in this step can vary depending on how automatic the DT-based control of and intervention into the real-world can be.

7.3.13 General principles

Purpose-driven: Digital twinning is inherently a purpose-driven process (this assertion is further supported later in Section 8.2.3.9). As Brilakis *et al.* (2019) highlighted, 'digital twins should be driven by purpose … (therefore) … different use cases will require different update methods and different levels of detail.' Consequently, any methodology developed to guide the implementation of a DT should be motivated by the specific objectives that led to the intervention and the creation of the DT.

Pluralism: This is a fundamental principle that should be embraced throughout the implementation of DM2. It is rooted in critical realism's assumptions of an independent ontological reality, epistemic relativism and judgemental rationalism (Sections 2.2.1, 2.2.2 and 2.2.3, respectively). From a critical realist perspective, reality is presumed to exist independently of our knowledge of it, and we can only access this reality imperfectly. Thus, it is advisable to adopt a pluralistic approach by employing a variety of

- data sources, including both secondary and primary data that represent diverse perspectives from various stakeholders involved (see *appreciation* in Section 7.3.3)
- data types and collection methods, encompassing both quantitative and qualitative approaches (e.g. interviews, documentation, sensor data etc.) (see *appreciation* and *digitisation* in Sections 7.3.3 and 5.4.3.6, respectively)

- data analysis methods, whether qualitative (such as *retroduction* and *communication*, in Sections 7.3.4 and 7.3.9, respectively) or quantitative (see *analysis* in Section 7.3.8)
- data timestamps, by collecting 'longitudinal data' over time (see *mirroring* in Section 7.3.7).

Pluralism may address the 'uniqueness' and 'limitedness' of human observations discussed in Section 3.7.4.1. McEvoy and Richards (2006: p. 72) suggest that CR allows for the utilisation of multiple methods to overcome the flaws of both 'uniqueness' and 'limitedness', arising from human incapacity to perceive all aspects of a complex phenomenon.

> *'From a positivist perspective, the goal of completeness may be to reveal different aspects of a phenomenon, whereas for an interpretivist it may be to provide a wider range of perspectives. Both these goals are compatible with a critical realist perspective. Quantitative and qualitative methods can be employed to reveal different facets (limitedness) of the same reality and also to examine reality from different perspectives (uniqueness).'*

The 'limitedness' of human observation highlights the extreme complexity and multi-dimensional nature of the real world, as much as it reflects human inadequacies. From this standpoint, Wynn and Williams (2012: p. 803) advocate for the use of triangulation and multiple methods to navigate the multi-faceted reality composed of various structures. They argue that 'different structures call for different means of developing knowledge about them and their properties, which necessitates the use of diverse methods and perspectives.'

In implementing pluralism, there are numerous ways to combine and integrate the different methods available. Varga (2018) offers a comprehensive review of the many mixed-methods research designs proposed in the literature. Table 7.2 lists the key designs and explains how each relates to DM2.

Non-sequential: The steps of DM2, as illustrated in Figure 7.1, are highly interdependent, and their presented order is largely arbitrary. Therefore, the sequence in which DM2 steps are undertaken is flexible, without a fixed or strict order, and it is likely that one will need to loop back and iterate between phases during the same intervention.

Critique: The principle of critique is derived from the critical realist assumption that life has an ethical dimension (Section 6.2.3.2). It calls for adopting a critical stance at every step of DM2 to explore possible ways to enhance capabilities and eliminate unnecessary constraints, thereby fostering the well-being of both humans and nature. Table 7.3 provides examples of critical questions that may be raised at different DM2 steps.

Table 7.3 Examples of the critical dimension to every step of DM2 (continued on next page)

DM2 step	Examples of Critique
Appreciation	Are all marginalized and underrepresented groups involved in appreciating the problem situation?
Retroduction	What are the underlying generative mechanisms limiting the capabilities or different stakeholders?

Table 7.3 Continued

DM2 step	Examples of Critique
Digitisation	Are there barriers, such as digital divide, limiting accessibility to key digital datasets and leading to mis- or underrepresentation of some groups or structures and giving rise to a biased digitised version of reality?
Mirroring	Do mirroring techniques or technologies used violate privacy, security, or cultural and social norms?
	Do less powerful groups know about their data rights and are these rights protected? For example (OASIS, 2013): others can use my data only with my consent, I can check who uses my data, I can manage my own data or choose someone to do it or me, I see personal benefits in exchange for consent to data sharing and so on.
Analysis	Are the analytical techniques or technologies unethical, or do they have the potential to cause discrimination, intrusion, interpretability or transparency issues?
Communication	Are the means of communication used suitable for all project stakeholders, including experts and non-experts, to comprehend and understand DT outputs?
Planning	Is the proposed plan designed to alleviate constraints and enhance capabilities?
Virtual implementation	Were all proposed alternative plans virtually implemented? Were the outcomes of the *virtual implementation*, including the expected systemic consequences of the plan, published and shared with all stakeholders transparently?

7.4. Summary

This chapter advances the development of the methodological element of the DATUM, focusing on uniting its methodical and philosophical components. It builds on two key outcomes of previous chapters. First, the DTUCS, which comprises a multi-dimensional classification framework, a taxonomy of DT uses, and a UML-based approach for modelling DT use case scenarios. Second, the philosophical foundation, based on CR, addressing the dilemma of pluralism in digital twin research for urban management. However, these two components lacked coherence, with DTUCS lacking philosophical grounding and CR being too abstract for practical use. To bridge this gap, this chapter introduces the data-driven multi-method (DM2) methodology, formulated by reviewing CR-informed methodologies and adapting them for digital twin processes. By integrating methodical precision with philosophical depth, this chapter ensures that DATUM offers a cohesive framework, providing both practical tools and a strong theoretical foundation for practitioners. It concludes by illustrating the combined elements of the DATUM (Figure 7.2).

Figure 7.2 The DATUM (author's own)

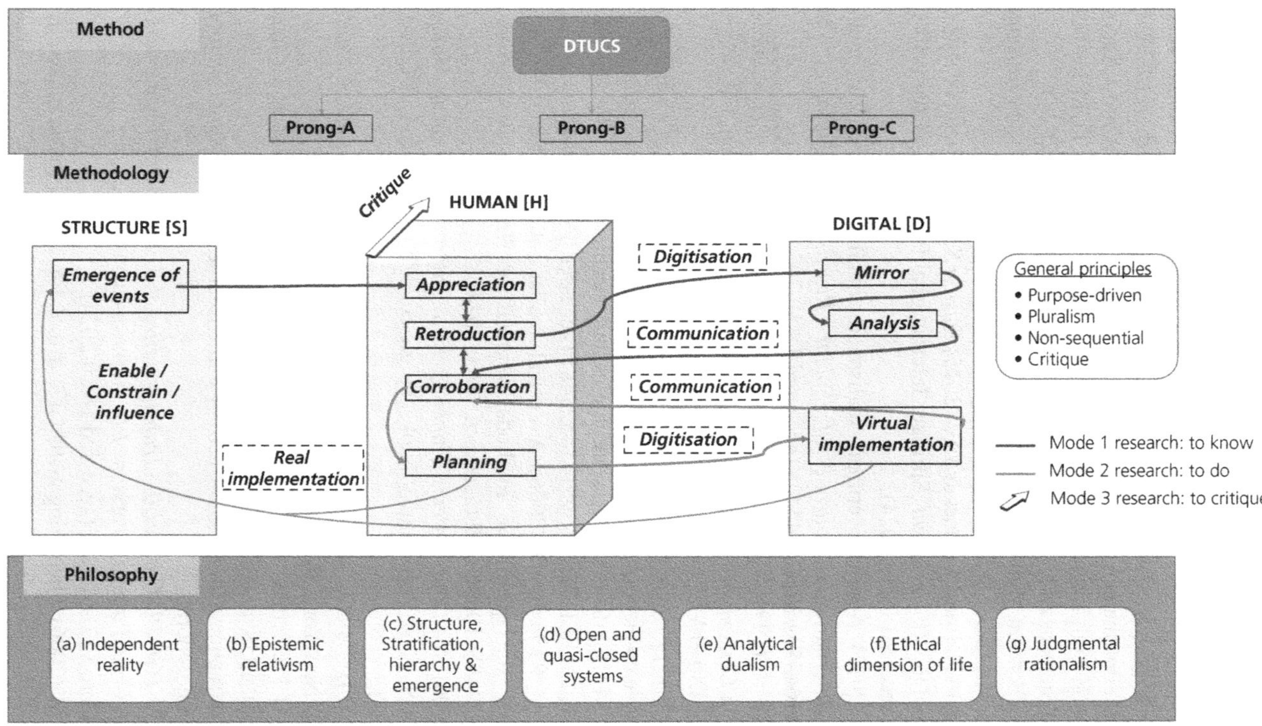

REFERENCES

Ackroyd S and Fleetwood S (2000) *Realist Perspectives on Management and Organisations*. Routledge, New York, NY. https://books.google.co.uk/books?id=CKG9yMsdPpQC (accessed 18/10/2024).

Bhaskar R (1975) *A Realist Theory of Science*. Leeds Books Ltd, Leeds, UK.

Bhaskar R (1993) *Dialectic: The Pulse of Freedom*. Verso, London, UK. https://books.google.co.uk/books?id=dIci5ETjznwC (accessed 18/10/2024).

Bhaskar R (1994) *Plato etc: Problems of Philosophy and Their Resolution*. Verso, London, UK. https://books.google.co.uk/books?id=8xePAgAAQBAJ (accessed 18/10/2024).

Brilakis I, Pan Y, Borrmann A *et al.* (2019) *Built Environment Digital Twining* [Report]. International Workshop on Built Environment Digital Twinning presented by TUM Institute for Advanced Study and Siemens AG. 10.17863/CAM.65445 (accessed 18/10/2024).

Danermark B, Ekstrom M and Jakobsen L (2005) *Explaining Society: An Introduction to Critical Realism in the Social Sciences*. Routledge, New York, NY.

Dresch A, Lacerda DP and Antunes Jr JAV (2015) *Design Science Research: A Method for Science and Technology Advancement*. Springer International Publishing, Cham. 10.1007/978-3-319-07374-3 (accessed 18/10/2024).

Fletcher AJ (2017) Applying critical realism in qualitative research: methodology meets method. *International Journal of Social Research Methodology* **20(2)**: 181–194. 10.1080/13645579.2016.1144401 (accessed 18/10/2024).

Gibbons M *et al.* (1994) *The New Production of Knowledge: The Dynamics of Science and Research in Contemporary Societies*. SAGE Publications, London, UK. https://books.google.co.uk/books?id=bYCZerx0d6oC (accessed 18/10/2024).

Jackson MC (2019) *Critical Systems Thinking and the Management of Complexity*. Wiley. https://books.google.co.uk/books?id=1jmNDwAAQBAJ (accessed 18/10/2024).

Kitchin R (2014) Big data, new epistemologies and paradigm shifts. *Big Data & Society*, April–June, pp. 1–12. 10.1177/2053951714528481 (accessed 18/10/2024).

McEvoy P and Richards D (2006) A critical realist rationale for using a combination of quantitative and qualitative methods. *Journal of Research in Nursing* **11(1)**: 66–78. 10.1177/1744987106060192 (accessed 18/10/2024).

Mingers J (2006) *Realising Systems Thinking: Knowledge and Action in Management Science*. Springer US, New York. https://books.google.co.uk/books?id=C-UcDQEACAAJ (accessed 18/10/2024).

Mingers J (2014) *Systems Thinking, Critical Realism and Philosophy: A Confluence of Ideas*. Routledge, New York, NY.

Næss P (2004) Prediction, regressions and critical realism. *Journal of Critical Realism* **3(1)**: 133–164. 10.1558/jcr.v3i1.133 (accessed 18/10/2024).

National Infrastructure Commission (2017) *Data for the Public Good*. National Infrastructure Commission Report.

OASIS (2013) *TGF-PL-Core-v1.0—Transformational Government Framework (TGF) Pattern Language Core Patterns Version 1.0*. http://docs.oasis-open.org/tgf/TGF-PL-Core/v1.0/os/TGF-PL-Core-v1.0-os.html (accessed 18/10/2024).

Pavlovskaya M (2006) Theorizing with GIS: a tool for critical geographies? *Environment and Planning A: Economy and Space* **38(11)**: 2003–2020. 10.1068/a37326 (accessed 18/10/2024).

Price L and Martin L (2018) Introduction to the special issue: applied critical realism in the social sciences. *Journal of Critical Realism* **17(2)**: 89–96. 10.1080/14767430.2018.1468148 (accessed 18/10/2024).

Sayer RA (1992) *Method in Social Science: A Realist Approach*. Psychology Press, Hove, UK. https://books.google.co.uk/books?id=kr8QvOpM2RwC (accessed 18/10/2024).

Sayer RA (2011) *Why Things Matter to People: Social Science, Values and Ethical Life*. Cambridge University Press, Cambridge, UK.

Simon HA (1977) The organization of complex systems. In *Models of Discovery: And Other Topics in the Methods of Science*. Springer Netherlands, Dordrecht, pp. 245–261. 10.1007/978-94-010-9521-1_14 (accessed 18/10/2024).

Simon HA (1996) *The Sciences of the Artificial*, 3rd edn. MIT Press, Cambridge, MA, USA.

Tomko M and Winter S (2019) Beyond digital twins – a commentary. *Environment and Planning B: Urban Analytics and City Science* **46(2)**: 395–399.

Varga L (2018) Mixed methods research: a method for complex systems. In *Handbook of Research Methods in Complexity Science: Theory and Applications* (Mitleton-Kelly E, Paraskevas A and Day C (eds)). Edward Elgar Publishing, pp. 545–566.

Weick KE (1989) Theory construction as disciplined imagination. *Academy of Management Review* **14(4)**: 516–531. 10.5465/amr.1989.4308376 (accessed 18/10/2024).

Wilson M and Greenhill A (2004) Theory and action for emancipation: Elements of a critical realist approach. In *Information Systems Research* (Kaplan B, Truex DP, Wastell D *et al.* (eds)), Springer, Cham, pp. 667–674.

Wynn D and Williams CK (2012) Principles for conducting critical realist case study research in information systems. *MIS Quarterly* **36(3)**: 787–810. 10.2307/41703481 (accessed 18/10/2024).

Ramy Elsehrawy and Bimal Kumar
ISBN 978-1-83662-235-2
https://doi.org/10.1108/978-1-83662-234-520251010
Copyright © 2025 by Ramy Elsehrawy and Bimal Kumar. Published under exclusive licence by Emerald Publishing Limited

Chapter 8
Putting DATUM to the test

8.1. Introduction

As discussed in the previous chapters, there is a pressing need to systematise and unify the emerging paradigm of digital twin for urban management (*DT for UM*). To address this, the artefact 'DATUM' was introduced in Chapter 4, comprising three core components: philosophical, methodological and methodical. Chapters 5, 6 and 7 fully developed each of these elements. However, DATUM remains, at this stage, a theoretical construct. To demonstrate that an artefact can fulfil its intended purpose, it must be tested in practical application. Therefore, the objective of this chapter is to evaluate DATUM, as outlined earlier in Figure 2.6, against the evaluation criteria set forth in Figure 4.8.

First, in Section 8.2, the 'abstract' research phase evaluates the philosophical component of DATUM, which is grounded in critical realism (CR), assessing its capacity to unify various DT practices under a single philosophical framework. This evaluation is conducted through focus group discussions involving DT researchers and practitioners. Section 8.3 applies 'extensive' research, using data from multiple DT case studies to evaluate the generalisability of DATUM's methodical component (i.e. DTUCS). Lastly, in Section 8.4, all three elements of DATUM are assessed as a unified whole through 'intensive' action research, involving two distinct DT projects. This final step examines DATUM's overall usability and practical effectiveness in addressing real-world challenges.

8.2. Evaluation of philosophical element

8.2.1 Introduction

As introduced earlier in Section 2.7.3.2, this stage of the evaluation process aims to assess the philosophical component of DATUM through abstract research. The objective is to determine how effectively this philosophical foundation, rooted in CR, can unite the various worldviews held by practitioners within the field under a single philosophical framework, addressing any conflicts and reconciling inconsistencies. The findings from this evaluation, including an exploration of the prevailing worldviews and their underlying philosophical assumptions, are then examined in relation to DATUM's philosophical core, which is grounded in the principles of CR (Table 6.1). This comparison seeks to determine whether CR provides a pluralistic and cohesive foundation capable of accommodating these diverse philosophical perspectives.

8.2.2 Methodology

8.2.2.1 Philosophical stance

It is essential for all researchers to declare their philosophical position or worldview and understand how this influences the methodologies and methods they choose (Creswell, 2009). This argument has been a recurring theme throughout this book (see Section 6.1). It is based on this reasoning that the exploration of the philosophical worldviews held by members of the field is considered vital. Similarly, it is on these grounds that the philosophical stance guiding the focus group discussions here will be clearly stated. Previous research links focus groups to phenomenological studies (Bradbury-Jones *et al.*, 2009), which is primarily a qualitative approach. Phenomenology, emerging as a form of inquiry in the early twentieth century (Barkway, 2001), focuses on uncovering and describing the essence of a phenomenon as consciously experienced from a first-person perspective (Spiegelberg, 1975). Bradbury-Jones *et al.* (2009) provide an overview of two main phenomenological approaches: Husserl's descriptive phenomenology and Heidegger's interpretive/hermeneutic phenomenology. Husserl's approach centres on phenomenological reduction, or 'bracketing', a method used to uncover the essence of a phenomenon by distancing oneself from everyday distractions. In contrast, Heidegger's approach emphasises understanding rather than mere description, rejecting the idea that observers can detach themselves from reality, as they inevitably impose their own interpretations to make sense of the phenomenon.

However, the focus groups in this book are conducted from a critical realist standpoint. A critical realist approach seeks to explain phenomena in terms of unobservable causal mechanisms operating beneath the surface (Pawson and Tilley, 1997). The phenomenon under investigation here is the current DT practices within the field of UM, where DTs are employed to tackle a range of urban challenges. According to the assumptions made in this book, the unobservable mechanisms driving these various DT practices are the philosophical worldviews of the practitioners (Section 3.8.4). Moreover, as outlined in Section 4.5.2.2, when practitioners depart from *typical practices* associated with the established paradigm, they begin to adopt *atypical practices*, often driven by agential choices. The generative mechanisms shaping these *atypical practices* are argued to be the philosophical worldviews that practitioners hold, whether consciously or subconsciously.

Thus, from a critical realist perspective, the purpose of the focus group discussions is to explain the observable events (i.e. the current DT practices) by identifying the generative mechanisms behind them (i.e. the philosophical assumptions and worldviews of the DT practitioners). In other words, the focus groups aim to explore 'what is it about the practitioners' worldviews that drives them to use DTs in UM as they do?' By revealing the participants' philosophical assumptions, one can then assess how compatible these views are with the core principles of CR, and, consequently, how effective CR is as a unifying philosophical foundation for DATUM and the emerging paradigm *DT for UM*.

8.2.2.2 Selection of method

The decision to conduct focus groups to evaluate the philosophical element of DATUM was driven by an understanding of the current state of the field and the recognised strengths of focus groups as a research method. After reviewing relevant studies, it became apparent that many within the discipline show limited concern for their own philosophical positions (see Section 3.8). For this reason, focus groups were deemed an appropriate method for exploring these positions (Crinson, 2001). Unlike interviews, which often focus on extracting direct answers to specific questions, focus groups use questions as prompts or thought-provoking statements, encouraging participants to delve into deeper ideas (Krueger, 2014). In this way, focus group discussions allow for the 'ontological depth' (Crinson, 2001) needed to reveal participants' philosophical worldviews, especially for those less inclined towards philosophical discussions. The dynamic nature of these conversations enables participants to confront and challenge each other's views, ultimately helping them uncover and articulate their own philosophical positions.

8.2.2.3 Recruiting

The study followed a 'single-category design' (Krueger, 2014), involving a purposeful sample of homogeneous groups of participants identified as DT practitioners and researchers. These individuals were primarily focused on DT applications and real-world implementation within the context of UM. The design was driven by the study's aim, which did not require dividing participants into distinct categories, such as industry professionals or applied researchers from academia. Both groups were seen as members of the same discipline, with shared interests in building and utilising DTs for urban management, and there was no need to compare or contrast them based on any particular characteristic.

Two 90-minute focus groups were conducted, each involving six participants, after which 'saturation' was largely achieved. Saturation refers to the point where no new insights are emerging (Krueger, 2014), meaning no additional philosophical worldviews were expressed. Given the restrictions imposed by the pandemic and the fact that participants were located in different countries at the time of the study, both focus groups were held virtually by way of online video calls. Ethical approval was obtained prior to data collection, with informed consent received from all participants.

8.2.2.4 Questions

To support a purpose-driven analysis, the fundamental principles of CR (Table 6.1) served as the guiding framework for shaping the focus group questions, activities, and overall discussion pathway. This included open, key, and concluding questions (refer to Table 8.1), with time allocated to each question based on its complexity, category, and the level of depth desired from the participants.

8.2.2.5 Analysis

To ensure rigour and reliability, the analysis followed three guiding principles and a methodology informed by critical realism for analysing focus group data, as derived from Kreuger (2014) and Crinson (2001). The key principles are as follows:

Purpose-driven: The analysis is guided by the purpose behind conducting the focus groups.

Verifiable: All findings are clearly traceable, allowing any other researcher to reproduce the analysis through a well-documented trail of evidence. As a result, conclusions and reflections are kept distinct from the findings themselves.

Systematic: The process must be prescribed, deliberate, and planned. In line with this, the systematic critical realist approach outlined by Crinson (2001) and depicted in Figure 8.5 was adopted.

TRANSCRIPTIONS

The discussions were video-recorded and automatically transcribed, with auto-transcripts cross-checked against a manually created 'abridged transcript' to enhance data validity. An abridged transcript includes only the relevant parts of the conversation, excluding introductions, initial questions, excessive moderator instructions, and irrelevant comments (Krueger, 2014). Additionally, any comments made by participants by way of the chat box during the virtual sessions were also included.

INDEXING

This step involved assigning non-exclusive index codes to segments of the conversation, covering various analytical topics raised during the discussions (Crinson, 2001). Avoiding early exclusive coding ensures that the analysis captures the full range of participant insights, preventing propositions from being based on just select parts of the data. Non-exclusive coding also highlights when participants contradict themselves or each other, offering insight into important dynamics that might otherwise be overlooked.

Table 8.1 Focus groups' questions

No	Question category	Question	Figure	Purpose	CR Principles	Time (mins)
1	Opening	Would you tell us your name and what you are currently working on?	—		—	7
2	Key	Rank these 4 pictures from most to least important DT practice.	Figure 8.1	To what extent participants are pluralist practitioners.	[Pluralism]	13
3		Do you believe a DT mirrors reality or the developers' views of reality?	—	Explore participants' ontological position.	[P2] [P3] [P6]	12
4		'A DT model can provide insights beyond what is currently seen' (National Infrastructure Commission, 2017: p. 63). Which of the following 3 sentences you think best describes a 'DT insight': 1- A failure in substation X detected 2- There is a significant increase in water consumption at district X 3- When number of covid-19 cases rise in district X, daily traffic volume increases on road Y 4- Other sentence	Figure 8.2	Explore participants' views of observable regularities and role of unobservable causal mechanisms. *Choices #1 and #2 represent observable events in domain of empirical. Choice #3 represents a possible mechanism in domain of real.	[P1] [P4]	12
5		practitioner X: 'I believe city dynamics including citizens behaviours and patterns are stable, repeated and 'lawlike'. My DT uses high-tech like AI to analyse the big data collected, uncover city dynamics laws and use these laws to predict the future. I deliver what people need.'	Figure 8.3	Explore participants' views on openness of the world (extrinsic: interconnectedness/ intrinsic: human nature: agential free-will and determinism).	[P5] [P7]	12

		Practitioner Y: 'I believe city dynamics including citizens behaviours and patterns are determined by the free will of people and therefore, they are unpredictable and do not conform to laws. My DT uses advanced visualization tech. and user-friendly open-access platforms for people to create, simulate and visualise different future scenarios. I deliver what people want.' Imagine you were assigned to create a DT for your city, what would your approach be?		*Strictly, practitioner X represents a deterministic view, while Y represents a voluntaristic view.		
6		Practitioner X: 'I believe ethics are universal and fixed. My DT is ethical because it follows the 'code' of ethics.' Practitioner Y: 'I believe ethics are culture-dependent, continuously evolving and changing. My DT is ethical because it goes with the existing norms of the community within which my DT operates' Imagine you were assigned to create a DT for your city, what would your approach be?	Figure 8.4	Explore participants' views on the ethical dimension of life – is it objective and real or subjective and a matter of culture or taste.	[P8]	12
7	Closing	Of all the things we've talked about today what to you is the most important thing that has been said?	—		—	12

CR principles: [P1]: Structure, hierarchy and emergence; [P2]: Ontological realism; [P3]: Epistemic relativism; [P4]: Stratified reality; [P5]: world is open with occasional quasi-closed systems; [P6]: Judgmental rationalism; [P7]: Analytical dualism; [P8]: ethical dimension of life

Figure 8.1 Figure associated with focus group question (author's own)

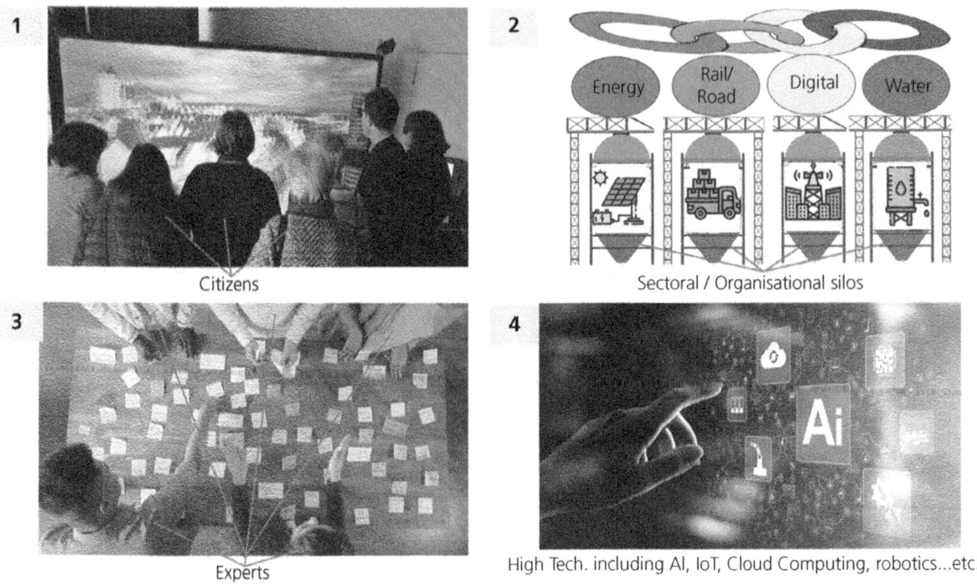

Figure 8.2 Figure associated with focus group question (author's own)

'A digital twin model can provide insights beyond what is currently seen' (Data for public good, NIC)

This is a City Digital Twin

(including big urban data from infrastructures systems, natural environment, ...etc.)

Imagine this a city DT, this guy has gained a 'DT insight', which of the following 2 sentences (if any) you think best describes a **DT insight:**

(a) *'There is a significant increase in water consumption at district X'*

(b) *'When number of COVID-19 cases rise within district X, daily traffic volume increases on road Y'*

(c) *YOUR SENTENCE?*

Figure 8.3 Figure associated with focus group question (author's own)

I am DT practitioner X	I am DT practitioner Y	I am...
I believe city dynamics including citizens behaviours and patterns are stable, repeated and 'lawlike'.	I believe city dynamics including citizens behaviours and patterns are determined by the free will of people and therefore, they are unpredictable and do not conform to laws.	I believe city dynamics are...
My DT uses high-tech like AI to analyse the big data collected, uncover city dynamics laws and use these laws to predict the future.	My DT uses advanced visualisation tech. and user-friendly open-access platforms for people to create, simulate and visualise different future scenarios.	My DT is...

Figure 8.4 Figure associated with focus group question (author's own)

I am DT practitioner X	I am DT practitioner Y	I am...
I believe ethics are universal and fixed	I believe ethics are culture-dependent, continuously evolving and changing	I believe ethics are...
My DT is ethical because it follows the never-changing 'code' of ethics.	My DT is ethical because its goes with the existing norms of the community within which my DT operates	My DT is ethical because...

INTERPRETATION

Interpretation focused on understanding and conceptualising 'themes' based on participants' perspectives, without making any inferences. Several analytical features were considered during interpretation, which may point to key underlying mechanisms or beliefs, including

- spontaneous and inconsistent comments, where participants may revise their opinions due to newly gained insights
- variations in word usage, where context and overall comments help clarify different meanings or synonyms.

Additionally, criteria used to assess the weight of ideas included:

Frequency: How often something was mentioned.

Specificity: Whether comments were detailed and specific.

Emotion: The intensity or enthusiasm displayed by the speaker.

Extensiveness: How many different people shared the same idea.

Importance: The relevance of the idea to the study's purpose. As Kreuger (2014: p. 355) noted, 'sometimes a really key insight might have been said only once in a series of groups. You have to know enough about what you are studying to know a gem when it comes along ... cutting-edge thinking may have only been voiced once in a series of groups, but it may be crucially important to the study.'

THEORISATION

This step involved applying existing theories to the interpreted themes to provide context. For this book, theories concerning philosophical assumptions (Figures 3.14 and 3.15) helped create context for the discussions and prepared for the subsequent step of retroduction.

RETRODUCTION

In this phase, the aim is to infer the necessary generative mechanisms behind the observed phenomena. It was assumed that the worldviews held by participants, often implicit or unconscious, were the causal mechanisms driving the current atypical practices within the discipline during its pre-paradigm phase. Retroduction seeks to uncover these philosophical worldviews and assess whether CR can accommodate and unify them within its philosophical framework.

If the participants reached a consensus on a worldview that contradicted CR's principles, it would indicate a failure of CR to provide a sufficient level of philosophical unification – for example, if all participants agreed that reality is objective, value-free and directly accessible by DTs, this would conflict with CR's principle of epistemic relativism. Similarly, if participants concluded that DTs could make precise predictions about future events within urban environments, this would oppose CR's understanding of the world as an open system, occasionally operating within quasi-closed systems.

8.2.3 Findings

This section provides a qualitative analysis of the focus group data, covering the 'indexing' and 'interpretation' stages of the critical realist-informed approach (Figure 8.5). The results are organised into the following ten analytical themes.

8.2.3.1 Theme [T1]: Inconsistencies

The non-exclusive coding method of indexing (Crinson, 2001) helped draw attention to several inconsistences in the discourse – for example, participants frequently used profoundly different terms, such as *'reality(s)'* and *'view(s) of reality'*, interchangeably as synonyms, assuming that both imply the same meaning. Other inconsistences arose from participants changing their minds, views or statements, thus contradicting one another or even themselves – for instance, one participant quickly advocated the approach Y to developing DTs in response to [Q5]. However, a few minutes later, the same participant typed, *'Y with a little bit of X'*. Another instance involved a participant initially agreeing – with a 'thumbs up' reaction – to another participant's statement that

Figure 8.5 CR-informed analysis of focus group data. Adapted from: Crinson (2001)

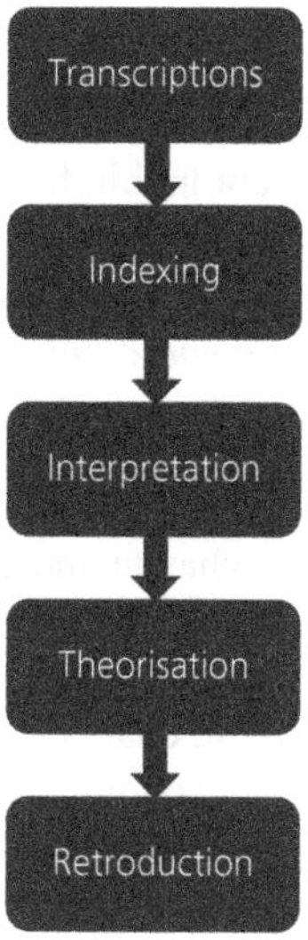

'the digital twin itself is neither ethical nor unethical; it's the way it is used that matters'. Nevertheless, a bit later the same participant said, *'I think that design* (of a DT] *as well could be ethical or unethical'*. While not a general theme, some participants occasionally provided answers or made statements that they could not support with a sound explanation or justification when challenged by opposing opinions or follow-up questions – for instance, they might argue that their claim is true '*...because it makes more sense*'.

8.2.3.2 Theme [T2]: Advocacy of pluralism

Throughout the discussions, all participants emphasised the importance of pluralism in DT practice, which entails the adoption of multiple approaches to developing and implementing DTs. They were presented with four images (Figure 8.1), each illustrating a distinct approach to DT implementation. These four approaches (i.e. tech-driven, cognitive, disruptive and humanistic), as conceptualised in Chapter 6, are underpinned by four different philosophical paradigms (i.e. functionalism, interpretivism, radical structuralism and radical humanism), respectively. Participants advocated for a pluralistic approach, albeit from four distinct viewpoints.

Incommensurability of methods

Within this context, a few participants articulated that relying solely on one DT approach is challenging due to the incompatibility of different DT approaches. One participant said

> *'I just wanted to comment that it was really hard to rank this. It's almost impossible because they are not sort of ... They are not of the same type.'*

Moreover, each approach is unique, offers a different perspective, and achieves a different purpose – for instance

> *'some developers go for very photorealistic models and sorry visual models, and that's not necessarily the best way to convey a message or a certain kinds of data.'*

Biases of people

More participants emphasised the need to use more than one approach to account for the biases and diversity of perspectives and views of people involved in the problem situation – for instance, one participant asserted, *'there will always be biases and should be aware of those biases'*, echoing the sentiment of another participant who suggested, *'obviously there's going to be bias and point of views involved.'* As a result, one participant highlighted the need for collaboration, stating, *'we need each other, work together, get a more unified or closer picture to what reality really is.'* In a similar vein, another participant emphasised, *'the most important thing is to spend more time without having to identify things very quickly and not letting one person to identify what a digital twin will do ... That approach needs to be more diverse and longer and so on.'*

Dynamic nature of urban environment

Other participants highlighted the relentless change and evolution of the complex urban environment over time, as well as the continuous evolution of our experiences, knowledge, technology, and the various social systems and organisations. Consequently, they emphasised the need for adaptability and flexibility in response to these changes. In other words, a DT practitioner might need to employ different approaches depending on how urban dynamics unfold and how the urban environment evolves.

> *'I see this as an evolutionary thing', one participant remarked, and another concurred by saying, 'I agree with [the former's name] way of thinking!'*

One more argued that nowadays the four approaches have relatively different importance, yet this ranking is expected to change over time: *'at this moment of time is 1, you know, 2, 4 and 3. Really ... that's I [sic] think it will change'*.

Multi-dimensional and socio-technical nature of urban environment

However, more participants emphasised the multi-faceted and multi-dimensional nature of complex urban problem situations. They highlighted that these situations often encompass various aspects, making pluralism – equal consideration of distinct DT approaches – vital for addressing the diverse dimensions of such complex problems – for instance, one participant argued that *'a digital twin for an individual asset'* and another DT at *'this level of urban level, city level'* will require entirely different DT approaches to develop and implement.

Another participant pointed out, *'I think it depends on the application, and this is why I ranked them equally important'*.

Another participant believed that although the humanistic DT approach should be the central one, other approaches should complement it: *'citizens come first because people ... It has to be all about people ... and nature! ... And then I put all the others together, equal second and refused to make a distinction between them'*.

A different participant asserted that *'may be an integrative view would be the best approach'*.

Furthermore, the majority of participants demonstrated their advocacy for pluralism by emphasising the need to integrate both quantitative and qualitative methods in practice while developing DTs. For instance, one participant recognised that some problems related to technical systems, such as infrastructure systems, can be addressed using quantitative methods, while other social aspects of the urban environment, where citizens are directly involved, require qualitative approaches.

> *'When you talk about infrastructure like power networks, road networks, etc, they are systems that have a purpose, right? You can measure them by providing power to people, water utilities. Cities do not have a purpose in that sense, therefore a digital twin of a city doesn't have a purpose in the same sense. Cities are supposed to be messy, chaotic, places where different people lead different lives and everyone kinda muddles along and the smart cities narrative and digital twins narrative tries to control and quantify something that shouldn't really be quantified which is why none of us would actually want to live in a smart city because it would be bloody depressing.'*

One participant, reflecting on their experience in developing DTs, argued that quantitative methods proved more helpful when analysing data related to city dynamics and infrastructure performance for supporting urban planning applications. Conversely, qualitative methods were more suitable when capturing data about end users' satisfaction levels at an airport to evaluate the '*social performance*' of the asset.

When comparing strictly quantitative and qualitative approaches, all participants concurred that it is rare for a DT practitioner to exclusively employ one of these two extremes. Instead, they agreed that a hybrid integrative is typically the most effective way forward.

> *'I think there should be an area in between ... and I think there should be a process when designing the system which allows to be in between ... I think one of the problems is that we are pushed to make decisions to go either one of these two ways.'*

Another participant emphasised that '*city dynamics are organic*' and, as a result, '*most are predictable, but some are not*'. Therefore, the same participant continued, '*my DT uses a combination of AI analytics and open creative systems.*'

Another participant regarded quantitative and qualitative methods as more valuable when integrated together, complementing each other. Whereas the former can assist in identifying patterns and regularities in the urban environment, the latter can help in understanding these dynamics as they continuously evolve.

> *'And so rather than just say, you know, that was the pattern last time, and therefore it's going to be exactly the same this time, I think it's much more interesting and useful to understand the patterns as they change.'*

8.2.3.3 Theme [T3]: DT is 'both' objective and subjective

A few participants believed that a DT is closer to being a twin or a mirror of its developer's views. One participant expressed this by stating, '*reality is a co-construction. So I guess then accordingly, digital twins would be a co-constructed view of reality.*'

However, some of those who initially appeared to be more convinced that DTs mirror the views of developers seemed to change their perspective as the discussion evolved. This change of perspective, as well as the significance and implications of other detected inconsistencies, different words used synonymously or change of minds are discussed in theme [T1] – for instance, one participant stated, '*...there are 'multiple realities' and you're right to point out that there might be a distinction between the developer's reality and somebody else's...*'

Nonetheless, the same participant implicitly acknowledged elements of an objective, independent 'reality' – '*but that's OK because it's OK to have 'different views of reality' and they can all be important.*'

Again, the same participant implied the same realist account a bit later by referring to a reality out there, distinguishing between a DT, on the one hand, and what a DT attempts to model with varying levels of precision, on the other hand.

> *'the more practical thing to be considering is about precision. It's about fidelity. So, you know, what is the fidelity of the model relative to 'what it's trying to model'.'*

Also, another participant initially argued that

> *'in terms of deciding which reality is going to be presented in this twin, it really depends on who you ask and also, it depends on when you ask them because everyone can have their own worldviews and opinions and desires ... I believe that it really mirrors the reality, not only of the developer ... but that it's mainly whoever is involved in the development process.'*

However, a few minutes later the same participant tried to explain that a DT may, in fact, encompass both objective and subjective elements of a phenomenon under investigation. The objective element pertains to the various facets of reality as examined from the so-called 'scientific' perspective, while the latter is more concerned with examining the same reality but from the different viewpoints of social actors involved. This insight was abruptly pointed out.

> *'You know, I just when I heard you talk, I started to think about this in constructing a digital twin or developing it, that there are both objective and subjective worldviews or realities to be represented. So, from a modelling perspective, there are certain elements in an urban digital twin that you know, could and should be scientifically based and that should be objectively developed, as far as you can, obviously, you want to sort of core the phenomenon and understand how that changes. But then you have the subjective element which goes into deciding what goes into the twin. What type of models do we include? What type of data do we include and what type of functionality do we provide to the users?'*

The argument above indeed resonates with a comparison made by another participant during a separate focus group discussion. This comparison involved quantitatively captured data of city dynamics, such as *'births and deaths'*, and qualitative data concerning the *'social performance'* of a specific asset, representing distinct worldviews of the same phenomena.

The strong theme that emerged from the discussions was the understanding of a single existing reality that individuals perceive differently. Most participants contended that a DT ultimately seeks to mirror this reality, albeit from different perspectives, and, at best, it can only capture it partially and with varying levels of fidelity.

The distinction between the independent reality, and our individual views, knowledge, and understanding of it was implicitly made at different occasions throughout the discussion, namely

> *'two different stakeholders could use the twin trying to explain the same phenomena or assess it and come up with different results, right ... It depends on how you scope, the way you perform an analysis, where you include and exclude, where you emphasise in terms of visualisation.'*

Some participants used the word *'both'* to indicate the coexistence of these two distinct things: the objective reality and the various interpretations of reality along with opinions on how it should evolve – for instance, one participant stated *'it's both, and I don't think that's necessarily a bad thing ... we've talked before about this ... whether a digital twin is a platonic digital twin or Aristotelian digital twin' ... 'The opinions, you know, the what-could-be is quite important about as much as the what-is currently depending on how you're using the two sets of things.'*

Another argued

'it has to be both (objective and subjective], so that there isn't one or the other ... you are modelling reality but also you have the biases ... and that will never disappear ... there will always be biases and should be aware of those biases.'

One more participant referred to biases and presence of various points of view that lead to partial or flawed mirroring of this reality to varying extents.

'in question of whether a digital twin mirrors reality, its more of a to what extent how we manage to capture reality in the models that we have ... obviously there's going to be bias and point of views involved in that, but I think it's almost like there's a stage in between.'

The expression *'a stage in between'*, used in the comment above, appears to imply the same stance taken in the earlier comments using the word *'both'*. Simply put, it refrains from adopting either of the two extreme theoretical assertions. It neither fully accepts the assertion that a DT is a truc and pure mirror of reality, nor does it assume that a DT is nothing but a representation of its developer's opinions. Therefore, this stance obviously rejects the idea of dissolving one into another. Put bluntly by another participant:

'Ultimately there is one reality and there's people experiences ... we all have a limited chunk of reality.'

In the same context, a participant suggested that a DT is a flawed representation of reality, and this is because of two reasons. First, it is because developers are the ones who *'decide where to put the focus'* – for example, a strong preference for visual representations or graphical models might not necessarily be *'the best way to convey a message or certain kinds of data. It might even distract you from the important thing you're trying to communicate'*. The second reason is the tendency to use the available models and not necessarily the right models – for example, nature, which is a part of reality, is usually underrepresented because, as the participant mentioned, *'we don't have good models for assessing green infrastructure, biodiversity and so on. So we tend to not do that only because we don't have the models'*. Another participant agreed with that latter argument, confirming that '*we have to work with what we have*'.

Due to these flaws, a consensus emerged through discussions about the importance of involving various stakeholders in the development of a DT, starting from the very early stages of the process.

'what is the aim of building a digital twin, and trying to get answers to that question from different stakeholders so we can see the different viewpoints ... for me, the most important thing is to spend more time without having to identify things very quickly and not letting one person to identify what a digital twin will do ... That approach needs to be more diverse and longer and so on.'

In support of the same idea, another suggested that

'we all have a small piece of reality in our hands ... until we get to that one reality, and eventually a unified comprehensive city digital model ... we need each other, work together, get a more unified or closer picture to what reality really is.'

8.2.3.4 Theme [T4]: DTs are evaluable

Many participants confirmed, at least implicitly, that a DT can be judged, evaluated and compared with its rivals, including other DTs built to capture the same phenomenon albeit in different ways. The extent to which a DT can capture the reality 'out there' is one key criterion upon which it can be assessed, namely

> *'it's more of a to what extent how we manage to capture reality in the models that we have.'*

In the same sense, while it has been acknowledged through discussions that a DT can never perfectly mirror reality, it was implied that the accuracy of a DT and its associated tools in describing the world can still be evaluated.

> *'you are modelling reality but also you have the biases ... and that will never disappear. We may get more accurate on the languages we can use to describe the real world, and the tools we use to do the describing, but there will always be biases and should be aware of those biases.'*

Others emphasised the extent to which a DT can fulfil its intended purpose or deliver the required outcomes as a central criterion for evaluating it.

> *'I think that what defines a digital twin being better than another is to do with how, how well they achieve their purpose.'*

Likewise, another participant confirmed that

> *'the outcome should be the benefit for people, for society, for the nature and to the extent that you're able to use this twin to arrive at decisions that lead you to those outcomes, then I would say, yes, you could. Objectively speaking, say that's better, that's the objective of the twin.'*

However, participants appeared to have difficulties in making similar assertions about the possibility or necessity of judging DTs when it comes to ethics and ethical considerations. They seemed to be very conscious of not implying that ethics, which are culturally defined and shape how a DT is developed and used, judged, praised or criticised when compared to its rivals in different cultures. These rivals include other DTs built to capture similar urban phenomena or achieve similar purposes but in different cultures. Further discussion on this topic will be presented in the 'DT and ethics' theme [T8].

8.2.3.5 Theme [T5]: DT insights

The question [Q4] (i.e. Which of the following three sentences you think best describes a 'DT insight'?) primarily aimed to investigate the phenomenon of gaining DT insights. Most of the participants who provided a direct response to this question suggested that option (b) (i.e. When number of covid-19 cases rise in district X, daily traffic volume increases on road Y) is the one that best represents a 'DT insight' from the choices provided. However, what is of greater concern to this discussion is why they believe so. Some participants struggled with providing a detailed explanation of how they understand a DT insight or give a clear justification for their answers – for example, *'...because it makes more sense'* (see theme [T1] above).

A few based their views on an assumption that links gaining insights from a DT to acquiring new information. Hence, to know if a DT has offered insights, one participant emphasised, one would need to ask '*what information has been unlocked here? Is this (information) new?*'

More participants, however, associated DT insights with gaining new actionable knowledge – that is, knowledge that can aid in taking actions. Accordingly, a DT practitioner would need to ask

> *'why do we need it [acquired information] and how can we go back and do something in the urban environment, like there's some kind of feedback to the physical world.'*

Another participant agreed, picturing insights as a necessary midway step between collecting data and unlocking the value within this data by making better decisions.

> *'So insight, I see as being one of those things that you get partway through the value chain and then you use the insights to make better decisions. So that, that to me is a really key thing, you know, from data to insight to decisions.'*

This same participant imagined a scenario of how a DT insight can be of great value to decisionmakers.

> *'the type of thing I would want them to be doing is recognising that the connected digital twins are telling them that there is likely to be extreme flooding in such and such a district because it's connected into the weather digital twin as well, and as a result of that extreme flooding – that hasn't happened yet – they can identify the assets which would be badly affected and how those assets would then affect others ... and they can make operational decisions to save lives in advance of the problem occurring.'*

Another participant expressed a similar opinion but with one additional dimension. She highlighted that, in addition to providing actionable knowledge or '*something that we could react upon*', a DT insight should also guide us towards determining *'the reason'* that led to a particular observed event happening in the first place.

From a hierarchical view of reality rooted in systems thinking, some participants argued that *'if you think about these things hierarchically ... certain patterns can emerge at the higher levels of the system'*. From this perspective, a DT insight is realised not when something is merely observed or experienced but, as one stated, when *'there is some sort of inference being made'* about the rather unobservable deeper levels of the systems.

An emergent theme began to dominate the discussion, assuming that a DT is exclusively employed for problem-solving. However, one participant expressed her desire to *'challenge this idea that is problem-solving tasks alone that you can use these twins for'*. She felt compelled to emphasise the exploratory capabilities of a DT. Consequently, DT insights can assist us in comprehending complex phenomena and structuring problem situations without necessarily intending to do so, although they can later be beneficial in addressing specific well-defined problems.

> *'you may also use it (DT) to gain insights in terms of problem exploration. So you're not really sure what the problem is, but you're using the twin and its models and everything to understand potential future adverse effects or problems and understand how things relate. So you're actually using it to gain insight in the form of how something is currently working and why things are happening ... you may understand something about the societal, technological, natural systems and how they interact. And you may apply that later on to know what type of measures you should do should a problem arise.'*

In response, the participant within the same group, who earlier emphasised the value of DT insights in the value chain from '*data to insight to decisions*', concurred that a DT is not solely applicable to problem-solving. He implied that the concept of *'decision-making'* encompasses a broader field that includes exploration alongside *'problem-solving'*:

> *'Just to be clear on my position, I'm not so much talking about 'problem-solving', more about 'decision-making, and yes, some of those decisions may be in relation to problems that need to be solved.'*

8.2.3.6 Theme [T6]: Predictability of urban dynamics

Everyone agreed that the urban environment is complex, dynamic and constantly changing. This was one of the reasons for advocating pluralism in DT practice, as discussed in theme [T2] above. For the same reason, some participants suggested that the idea of prediction is irrelevant to certain aspects of the urban environment.

> *'When you talk about infrastructure like power networks, road networks, etc, they are systems that have a purpose, right? You can measure them by providing power to people, water utilities. Cities do not have a purpose in that sense, therefore a digital twin of a city doesn't have a purpose in the same sense. Cities are supposed to be messy, chaotic, places where different people lead different lives and everyone kinda muddles along.'*

Another participant agreed to this, *'I agree completely with [the former's name] point.'*

However, others argued that urban problems are inherently different from those in the natural sciences, where natural laws govern the behaviour of systems. In contrast, urban planning involves 'wicked problems' that lack natural laws to guide predictions.

> *'our particular application is urban planning and this is a wicked problem. So what distinguishes this from, let's say, the more technological or physical problems that you solve is that you don't have natural laws that sort of guides what happens if you perturbate a system.'*

The absence of natural laws governing the urban environment restricts the ability to predict outcomes in terms of time and space, particularly with non-enduring patterns. One participant pointed out that

> *'from what I've seen so far, I think there are epochs of stability where you could say that there is a certain pattern of behaviour, but then you have disruptions and they change. But I think that there are subcomponents or certain parameters that are fairly stable in certain epochs or time epochs... I think both... there are some chaotic elements, something we cannot predict, but I do believe that for certain subsets of parameters in a certain time epoch, you could in fact assume that there is some.'*

The same participant drew upon her personal experience with a DT used for a road network owned by a public authority to illustrate how the DT's predictive capabilities are compromised by the ever-changing broader environment, namely

> *'for instance ... public roads authorities have a very big regional transportation model to understand current and possible future transportation systems and measures... that usually actually works to some extent, plus minus some uncertainty margin. But it's OK. It's good enough for making decisions... But then you have either over a long period of time, where things change like unforeseen technology or unforeseen phenomena that really changes everything overnight and COVID is, for instance, one of them.'*

Additionally, the changes in citizens' behaviours due to their natural reflexivity further compound the challenges associated with making predictions. The same participant continued:

> *'People think and act differently. Such as maybe now since 2015, I would say we've started to think very much about sustainable development, about our individual behaviour, and we're slowly changing ... Then a model from before that epoch could not have predicted that.'*

When examining citizens' behaviours more closely, it became apparent that there was a consensus that individuals possess free will. Nevertheless, their actions are influenced by external conditions, motivating them to behave in specific ways. Consequently, by virtue of such influences, people tend to exhibit certain observable patterns of behaviour for a certain period of time. However, given the immense complexity and constant change of external conditions, individuals may need to exercise their free will to make different choices. Consequently, regardless of the patterns they exhibit, these patterns will never remain constant over an extended period of time, making long-term predictions more challenging.

> *'It is clear as per Y that people have free will, but people use their free will in ways that forms patterns and it is observable, but in different circumstances they use the free will and come up with different, you know, different patterns ... So for example, the behaviour of citizens pre-COVID to post-COVID is very different that the same citizens with the same free will, but they might do different things ... So there's always going to be some kind of external influence that will affect our decisions. And the combination of external influences means that there's, you know, it's vastly complex how different people will make decisions but there are still observable patterns... I'm also recognising that that free will lead to patterns.'*

In the same sense, another participant believed that '*city dynamics are organic – on that basis, most are predictable, some are not.*'

A similar assertion – that some dynamics can be predicted to a certain extent while others are significantly more unpredictable – was proposed by another participant, albeit from a systems perspective.

> *'I think if you think about these things hierarchically ... certain patterns can emerge at the higher levels of the system, that can be at an aggregate level across the population, can be relatively confidently predicted from the interactions at the lower levels. And equally, there'll be other things which are tending towards the randomness, chaos, things that are difficult to predict from the interactions at the lower levels, and there'll be some things whether the extent of free will and choice or changes that happen over a longer period of time make it harder to predict these kinds of things.'*

In a nutshell, there seemed to be a growing consensus that while urban patterns can be observed and are sometimes crudely predictable, the dynamic and ever-changing urban environment, along with autonomy of individuals, prevent these patterns from being enduring, and thus from being consistently predictable, especially with high certainty. As a result, some participants argued that it might be more valuable to utilise these patterns for a deeper understanding and explanation of how urban structures operate, rather than relying on them for predicting the occurrence of future events.

> *'rather than just say, you know, that was the pattern last time, and therefore it's going to be exactly the same this time, I think it's much more interesting and useful to understand the patterns as they as they change.'*

8.2.3.7 Theme [T7]: The dialectic between urban structures and people

The theme [T6] above, which delved into the predictability of urban dynamics and the influence of citizens' behaviours on the emergence of regularities, naturally promoted participants to reveal their worldviews regarding the relationship between urban structures and people. Initially, a clear distinction was made between citizens as individuals and society as a whole.

> *'I'm making an intentional distinction between people as individuals and society, as a collection of individuals.'*

Most of the participants used the phrase *'somewhere in between'* to indicate what they perceived as a false dichotomy between determinism, in which the urban environment dictates how citizens behave, and voluntarism, where citizens freely shape the environment according to their desires, regardless of its existing state – for instance, in response to question [Q5] (i.e. Imagine you were assigned to create a DT for your city, what would your approach be?), as shown in Figure 8.3, one participant said, '*I think my position would be somewhere in between the two.*'

Describing a position that falls *'somewhere in between'*, another participant noted that *'it is clear that people have free will'*. However, he later asserted that *'there's always going to be some kind of external influence that will affect our decision'*. Therefore, based on both assumptions, it is inferred that *'people use their free will in ways that form patterns, and it is observable; but in different circumstances they use the free will and come up with different, you know, different patterns'*. From the same standpoint, the same participant further elaborated.

> *'So I think what becomes important is, is not to think that humans are some kind of automatons and there will always follow the same laws, but also not to think that it's a complete free will and therefore you can make no sense of it. And it's neither one nor the other.'*

In terms of how urban structures can influence the actions of people, another participant suggested that such structures do not only impose constraints but can also act as enablers.

> *'So, people getting the electricity, water, transport services and mobility services that they need in order to make their own decisions and less about inadvertently or deliberately influencing the movement of people themselves.'*

8.2.3.8 Theme [T8]: DT and ethics

When shifting the discussion to ethics, participants began contemplating and discussing what ethics are about in context of digital twinning – for instance, one at first wondered, *'is it about outcome or action?'*, and the first response to this question by another participant was *'not sure!'*

However, as the discussions progressed, participants seemed to shed light on three different dimensions in the context of digital twinning. The first is concerned with DT outcomes, the second focuses on the DT practices themselves, including the ways DTs are used to support specific decisions or interventions, while the third involves the means used to develop the DT in the first place.

Primarily concerned with DT outcomes, one participant commented.

> *'I believe ethics are about outcome not intention. My DT is ethical because it does not cause undue harm to the populace or minority groups.'*

Similarly, another participant pointed out that, from an ethical point of view, it is unacceptable to have a DT that would end up marginalising some groups.

> *'What you can't have is digital twins kind of benefitting society on average, but some people being left behind. So, I think people and society and nature, so that's that for me is like the starting point.'*

It was also emphasised that outcomes should not be only about avoiding harm but also enhancing citizens' capabilities. An ethical DT, from this perspective, is one that promotes the empowerment of individuals. It is a DT that can assist citizens reclaiming their freedom by facilitating interventions that provide them with what they need, including infrastructure provisions and services as enablers rather than constraints. As a result, people can make their own decisions freely.

> *'It's both more practical and more ethical to base our digital twins on infrastructure that provides people with what they need ... So, people getting the electricity, water, transport services and mobility services that they need in order to make their own decisions and less about inadvertently or deliberately influencing the movement of people themselves.'*

However, another participant emphasised the significance of DT practices and their pivotal role in shaping the DT outcomes.

> *'It will be possible, for example, to have exactly the same digital twin, which in one country is used to control citizens in quite an unpleasant kind of way and in another country, is used to help citizens to be more free and to have more choice and to get better outcomes for themselves and their children and the nature around them.'*

Others believed that ethics are not only about outcomes, but also the means, including the processes through which a DT is developed or built, which can have significant ethical implications. As one participant explained.

> *'I also think that you can distinguish between sort of the outcome that came from the use ... and then the other component is the way it is being used, so the process, is that an ethical process? The way you are designing this, who can access the system, how they can interact with it ... that cannot happen in a vacuum. It is a process that surrounds it. It's a decision making process, it's a political process or an administrative process, and I think that that design as well could be ethical or unethical ... we are experiencing this sort of political interests in this twin (a DT the speaker is working on developing it with local authorities) and the acknowledgement and understanding that this is a very powerful tool because you can, you can use this to communicate your worldview, It is possible to manipulate.'*

Another participant provided more detailed examples of how ethics or ethical considerations are inherently and inevitably intertwined with some of the modelling processes and methods used while building a DT.

> *'The twins we're going to build, or any kind of models, there are some decisions about what to include and what not to include, even when it's trying to be a twin, if you're doing white box modelling or black box modelling, if they're getting verifiable results that replicate reality, but not necessarily for the same reasons that occur in reality. There will be decisions made by the people who built that model about what to include, what not to include, assumptions about what's causing things and those could encode, if you know, perhaps, if you're not really thinking about it, or perhaps it's unavoidable, those things could encode aspects of the*

> *ethics, the values, the morality of the people who made them. And that may mean that those things aren't transferable. The framework for building them could be transferable, but, you know, a digital twin, they can get one from one place, can't just be forced into another, one element to another, if it's involving people and the construction of it has embedded assumptions about modellers that might not be translated, you know.'*

Despite participants focusing on different aspects of ethical digital twinning, it was generally acknowledged that all these dimensions matter. One participant pointed out, *"the purpose of technology is what it does' (a statement made earlier by someone else), was probably more what I wanted to say and how you use it is another thing, but it's a very important thing'.*

As the discourse about ethical digital twinning continued, participants were prompted to unveil their deeper worldviews and philosophical understandings of the notion of ethics and its nature. Over time, it became evident that DT practitioners would not genuinely embrace ethics without a clear understanding of what ethics entails. To this end, all participants agreed, albeit to varying degrees, that ethics and culture are interconnected, as the latter define the former, with the understanding that culture naturally changes and evolves over time. As one participant articulated, *'my view is ethics are culture-dependent ... and it will evolve over time'*. Consequently, he believes that cultural or societal shifts will inevitably reshape our understanding of ethics – for example, he continued, *'digital twins and generally use of data, quicker use of data, quicker decision-making is a cultural shift ... it's an industrial revolution, so, there would be a massive shift, ethical shift, to what is good, what is considered as good, and digital twins contribute to that shift as well.'*

Similarly, another participant shed light on the interdependency between ethics and cultural variations over time, as well as the relationship between ethics and the society within which ethical standards are defined.

> *'I happen to think that ethics isn't fixed ... is ethics fixed or does it change? I'm in the camp of it changes because society changes. So, if you look within one particular culture, I think the ethics can potentially change with time ... I think we can see that very clearly in terms of what is acceptable in society now compared to 20 years ago. But also there's a cultural context which changes with place, so what is considered ethical in Europe is different, different from what is considered ethical in in the far east. ... I think there's a cultural context that you can't get away with.'*

Although all participants agreed that ethics are culture-dependent, some held strictly relativist views while others adopted a slightly more realist stance. The relativists were evidently indifferent to influence of various cultures. For them, since cultures themselves define ethics, they believed that they were not in a position to judge any culture, and consequently, any DT that stemmed from it, in terms of its ethical nature.

> *'In one society it may be chosen, you know, in a very democratic kind of way that the most important thing is people, forget nature, we're just going to kind of optimise the outcomes for people. And in another society, which in my mind would maybe the society I would want to live in, the decisions which would benefit both people and nature.' ... And it's I don't think it's to say that one is better than the other or one is right, it's just to say that they're different.'*

Another participant confirmed the same.

> *'The ethics are culture dependent, we can talk about the social values of different places and with those, as [participant's name] said, its not saying any are right or wrong, It's just that they're different from one another and they can vary from place to place and over time, so that dynamic can change.'*

It is worth mentioning, however, that some of the relativists exhibited signs of inconsistency. They insisted on refraining from claiming that any culture or society can be considered more ethical than another. Nevertheless, they repeatedly implied that some principles or actions, like choices made *'in a very democratic kind of way'*, or *'issues of ... discrimination'* are universally clearly known as ethical or unethical – for example, the participant who earlier said that *'it will be possible, for example, to have exactly the same digital twin, which in one country is used to control citizens in quite an unpleasant kind of way and in another country, is used to help citizens to be more free and to have more choice and to get better outcomes for themselves and their children and the nature around them'*, implying that one is used in an ethical way and the other is not, was the same one who also stated few minutes later, when comparing between two DTs in two different societies, *'I don't think it's to say that one is better than the other or one is right, it's just to say that they're different.'*

The fewer realist participants, however, argued that no matter how cultures vary over space and time, there will always be some fixed principles of ethics that all humans agree upon in every context and era.

> *'My concept of ethics when it comes to think about it is a bit like a Venn diagram ... there's a big, strong central bit but around the outside they change.'*

8.2.3.9 Theme [T9]: DT is purpose-driven

This is one of the most prominent emerging themes. All participants emphasised the significance of the purpose of a DT is to such an extent that by the end of the discussion several participants thought that this was the most important thing mentioned in the meeting.

> *'Something which I and I feel was important and part of what makes me continue to think it's important is what felt like a lot of people around the call kind of coalescing and agreeing on it, which was that purpose point. I think many, many people, one way or the other talked about the purpose of digital twins and how that then defines so much else.'*

8.2.3.10 Theme [T10]: Philosophical awakening

This theme came to the forefront towards the end of the discussions. It was after the participants had been presented with several open-ended, philosophically oriented questions that demanded detailed explanations that they began to feel enlightened about the idea of revealing philosophical worldviews. Someone noted that the questions were 'surprisingly difficult', whilst another said

> *'I think I've been exposed to some very interesting insights from everyone who talked, and I feel because we're also struggling very much with your questions, I think they're very good, by the way.'*

8.2.4 Theorisation and retroduction

This section involves the theorisation and retroduction steps of the focus groups method (Figure 8.5). These steps are essential for evaluating the capacity of CR to provide a unifying philosophical foundation for the implementation of DT for urban management. Using retroduction, the mechanisms or causal factors responsible for generating the worldviews articulated throughout the discussions and for the emergence of the identified analytical themes are inferred. Consequently, these mechanisms, being the underlying philosophical assumptions, are leveraged to evaluate CR in terms of its ability to encompass these diverse philosophical worldviews in a unifying manner.

One of the principles guiding the analysis of focus group data is 'verifiability' (Krueger, 2014). According to this principle, all conclusions must be traceable back to the collected data through a clear trail of evidence. To render this connection explicit and transparent, the conclusions below are presented in a tabular form (Table 8.2). This format creates a direct link between (a) analytical themes, (b) retroduction or inferences made, and (c) the evaluative arguments for assessing CR in terms of its capacity to unify the various philosophical worldviews expressed by the participants, underpinning their vision of DT practice in context of urban management.

8.2.5 Implications of critical realism for digital twin practice

CR demonstrated the capacity needed to provide a philosophical groundwork for this new paradigm. Thanks to its inherently pluralistic nature, CR seamlessly accommodates the various DT practices discussed and the philosophical assumptions underpinning them, fostering a unified and complementary approach. Consequently, this paper lays the foundation for future DT practices and research in context of urban management to generate scientifically rigorous insights explicitly grounded in a robust theoretical framework, enabling the evaluation and critique of findings and knowledge produced.

In addition to providing a meta-theoretical foundation for DT practices, CR promises further practical insights and guidance to DT practitioners. The principle of epistemic relativism and the concept of laminated totality, as described in CR, present a compelling framework that can inspire DT practitioners to embrace a multidisciplinary approach when addressing complex urban problems and making crucial decisions in urban development. Laminated totality highlights the interconnected layers of reality that shape urban environments, acknowledging the intricate web of social, economic, cultural and environmental factors in play.

By recognising that urban systems comprise diverse and interdependent dimensions, DT practitioners are encouraged to move beyond mere technological enthusiasm, transcend disciplinary boundaries and collaborate with experts from various fields while developing the DT or when interpreting its outcomes. This approach fosters a more comprehensive understanding of urban challenges, enabling the formulation of holistic strategies that consider both immediate concerns and long-term consequences. Embracing epistemic relativism and laminated totality, DT practitioners can cultivate a more inclusive and adaptable digital twinning process that reflects the complexity of real-world urban dynamics and strives to create sustainable and resilient cities for the future.

The concept of stratified reality provides DT practitioners with a valuable framework for interpreting observed urban patterns and correlations within the urban environment. By acknowledging that the urban landscape consists of multiple layers, including empirical data, underlying mechanisms, and contextual influences, this perspective empowers DT users to go beyond surface-level observations revealed by DTs. Critical realist DT practitioners can delve deeper into the root causes of observed phenomena, discerning the intricate interplay of social, economic, political, and environmental factors. This necessitates the interdisciplinary approach advocated above and the involvement of expert knowledge.

Moreover, this comprehensive understanding of stratified reality can guide DT practitioners when developing the conceptual design of the DT. Recognising the underlying structures and mechanisms shaping urban dynamics can inform the early mapping of the urban system under focus. This, in turn, helps identify the relevant datasets required to feed into the DT or specify the operating DTs

Table 8.2 Retroduction from focus group findings and evaluation of CR as underpinning philosophy for the paradigm *DT for UM* (continued on next page)

Theme	Retroduction	Evaluation of CR
[T1] Inconsistencies / [T10] Philosophical awakening	Several inconsistencies suggest that some participants may have a lesser concern for philosophical underpinnings and possess an underdeveloped awareness of their own philosophical worldviews. Many participants found value in, and some were even surprised by, the exposure to deeper philosophical debates.	The themes [T1] and [T10] exemplify how CR can facilitate a greater focus on the philosophical foundations that shape individuals' worldviews. [T1], which includes the detected inconsistences in the discussions, indicates that DT practitioners typically do not readily declare or even reflect upon their own philosophical worldviews. However, as CR formed the basis for the focus group questions, theme [T10] illustrates how participants welcomed the opportunity to engage in philosophical debates. It demonstrates their appreciation for gaining insights into these foundational topics and their profound implications for practical and concrete practices.
[T2] Advocacy of pluralism	In this theme, participants recommended the use of multiple and mixed methods because they believe in the *(a)* incommensurability of methods *(b)* biases of people *(c)* multi-dimensional and socio-technical aspects of the urban environment *(d)* dynamic nature of the urban environment over time.	CR provides a solid ground for DT practitioners who advocate for a mixed-method or *'integrative'* approach, encompassing both quantitative and qualitative methods. CR aligns with the advocacy of pluralism, as argued by many critical realists (Danermark *et al.*, 2019; J. Mingers, 2006; Sayer, 1992). CR's key principles are in line with the main reasons mentioned by participants during the discussions to justify their support for pluralism (a, b, c, and d): According to CR's notion of epistemic relativism [P2], all views and theories are fallible, and every perspective is both limited and unique. Therefore, different DT methods, rooted different views, are (a) limited because each method is designed to serve a different purpose, and (b) unique because they are inevitably value laden, carrying values and assumptions. Consequently, adopting a pluralistic approach is considered best practice for a critical realist. CR posits that the world is open [P4], acknowledging its (c) multi-dimensional and (d) constantly changing nature. Hence, the use of various methods over time is essential as new events unfold and new structures emerge. Moreover, the recognition of the (d) multi-dimensional nature of the urban environment aligns with Bhaskar's concept of 'laminated totality' (2010), which views the world as a complex inter-meshing of various systems operating at different levels. This necessitates interdisciplinary research to comprehensively understand its diverse facets.

Table 8.2 Continued

Theme	Retroduction	Evaluation of CR
[T3] DT is 'both' objective and subjective	Participants in this theme believe that a DT possesses both objective and subjective aspects. They affirmed a DT is *'both'* objective and subjective, or at a *'stage in between'* as articulated by some participants. It seems that they view a DT as objective, not in the sense of being entirely 'value-free' as positivists may propose, but rather as a representation of structures and events with real ontological existence 'out there'. However, they also emphasise that a DT is subjective and fallible, as it invariably carries the *'biases and point of views'* of the individuals involved in digital twinning.	CR can readily accommodate the views expressed by participants in this theme. Much like any critical realist, the participants avoid committing the epistemic fallacy, which entails conflating ontology (i.e. reality) and epistemology (i.e. our fallible knowledge of reality). They contend that a DT is *'both'* objective and subjective, neither solely the former nor entirely the latter. As one participant pointed out, *'ultimately there is one reality and there's people experiences'*. The unifying potential of CR is most apparent when one can rely on both principles of ontological realism [P1] and epistemic relativism [P2] to recognise that a DT encompasses *'both'* objective and subjective dimensions.
[T4] DTs are evaluable	Participants believe that it is possible to judge a DT based on rational grounds and in terms of how accurate or useful a DT is compared to others. While arguing that a DT has a subjective side intrinsic to it (see T3 above), participants still believed that DTs can be judged. They assert one DT can be deemed *'better than another'* based on rational criteria such as *'the extent … to capture reality'* or *'how well they achieve their purpose'* and bring about *'benefit for people, for society, for the nature'*.	This theme aligns with CR's principle of judgmental rationalism [P7], which avoids falling into radical relativism. Radical relativism might assume that all DTs are essentially the same since they are subjective and value laden, resulting in indifference toward how the DT is developed or implemented. In contrast, participants contend that despite the inherent subjectivity of any DT, it is plausible for one to more accurately represent the external reality and bring about desirable changes. Furthermore, they believe that people can rationally assess and evaluate DTs accordingly. However, it is noteworthy that when it comes to evaluating how ethical a DT is, some participants take a more relativistic stance, which contrasts with their arguments in this theme and the concept of judgmental rationalism [P7]. This is further explored in theme [T8].

[T5] DT insights	Participants believe that DTs can help us gain knowledge about reality that guides our future actions. This knowledge can be either probabilistic or propositional. *(a)* Probabilistic: This knowledge is directly acquired when a DT recognises that a particular event of interest is *'likely'* to occur – for example, knowing that *'there is likely to be extreme flooding'* can guide and influence our actions and plans. The concept of using DTs for prediction is further explored in the following theme [T6]. *(b)* Propositional: This knowledge is indirectly inferred. It involves working out the reasons behind a particular event or to *'understand how things relate... how something is currently working and why things are happening'* in an explanatory sense. In other words, it involves thinking *'about these things hierarchically'*, recognising that *'certain patterns can emerge at the higher levels of the system'*. Consequently, there may be *'some sort of inference being made'* to postulate the involvement of certain causal mechanisms at the lower levels 'beyond what is currently seen' (National Infrastructure Commission, 2017: p. 63). These mechanisms contribute to the emergence of observed events or patterns.	The participants' understanding of DT insights and how they can aid decision-making aligns with two fundamental CR principles. First, acknowledgment of an independent reality [P1] that exists outside of our perceptions. Second, Assumption of a hierarchical and stratified reality [P3], where lower strata are unobservable. Consequently, recognising that knowledge about the mechanisms operating at these lower levels can only be fallibly inferred, or 'retroduced,' based on insights obtained from DT outputs. This knowledge about underlying generative mechanisms can help us comprehend and explain the reasons behind observable patterns at higher levels. Moreover, utilisation of this explanatory power of DTs to guide decision-making, potentially addressing the root causes of undesirable situations, could, in extreme cases, *'save lives in advance of the problem occurring'*.

Table 8.2 Continued

Theme	Retroduction	Evaluation of CR
[T6] Predictability of urban dynamics	Participants believe that *(a)* unlike natural phenomena, urban systems are not governed by fixed natural laws. *'urban planning … is a wicked problem … you don't have natural laws that sort of guides what happens if you perturbate a system'* *(b)* unexpected events might emerge, such as *'unforeseen technology or unforeseen phenomena that really changes everything overnight and covid is, for instance, one of them'* *(c)* people have free will and their thoughts and behaviours change over time – for example, *'since 2015 … we've started to think very much about sustainable development, about our individual behaviour, and we're slowly changing'* *(d)* the unique characteristics of the urban environment make it difficult to accurately predict concrete future events. Only crude predictions within limited time epochs are plausible. Therefore, *'it's much more interesting and useful to understand the patterns as they as they change'* instead of saying *'you know, that was the pattern last time, and therefore it's going to be exactly the same this time.'*	These five arguments, *(a)* to *(e)* align well with the principles of CR as explained below. Argument *(a)* corresponds to the view that urban systems are 'open systems' [P4]. This aligns with Archer's description of the vexatious fact of society (1995, 1), which, she argues, 'is different from natural reality whose defining feature is self-subsistence: for its existence does not depend upon us'. Similarly, Sayer (1992, 123) asserts that 'the social sciences deal with open systems but lack the advantage of their equivalents in natural science of having relevant closed system sciences on which to draw.' Arguments *(b)* and *(c)* further emphasise the challenges of maintaining closure in open systems, building upon the general idea presented in *(a)*. Argument *(b)* relates to the extrinsic condition for closure, where 'the relationship between the causal mechanism and those of its external conditions which make some difference to its operation and effects must be constant' (Sayer, 1992: p. 122). This condition is challenging to satisfy when unexpected external events, like Covid-19, disrupt the urban systems. Argument *(c)* relates to the intrinsic condition of closure, which focuses on maintaining stability within the system, such as people's thoughts and behaviours regarding sustainability. However, as Archer (1995, 70) points out, 'people by nature are reflective in thought and reflexive in action… which therefore makes attempted closure rather like locking the stable door on a horse who knows how to undo it'. Sayer (1992: p. 123) further highlights that 'one of the main reasons for the openness of social systems is the fact that we can interpret the same material conditions and statements in different ways and hence learn new ways of responding, so that effectively we become different kinds of people'. In *(d)*, the implications of *(a)*, *(b)* and *(c)* manifest as the challenging task of predicting future events within open systems in the urban environment. Næss (2004, 156–57) notes that 'a number of theorists belonging to the tradition of critical realism have rejected the possibility of prediction about social matters

	(e) Despite all the above, there could be some *'epochs of stability where you could say that there is a certain pattern of behaviour'*. In other words, while urban dynamics may exhibit unpredictable and unexpected events, they may also manifest some temporary and short-lived regular patterns.	… According to these authors, the implausibility of predicting specific concrete future events within the urban environment is a consequence of society's character as an 'open system', as distinct from the much more 'closed' conditions characterising the parts of reality focused on in the natural sciences.' Argument *(e)* is also consistent with a critical realist perspective. The idea of experiencing *'epochs of stability'*, where approximate spatially and temporally restricted regularities can be detected, aligns with the notion of quasi-closed systems and the demi-regs they may manifest.
[T7] The dialectic between urban structures and people	Participants believe that *'people have free will'*. Nonetheless, *'there's always going to be some kind of external influence that will affect our decision.'* Between the free will of people and the influence of external factors, participants believe that *'somewhere in between'* is a truer account of the relationship between citizens and urban structures. As one explained, *'So I think what becomes important is, is not to think that humans are some kind of automatons and there will always follow the same laws, but also not to think that it's a complete free will. And, therefore, you can make no sense of it. And it's neither one nor the other.'*	Once again, a critical realist stance can readily embrace the views expressed in this theme, which might have appeared to be challenging to reconcile from either a deterministic or voluntaristic perspective. It is in the face of this tension that the philosophy of CR, by virtue of the concept of analytical dualism [P5], provides an opportunity for the reconciliation of the agent-structure dichotomy. This theme also strongly relates to the concept of holistic causality,[1] where totality (i.e. the form and structure of the combination of constituent elements) causally determines the elements, whereas the elements collectively co-determine each other as well as the whole.

Table 8.2 Continued

Theme	Retroduction	Evaluation of CR
[T8] DT and ethics	The findings from this theme can be interpreted as follows. *(a)* Participants believe in the possibility of ethics. *(b)* Participants posit that, when employing DT for urban management, ethics primarily revolve around promoting emancipation and freedom while avoiding harm. *(c)* Participants acknowledge that ethics are contingent on culture. *(d)* Participants endorse a relativist stance on ethics, with the majority contending that it is challenging to assert that one culture or the DT derived from it is inherently more ethical or superior to another; instead, they see them as merely distinct.	CR readily aligns with the beliefs in *(a)*, *(b)*, and *(c)* as follows. *(a)* From a critical realist perspective [P6], ethics are conceivable because, as discussed in theme [T7], CR rejects deterministic views of humans and, thus, assigns them responsible for their actions. As articulated by the critical realist Porpora (2019: p. 276), 'CR offers a non-nomothetic view of causality that, in contrast with the nomothetic view, does not presuppose the determinism that undermines moral responsibility. CR thus rescues the very condition of possibility for ethics.' *(b)* CR lends support to this argumen – for example, in his works of dialectical critical realism, Bhaskar elucidates the purpose of social science, emphasising its concern for the emancipation of individuals and the empowerment to be and do without undue constraints. *(c)* This perspective resonates with CR (John Mingers, 2014). Sayer (2011: p. 136) consistently emphasises that 'human well-being is always culturally defined or conceptualised' and underscores how 'society shapes us'. *(d)* Participants in this context seemed to advocate for 'an emotivist account of ethics' or a form of 'cultural relativism that threatened the grounding for any non-arbitrary ethics' (Douglas V. Porpora, 2019: p. 277). They imply that ethics ultimately boils down to no more than 'expressions of personal or cultural taste'. This perspective arises from their belief that, contrary to CR's principle of judgmental rationalism, one should abstain from making truth claims.

[1]We appreciate the input of an anonymous reviewer in clarifying the implications of 'holistic causality' for theory and practice of digital twins.

However, we argue that participants are content with adhering to this relativist position largely out of respect for people's freedom and their right to select and shape their own cultures, which then become emblematic of their identities. It is understandable to argue in favour of people's and societies' rights and freedoms to define ethics (John Mingers, 2014). However, we assert that a relativist stance on ethics is flawed for two reasons.

First, from a critical realist position, Sayer (2011, p.136) critiques the epistemic fallacy inherent in arguments like *(d)*, where 'from the correct point that human well-being is always culturally defined or conceptualised – by way of particular cultural discourses – it is wrongly inferred that the condition that they define is itself successfully and exclusively determined by those discourses, so that cultural discourses can never be mistaken'. He further elucidates, 'we may want to argue that societies should be free to decide on their own conceptions of the good, but it does not follow that everyone – racist or non-racist, misogynist or feminist – has equally adequate conceptions.' (Sayer, 2011: p. 105).

Second, treating all cultural beliefs and the associated definitions and practices as equally valid can be seen as disrespectful. It involves a dismissive refusal to take seriously the claims or understandings posited by different cultures[2].

However, CR offers an account of ethics and ethical DT practices which can address these issues while still upholding the indisputable arguments in *(a)*, *(b)* and *(c)*. This account, best explained by Sayer (2011), is termed 'pluralist objectivism'. While pluralist objectivism acknowledges that 'different cultures provide different kinds and mixes of flourishing and suffering' (2011, p.135) and recognises that 'individuals vary and become adapted to different ways of being' (2011, p.136), it also maintains that 'well-being should … be thought of in terms of objective states of being which people strive to discover, achieve or create' (2011, p.134). In essence, it posits that 'there are many kinds of well-being, but that not just any way of life constitutes wellbeing.' (2011, p.135).

[2]This argument is inspired by Andrew Collier's (2004) discussions in *Realism, relativism and reason in religious belief*.

Table 8.2 Continued

Theme	Retroduction	Evaluation of CR
		Indeed, a pluralist objectivism perspective on ethics provides a suitable framework for some comments made in the focus groups discussions, such as thinking of ethics *'a bit like a Venn diagram … there's a big, strong central bit but around the outside they change'*. Similarly, it accommodates participants' recurrent tendency to assert truth claims when discussing ethical DT practices, such as respecting privacy of citizens or protecting their data rights.
[T9] DT is purpose-driven	Digital twinning is a purpose-driven process. Clearly, *'many people, one way or the other talked about the purpose of digital twins and how that then defines so much else.'*	This argument is also supported by CR – for instance, in Bhaskar's dialectical critical realism, one can derive 'ought' form 'is', and making judgments necessarily implies a commitment to action.

that need to be connected to create a group of connected DTs, offering a more accurate representation of the complex urban phenomenon under investigation. By integrating this perspective into their practices, urban managers can utilise DTs to formulate strategies that address not only immediate symptoms but also the deeper systemic issues. This results in more effective and sustainable urban development decisions that consider the multifaceted nature of cities and their complexities.

The implications of viewing the urban environment as an open system for DT practices are significant and fourfold. An open systems perspective entails recognising that urban dynamics involve continuous interaction between human behaviour, environmental factors, economic forces, social structures and more. Consequently, DTs can be developed to model these intricate relationships. Fuelled by data from sensors and various sources, DTs could facilitate the simulation of scenarios and predictions of urban changes over time. However, the same open-system perspective also encourages practitioners to exercise caution and humility when using the simulation and prediction capabilities of a DT, emphasising their role as aids for comprehending and navigating urban complexities rather than as definitive predictors. The openness, both intrinsically and extrinsically, of urban systems imply that DT predictions and simulations are inherently limited by the very nature of the twinned open systems. In short, by virtue of CR, urban planners would acknowledge that while DTs and urban data analytics can provide valuable insights, their predictions are not deterministic and should be interpreted within a broader contextual understanding.

Second, the open-system perspective encourages urban planners to prioritise the use of DTs for exploratory and explanatory purposes over predictive and prescriptive aims. Recognising the complex and dynamic nature of cities, planners are prompted to view DTs as tools that help unravel the multifaceted interactions shaping urban dynamics. Instead of seeking rigid predictions or definitive prescriptions, DTs are used to explore various scenarios, understand causal relationships, underlying generative mechanisms, and hidden system vulnerabilities or potential points of failure within the existing infrastructure system of systems. In other words, a critical realist DT practitioner would be more interested in understanding how a causal mechanism acts and how it affects an existing or proposed urban system than in trying to forecast specifics of future events.

Thirdly, by recognising the dynamic interactions within urban environments, DT developers would understand that the urban environment has an ever-evolving nature, and that the intricate web of factors influencing urban dynamics is subject to change. Hence, their open-system view of the urban environment would prompt them to be receptive to feedback and adaptive in how they maintain and adjust the DT and its underlying assumptions.

Fourthly, CR can aid in suggesting a suitable level of human involvement in the DT-based decision-making process. One of the key DT practical considerations involves determining the extent of human involvement in the DT's cyber-physical bidirectional loop. On one hand, the urban environment is recognised as an open system, necessitating a greater reliance on human expert judgments to support decision-making. However, in CR, it is also acknowledged that quasi-closed systems may emerge, identifiable through the demi-regularities they produce, making it comparatively less challenging for practitioners to make predictions, albeit to varying degrees. This, in turn, opens the door to the possibility of reduced human involvement and an increase in the DT's autonomy. Recognising both of these facts inevitably creates a practical tension concerning the appropriate level of human involvement or autonomy for a specific urban system's DT. While numerous factors should influence this decision, including the stakes involved and the consequences of decisions driven by the DT, it is argued that knowledge, defined by CR as the way causal mechanisms operate, can assist practitioners in determining the optimal level of human involvement. Once such knowledge is acquired, it can aid practitioners in making informed judgments about the stability of a quasi-closed system, thereby proposing a practically suitable degree of autonomy or frequency of human involvement in the cyber-physical bidirectional loop.

The concepts of analytical dualism and holistic causality empower practitioners who are developing and using DTs for urban management purposes to adopt a dual perspective. These critical realist principles provide DT practitioners with a comprehensive framework that encourages the integration of both top-down and bottom-up approaches in making informed urban development decisions.

In the top-down approach, by recognising that the totality of the urban environment's form and structure causally determines its elements, urban planners acknowledge the influence of wholes like physical, social and legal structures that impose constraints upon citizens and social interactions within the urban landscape. This understanding enables using the DTs to make coarse predictions about the influence of these systemic constraints and the resulting urban dynamics, guiding urban strategic interventions.

Simultaneously, recognising that elements co-determine the whole encourages the utilisation of open-data methods and the establishment of cross-sectoral connections between typically siloed urban datasets. This approach aims to capture the interdependencies between various system components and the resulting emergent behaviour they might produce under certain circumstances. Moreover, by acknowledging the significance of people's free will and the necessity to align urban development with the needs and aspirations of civic society, the bottom-up approach becomes paramount. This aligns with transdisciplinary practices 'that go beyond the academy to co-produce knowledge with urban stakeholders' (Karvonen *et al.* 2021, p. 8). As a result, with the aid of user-friendly platforms, realistic visualisations and immersive technologies, DTs can be used to engage the public and involve citizens, enabling urban planners to incorporate local knowledge, preferences and concerns into decision-making processes.

By embracing analytical dualism and holistic causality, urban professionals can harness DTs to bridge the gap between macro-level structural influences and micro-level individual agency, fostering a more balanced and effective approach to urban development that reflects the intricacies of both the built environment and the diverse communities that inhabit it.

As argued in Table 8.2, an objective pluralist account of ethics, promoted by Sayer (2011) and rooted in CR, offers a more reasonable and robust theoretical foundation compared to relativism. Moreover, it has significant practical implications for the use of DTs by urban managers. While both objective pluralism and relativism acknowledge the diversity of ethical perspectives, they differ in their underlying assumptions and implications for decision-making. Objective pluralism recognises that various ethical values, social norms and cultural variations can coexist and interact within different societies or even within the same society. However, it also seeks to identify common ethical principles, based on the concepts of flourishing and suffering, that transcend individual and cultural viewpoints. In other words, it does not assert that all values are equal or that there are no overarching ethical principles. Instead, it aims to navigate the tension between different values and find ways to engage in ethical discussions that respect multiple viewpoints without collapsing into a relativistic stance.

This would significantly influence how urban planners would use DTs to support decision making – for example, objective pluralist urban planners would first recognise the diversity of ethical perspectives involved when developing and using a DT of transportation networks. By acknowledging all ethical concerns of various groups, DT practitioners would strive to build a purposeful DT that reduces congestion and environmental issues while considering the potential impact on local communities and cultural practices – for instance, if the DT recommends rerouting traditional routes or changing the dynamics of a neighbourhood, the planner will engage with community members to understand their concerns and aspirations in order to accommodate their state of flourishing and suffering. Additionally, they would strictly adhere to overarching ethical principles defining the responsible use of DTs. This might involve implementing robust data protection, security, privacy, fairness and transparency measures.

In a nutshell, a critical realist DT practitioner seeks a middle ground that respects diversity while also recognising the need for principled ethical evaluations. These evaluations are universal to the extent that they should influence decisions with non-local impacts – for instance, decisions made in a city in the global north might cause suffering to cities in the global south due to planetary limits and scarcity of natural resources.

8.3. Evaluation of methodical element

8.3.1 Introduction

The worldviews and beliefs expressed by DT practitioners, as outlined in the abstract research above (Section 8.2), give rise to varying approaches in DT practice. As a result, members of the discipline tend to favour certain DT tools or methods during DT-based interventions. Given that DATUM seeks to unify the *DT for UM* paradigm, it should ideally present a standardised methodical framework that accommodates all forms of DT features and applications, regardless of practitioners' differing perspectives and preferences. Therefore, the 'extensive research' (Section 2.7.3.3) undertaken here evaluates the practical generalisability of DATUM's methodical framework (i.e. DTUCS). Specifically, it assesses how well DTUCS applies across different types of DT-based projects in the context of UM. This includes determining how effectively Prong-A describes the features of DT use cases, how Prong-B captures the full range of DT uses, and how Prong-C models the use case scenarios (UCS) across various DT projects.

8.3.2 Methodology

The 'extensive research' presented here utilises a multiple-case study method, drawing on both primary and secondary data. Case study research is a well-established qualitative approach (Yin, 2009) used to test theories or 'generate an in-depth, multi-faceted understanding of a complex issue in its real-life context' (Crowe *et al.*, 2011: p. 1) or to explore a specific "bounded system' … a program, an event, an activity, or individuals' (Creswell, 2009: p. 61). This method supports the development of 'more general theoretical statements about regularities in the observed phenomena' (Fidel, 1984: p. 274).

Stake (1995) identifies three types of case studies: intrinsic, instrumental and collective. An intrinsic case study focuses on a unique phenomenon, selected for its distinctiveness rather than its representativeness. Conversely, an instrumental case study seeks to understand a broader, more general phenomenon where the specific case selection is less critical. The collective case study 'involves studying multiple cases simultaneously or sequentially in an attempt to generate a still broader appreciation of a particular issue' (Crowe *et al.*, 2011: p. 2). By comparing the behaviour, similarities and differences of the phenomenon across various contexts, this approach helps to derive more generalisable conclusions. Yin (2009) suggests studying three to five cases to ensure reasonable generalisability and validation of a hypothesis.

Given the breadth of this research, multiple instances are examined. The collective or multiple case study method used here explores various DT projects from different organisations. However, the analysis remains at a concrete level, focusing on practical outcomes rather than engaging in philosophical or abstract discussions, which are covered in the abstract research (Section 8.2).

The nature of the extensive research, and the chosen multiple case study approach, requires the collection of data with sufficient reach and breadth across several DT projects. This data must also be diverse enough to ensure the findings are plausibly generalisable across the discipline. To this end, two data collection methods were used, detailed in Sections 8.3.3 and 8.3.4.

(a) *Archived case studies*: Secondary data from completed and documented DT projects were collected. A rich archive of DT case studies is available on the UK's Connected Places Catapult DT Hub[3]. This repository was excluded from the systematic literature review (Section 3.5), on which the development of DTUCS is based, to avoid duplication of data. Three cases were randomly selected from the DT Hub to assess the generalisability of DTUCS across different DT projects.
(b) *Synthesised case studies*: Primary data of potential DT use cases were gathered from a multi-city urban DT project in Germany. This project involved the design and creation of several connected urban DT use cases.

The aim of this research is to evaluate DATUM's methodical framework (i.e. DTUCS) in terms of its practical generalisability. The success of Prong-A in describing most DT features, Prong-B in capturing the majority of DT uses, and Prong-C in modelling the UCSs of various DT projects would demonstrate a satisfactory level of generalisability. Conversely, failure in these areas would indicate insufficient practical generalisability.

8.3.3 Analysis of archived case studies

8.3.3.1 Case study 1: City-scale digital twin prototype for Cambridge

This section outlines the first case study, structured according to DTUCS Prong-A's standard DT features and Prong-B's standard DT uses. The interactions between the actors and the DT are then modelled using the UML sequence diagram method, as recommended by Prong-C (Figure 8.6).

The focus of this DT is to 'explore behavioural insight for reducing car dependence by considering the socio-economic characteristics of various site users.' A GUC identified for this DT is to 'support city policy-making,' placing it firmly within the 'futures planning' *area of application* and addressing the 'initiation' *lifecycle stage*.

The DT *captured* GIS data of Cambridge's road network while monitoring automated number plate recognition (ANPR) sensor data, including travel direction, time of arrival and parking duration, for one week in 2017. This resulted in a *dynamic*, though *offline*, DT with a city-wide *spatial scale* and individual vehicle *spatial resolution*. A rule-based algorithm was applied to the data, enabling the classification of distinct car user groups through computational processes.

Currently, the DT operates at a sub-system *level of federation*, focusing solely on road networks. However, there are plans to scale up to a systems level by integrating multiple DTs related to

[3]https://digitaltwinhub.co.uk/

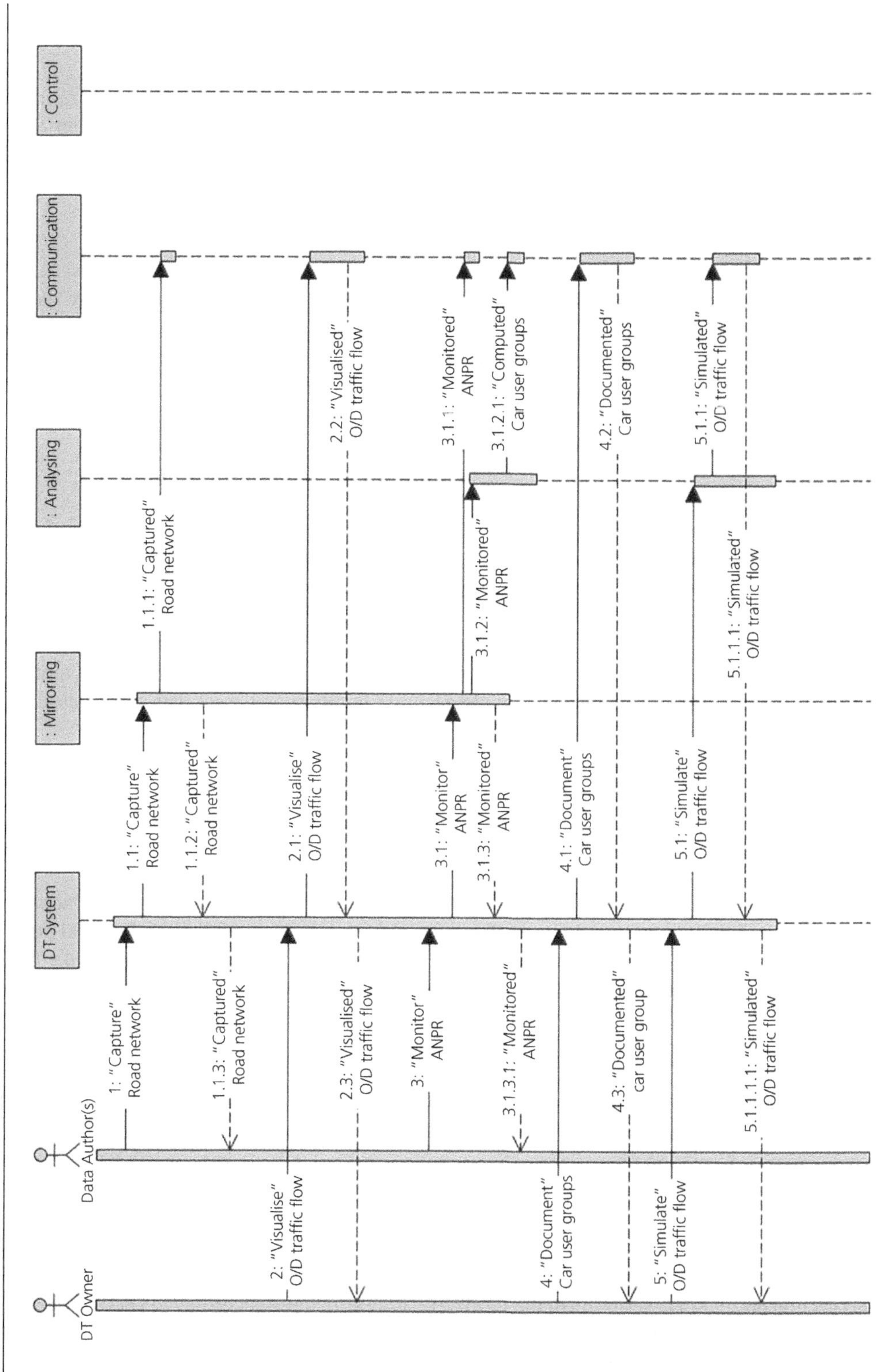

Figure 8.6 Case study 1 (City-scale digital twin prototype for Cambridge) modelled using Prong-C. (author's own)

transportation infrastructure systems, including 'roads, traffic signals, kerbside, bus/rail networks, and legacy systems.'

The DT is used to run *scenario simulations* by altering user-defined inputs, such as 'future employment, housing growth, and the associated spatial distribution.' These simulations help explore potential outcomes compared to users' expectations. This demonstrates how the DT operates across both *urban layers of* infrastructure assets – such as road networks – and socio-economic layers, including housing and employment variables.

Primarily, this DT is designed to *inform* policy-making and 'support human decisions ... hence does not include algorithmic decision making' (i.e. *actuate*). Origin–destination flow data and simulation results are *visualised* on a GIS platform, providing decision-makers with a clearer understanding of road network dynamics. The prototype, however, is not designed to support community *engagement*, as the 'user interface is oriented towards professionals in local authorities and academic users.' It is noted, though, that 'an extended user interface tailored to different user backgrounds is to be explored' in future iterations.

8.3.3.2 Case study 2: Coventry University digital campus

This section presents a description of the second case study couched in Prong-A standard DT features and Prong-B standard DT uses. The actor-DT interactions are then modelled using UML sequence diagram modelling method as recommended by Prong-C (Figure 8.7).

The DT of Coventry University is built through digitising over 110 individual buildings constituting the main Coventry campus. It is developed to 'manage building information more efficiently, reduce operational costs and provide accurate building and asset data for all estates and university stakeholders.'

The *GUC* inferred from the documented purpose can then be: 'reduce operational cost', which is most relevant to the *application area* of 'resource management' and the 'operation and maintenance' *lifecycle stage*.

The DT involved capturing BIM models of buildings at the Coventry campus and *monitoring* energy data and air temperatures by way of sensors, thus, *mirroring* both built and natural *layers of urban environment* in a dynamic online manner. Moreover, various building systems like CAFM and BMS were integrated and linked to the common data environment (CDE) as well, presenting a system *level of federation*. The DT, comprising a full campus, is spatially equivalent to a neighbourhood *scale* yet provides a higher *resolution* at an individual building space or zone.

Through *visualisations* of different building graphical and non-graphical information, the DT *informed* decisions including 'decisions on when to service assets, dates for planned preventive maintenance, information about energy consumption and reduction can be monitored, occupancy and space utilisation and access control management decisions'.

8.3.3.3 Case study 3: Smart energy digital twin for Bridgend County Borough Council (BCBC), Wales

This section presents a description of the third case study structured according to DTUCS Prong-A's standard DT features and Prong-B's standard DT uses. The user-DT interactions are then modelled using the UML sequence diagram modelling method as recommended by DTUCS-C (Figure 8.8).

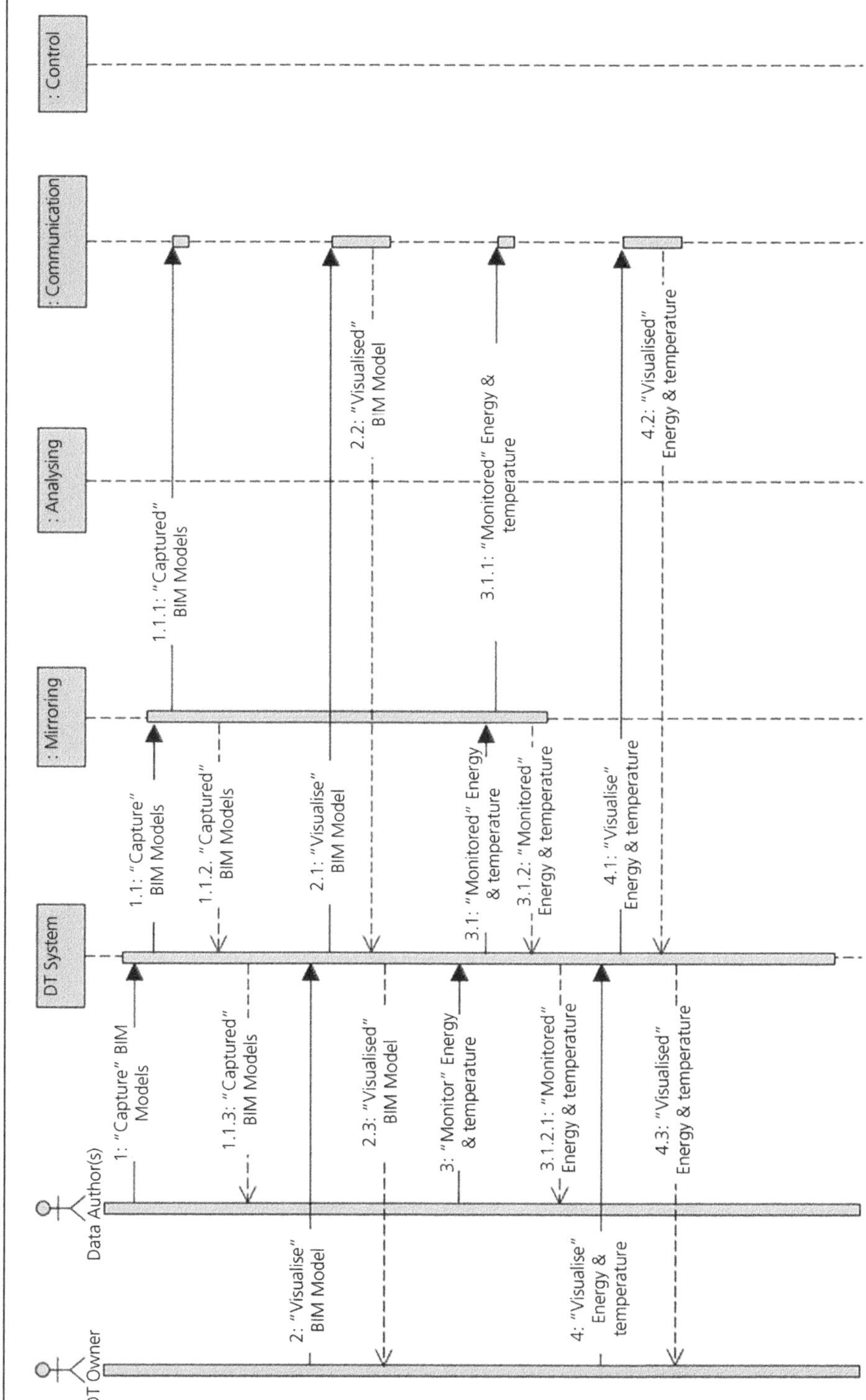

Figure 8.7 Case study 2 (Coventry University digital campus) modelled using Prong-C. (author's own)

Figure 8.8 Case study 3 (smart energy digital twin for Bridgend County Borough Council, Wales) modelled using Prong-C. (author's own)

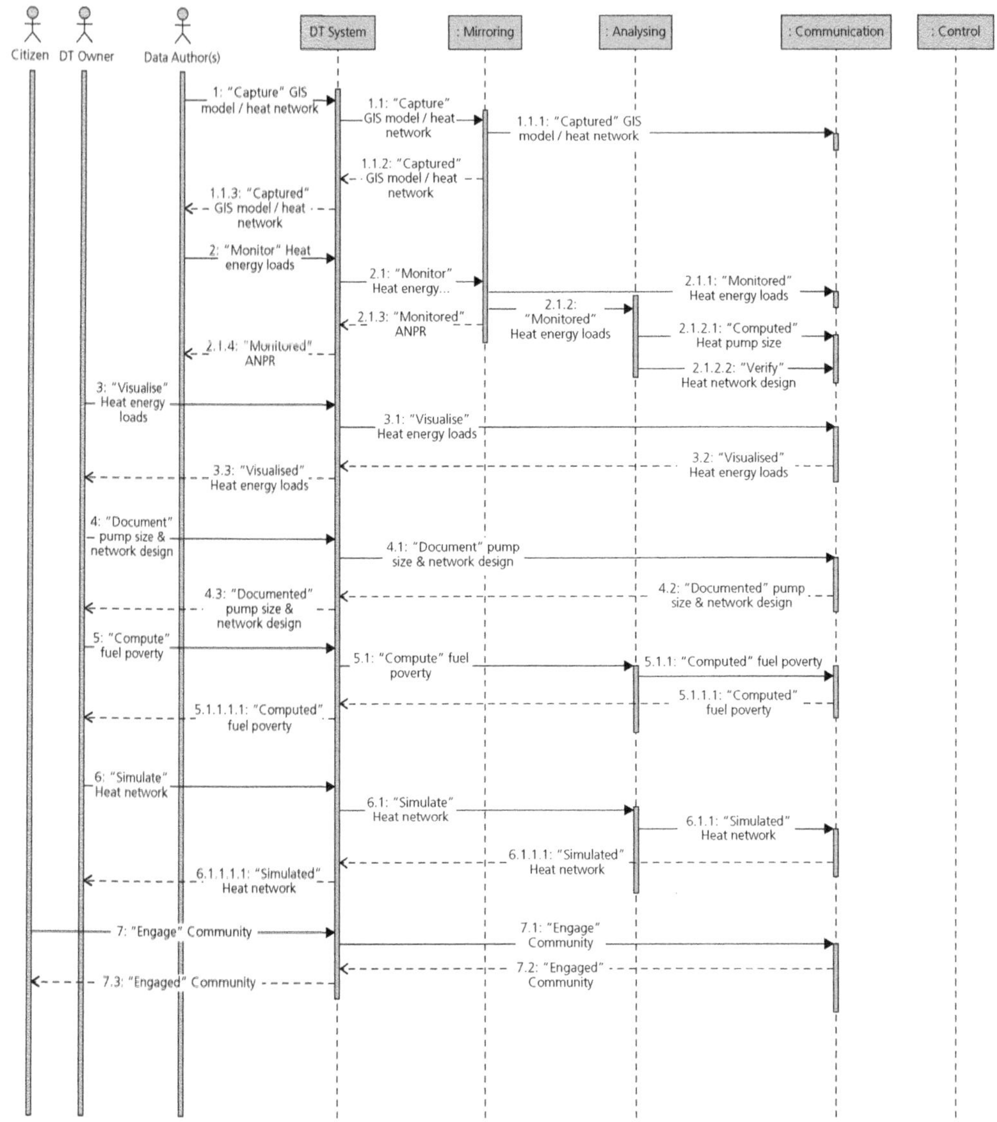

This case includes a district heat network DT that 'automates optimised plant, pipe sizing, and network routing based on peak load analysis using real property data in conjunction with established benchmarks'. Thus, a suggested GUC for this DT is 'optimise heat network design', which is relevant to 'resource management' *area of application.*

The DT is obviously serving the 'design' *lifecycle stage.* It *captured* the GIS model of the district of interest and other unstructured data sets. Also, it *monitors* the heat energy loads at district households. It, therefore, *mirrors* aspects of two *urban layers* – built and socio-economic environments – in a *dynamic online* manner. Moreover, datasets relevant to various heat energy network components like pumps, piping and outlets at households were integrated, demonstrating a system *level of federation.*

The DT, involving a heat energy network spreading over a full district, is spatially equivalent to a neighbourhood *scale* yet provides a higher *resolution* at an individual network component and household.

Data *mirrored* are leveraged in different ways. They are used to *compute* heat pump sizing and verify produced network design based on 'benchmarks'. Data were also exploited to *compute* fuel poverty indicators based on household income, household energy requirements and fuel price elements. Further, the DT can run *scenario-simulations* based on alternative input scenarios of energy usage profiles.

Communication with DT users is done through the *visualisation* of dashboard infographics (e.g. digital representation of a smart energy network that is automatically generated, flow metrics, health energy profiles, quarterly heat energy requirements) to provide analytics and insights. Another form of *communication*, in pursuit of transparency and community buy-in, is the use of interactive 3D web mapping platform for community *engagement.* Furthermore, DT developers are planning to use Unreal Gaming Platform for *immersive* interactions.

The DT *informed* decisions including 'decisions on when to service assets, dates for planned preventive maintenance, information about energy consumption and reduction can be monitored, occupancy and space utilisation and access control management decisions'.

8.3.4 Analysis of synthesised case studies

During the development of this book, DTUCS was published in a journal paper (Al-Sehrawy *et al.*, 2021). The publication had enough impact to draw the attention of key DT researchers and practitioners, including the DT toolkit team who created the DT toolkit on DT Hub and provided very positive feedback (e.g. In a meeting with the DT toolkit team, one team member said: *'I think that is brilliant. This is really good! ... I haven't seen anything better to be honest for what we're trying to do in the toolkit ... I think we should as a group seriously consider using the approach (DTUCS) here as a template for our DT roadmap.'* Another highlighted, *'this approach is far more detailed, our version of like the '3-bucket' (Figure 3.9-c) is just a very high-level view'*). The feedback underscored DTUCS's value as a comprehensive method for guiding DT development, showing its potential to become a foundational reference for practitioners across the DT community.

Similarly, the team working on the connected urban twins (CUT) project in Germany found DTUCS to be very useful. CUT is a five-year project with total value of 32.4 million Euros, involving 70 team members and 73 smart city pilot projects across the three cities of Munich, Hamburg and Leipzig.

At an initial stage of the project upon reaching out, one of CUT's project managers stated, '*we do have 2 days meeting ... to see if we all got the same understanding of your system and to check if it provides the expected value for our project*'. Few weeks later, the project team confirmed: '*Our workshop ... went really well ... We came to the conclusion that, considering your classification system and its categories, we can further develop a guidance to capture, document and communicate requirements of use cases in form of a questionnaire/checklist that will give us support to really comprehend a use case and its specifications at an early stage.*'

Subsequently, CUT's team used DTUCS to help specify the various DT use cases and required DT uses. A key task for CUT project team is to describe the DT uses employed for every DT use case implemented as a part of CUT project[4]. To carry out this task, the project team had to find a consistent and standardised framework to facilitate the process. Consequently, they drew on DTUCS Prong-B to complete this work. Figures 8.9 and 8.10 below show two examples of two different DT use cases that utilised DTUCS Prong-B to specify the DT uses required for delivering the DT GUC articulated. This process has indeed helped in severe and extensive testing and evaluation of DTUCS Prong-B against a wide range of different DT use cases.

8.3.5 Conclusion

The extensive type of research presented here evaluates the generalisability of DTUCS, the methodical component of DATUM. Given the nature and objectives of this evaluation, which focus on generalisability, DTUCS was applied across multiple case studies. These case studies fell into two categories: archival, which are real-world examples of completed DT projects, and synthesised, which are hypothetical case studies proposed by DT experts and urban managers.

When DTUCS was applied to three archival case studies, randomly selected from DT Hub, it demonstrated significant generalisability. It successfully captured and specified the diverse DT features and uses across all the case studies. Additionally, DTUCS Prong-C, which draws from UML, was straightforward to use for modelling the DT use cases described in each case study. However, some challenges arose. While these did not diminish DTUCS's overall generalisability, the manual task of extracting DT features and uses from textual documentation and modelling the use cases was labour-intensive and time-consuming. In some instances, crucial information, such as the stakeholders involved and the specific datasets used, was either missing or unclear in the documentation. Due to time constraints, Prong-C was only applied to three case studies. Although these cases were selected randomly to avoid bias, random selection alone does not ensure exposure to the full range of DT cases and scenarios. Extending Prong-C's application to more case studies of varying scales and applications would strengthen the evaluation.

In the synthesised case studies, the CUT project team used DTUCS Prong-B to identify all potential DT uses necessary for delivering the multi-city project's DT use cases. DTUCS Prong-B proved highly generalisable, providing the team with sufficient variety to account for all relevant DT uses. However, the team recommended a minor adjustment to further enhance DTUCS Prong-B's generalisability. They suggested adding a new specialised sub-use, 'pre-cast,' alongside 'forecast' and 'backcast,' under the specialised use 'predict'. According to the team, 'pre-cast' refers to the derivation of pseudo-data through analogy, such as inferring unknown private household energy consumption from known public household energy consumption.

[4]Until the date of writing this book, CUT project team applied DTUCS to seven different urban DT case studies (https://www.connectedurbantwins.de/) including the two examples provided in Figures 8.9 and 8.10 below.

Figure 8.9 Example 1 of a DT use case developed by CUT project team with DT uses specified based on DTUCS Prong-B (author's own)

As a user, I would like to analyse the current utilisation of all daycare centres.

Mirror	Analyse	Communicate	Control
Capture	Learn & Recognise	Visualise	Inform
• Daycare network (represented by KIVAN data)	N/A	• Map view of the city of Leipzig • Childcare centres as markers • Visualisation of capacity utilisation as a pie chart above the markers • Display of relevant key figures	• Support for the care of "needy" parents
Monitor	Simulate	Immerse	Actuate
• All data should be available at least weekly	N/A	N/A	N/A
Quantify	Predict	Document	
• Locations / buildings of the day care centres • Capacity per day care centre (if possible according to normal places and integrative places) • Number of occupied places	N/A	• Export of a utilisation report (PDF)	
Qualify	Qualitise	Transform	
• Care situation (illness, holidays, etc.)	N/A	• Export of the current occupancy rates as a list (CSV)	
	Compute	Engage	
	• Utilisation = occupied places / total places • Free places = total places - occupied places	N/A	

Figure 8.10 Example 2 of a DT use case developed by CUT project team with DT uses specified based on DTUCS Prong-B (author's own)

As a planner, I would like to simulate changes in the capacity of a daycare site.

Mirror	Analyse	Communicate	Control
Capture	Learn & Recognise	Visualise	Inform
• Daycare network (represented by KIVAN data) • Population forecast	N/A	• Map view of the city of Leipzig • Childcare centres as markers • Visualisation of capacity utilisation as a pie chart above the markers • Display areas with problematic capacity utilisation	• Support for the care of "needy" parents
Monitor	Simulate	Immerse	Actuate
• All data should be available at least weekly	• Simulation of closures • Simulation of partial closures, e.g. due to refurbishment measures	N/A	N/A
Quantify	Predict	Document	
• Locations / buildings of the day care centres • Capacity per day care centre (if possible according to normal places and integrative places) • Number of occupied places	N/A	N/A	
Qualify	Qualitise	Transform	
N/A	N/A	• Export of the current occupancy rates as a list (CSV)	
	Compute	Engage	
	N/A	N/A	

8.4. Evaluation of DATUM

8.4.1 Introduction

After evaluating both the philosophical and methodical components of DATUM, this section assesses DATUM as a complete framework in terms of its overall adequacy and usability. The value of this holistic evaluation is twofold. First, it enables a more thorough assessment of the methodological component, DM2. As DM2 is intended to bridge the philosophical and methodical elements, it is difficult to fully evaluate its effectiveness in isolation. Second, examining DATUM as a whole ensures that, when its elements are integrated, it can achieve the objectives for which it was developed.

This evaluation is conducted using intensive action research, as outlined in Section 2.7.3.1. The approach involves implementing DATUM in its entirety – philosophical, methodological and methodical components – to address a real-world problem. A satisfactory level of usability can be demonstrated by how well DATUM, grounded in its philosophical principles, helps practitioners identify stakeholders and accurately represent their differing worldviews. Additionally, a usable DATUM should facilitate the capture and understanding of the complex real-world system being studied. Finally, the ease with which practitioners can navigate the methodical framework, DTUCS, and the time required to understand its prongs and produce results will serve as another key measure of DATUM's overall usability.

8.4.2 Methodology

8.4.2.1 Introduction to action research

Action research is the process of 'learning by doing' (O'Brien, 2001) with the ultimate aim of 'improving practice' (Koshy, 2009: p. 15). It is a type of research that 'involves action, evaluation and reflection … based on gathered evidence … (is) situation-based … can involve problem-solving … (where) … findings emerge as action develops' (Koshy, 2009: pp. 15–16).

8.4.2.2 Model of action research

In this section, several popular models of action research and their distinguishing features are examined before introducing the model adopted in this book. Elliot (1991: p. 71) proposes a responsive and fluid model of action research, illustrating how planning, action, monitoring, and evaluation often overlap (Figure 8.11(a)). The model developed by Kemmis and McTaggart (2000: p. 564) introduces a 'spiral' structure with successive, self-contained cycles, each consisting of three essential steps: plan, act and observe, and reflect (Figure 8.11(b)). O'Leary's (2004: p. 141) model, shown in Figure 8.11(c), portrays action research as a cyclical process, where learning deepens and knowledge grows, leading to better understanding and improved actions in later stages compared to the initial ones. Table 8.3 below summarises the key and unique aspects of each of these three models.

Figure 8.12 presents the action research model developed and implemented in this study to assess the overall usability and adequacy of DATUM in practice. It depicts two cycles of implementation, each designed to expose DATUM, with DM2 at its core, to a unique problem situation. This approach evaluates the distinct steps and principles of DM2 in a variety of contexts. Each cycle begins with the 'plan' phase, where the designed artefact (DATUM) is applied to a selected UM problem. The next phase is 'act and observe,' where DATUM is implemented, and its performance is monitored. Finally, the 'reflect' phase involves evaluating DATUM against the evaluation criteria (Section 4.6.5) and introducing modifications or design changes as necessary. During the actual implementation of this model, reflection often occurred simultaneously with the 'act and observe' process.

Figure 8.11 Three existing models of action research. Source: (a) (Elliot, 1991) (b) (Kemmis and McTaggart, 2000) (c) (O'Leary, 2004)

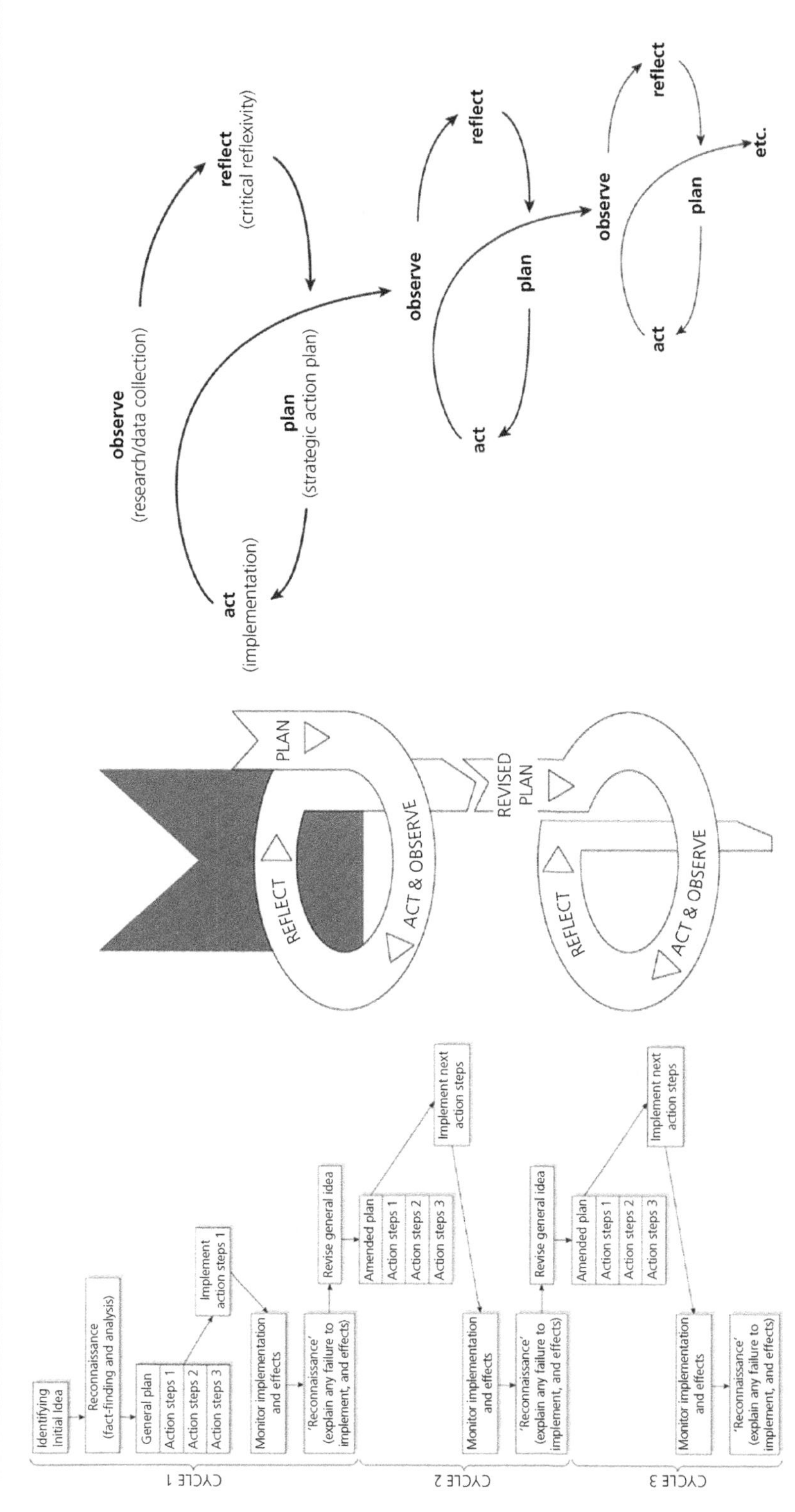

Table 8.3 Three existing models of action research and their key features

(Elliot, 1991)	(Kemmis and McTaggart, 2000)	(O'Leary, 2004)
This model explicitly shows how the process of action research is *neither neat nor rigid* – for example, it demonstrates how one can learn and revise the whole plan and the initial general idea while still in the middle of implementation.	*The spiral shape* instead of a fixed centric loop shows how the whole process takes place over time. To the extent that Winter and Munn-Giddings (2013) thought that this might give the impression that even the simplest processes may take longer time to complete. Like O'Leary's, this model also indicates how difficult it is to assume a rigid process of action research – for example, this model refrains from disentangling observing and acting, but present them as one task.	*The converging cycle* is a unique feature of this model. It indicates improvement of learning and practices within every cycle compared to the preceding one. Another unique feature of this model, the author argues, is *the shifting or the displacement of cycles*, with a completely new centre for each cycle. This implies that the contextual conditions within which each cycle is performed are never the same. Even if the same people collaborate to address the same problem twice in two successive cycles, their knowledge and understanding in the second cycle is improved compared to the first one.

The action research model is guided by five key principles, listed in Table 8.4, along with corresponding visual features as illustrated in Figure 8.12. The first three principles are drawn from existing models previously discussed. The fourth principle, regularity, emerges from the critical realist perspective adopted in this book, specifically the understanding of the world as an open system (Section 6.2.3.1) where problems are complex and multi-dimensional. From this viewpoint, creating a controlled, lab-like environment to implement and evaluate DATUM is unrealistic. Each problem presented for DATUM to address through 'intensive' action research will have different circumstances and conditions, meaning every situation provides a new and unique environment for DATUM to operate within. This implies that each problem will require focusing on certain aspects of DATUM while de-emphasising others, depending on the situation's nature.

This understanding has significant implications for the design and implementation of the action research model in this book. Given that problems in open environments are unique, every action research cycle dealing with these types of problems will have an irregular form – for example, Section 4.5.2.1 discussed how different UM problems may display varying levels of complexity. A problem exhibiting severe structural complexity may require greater emphasis on the *analysis* step of DM2 (Section 7.3.8), while a problem characterised by high levels of people's complexity will call for a broader *appreciation* (Section 7.3.3) to account for numerous and potentially conflicting viewpoints. Therefore, selecting a different type of problem for each action research cycle is essential to provide maximum exposure and rigorous testing of the artefact. In other words, the ideal problem situation for any given cycle is one that allows for evaluating the parts of DATUM that were least tested in the previous cycles.

Table 8.4 The four key principles and visual features shaping the action research model implemented in this book

Action research methodological principles	Corresponding visual feature
Implementation is undertaken over time (Kemmis and McTaggart, 2000).	Spiral (Y-axis).
Better understanding and improved practice (O'Leary, 2004).	Cycles converge (Scale).
Contextual conditions are unique for every cycle (O'Leary, 2004).	Cycles shift / displaced (X-axis).
Neither neat nor rigid (Elliot, 1991; Kemmis and McTaggart, 2000).	This principle is acknowledged and endorsed albeit not visually illustrated for simplicity.
Depending on the contextual conditions and the nature of the problem, learning is not equally gained or improved for all aspects of the tested plan. Some aspects are better implemented, more stressed upon, or better observed at each cycle.	Cycles have irregular shapes, with concave curves referring to aspects of the plan which were accurately implemented and thoroughly reflected upon, while the convex curves indicate the opposite. (regularity).

Figure 8.12 2-cycle action research model implemented in this book (author's own)

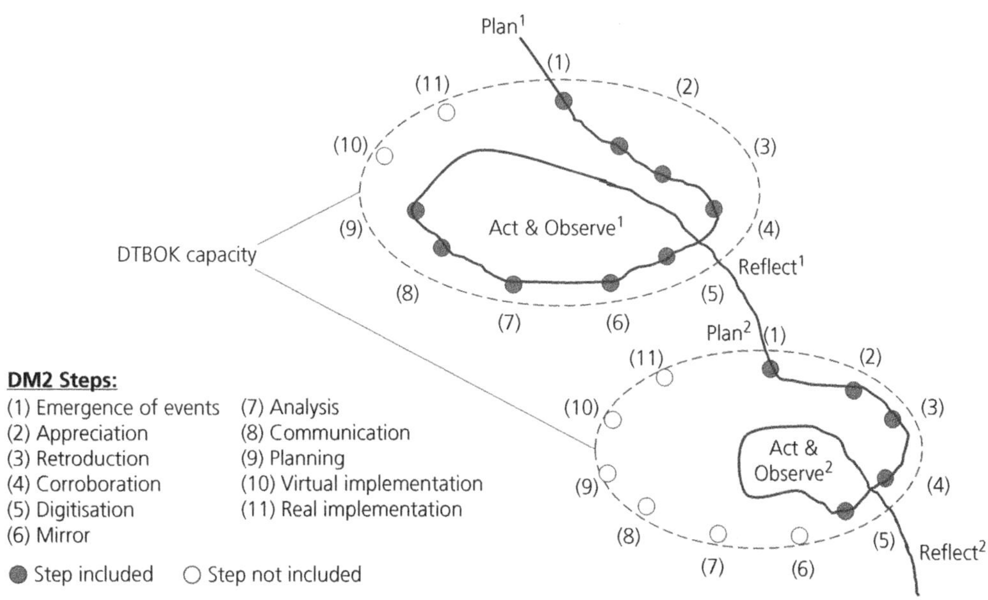

8.4.3 Cycle 1: DT of a building

8.4.3.1 Plan

The artefact being evaluated in the first cycle of the action research model (Figure 8.12) is DATUM, which encompasses its three core components: the philosophical, methodological and methodical elements, as originally designed and illustrated in Figure 7.2. The next stage of this action research cycle, 'act and observe,' is outlined in the following section and structured around the steps of DM2 – the methodological element at the heart of DATUM – applied to address the problem at hand. This approach clearly illustrates the specific actions undertaken, as recommended by DATUM.

8.4.3.2 Act and observe

EMERGENCE OF EVENTS [S]

In the initial phase of DM2, real-world structures generate new events, a subset of which begins to be observed and appreciated.

APPRECIATION [H], RETRODUCTION [H] AND CORROBORATION [H]

The issue under focus here is the indoor air quality within a prominent university building, referred to as BLD, located at the centre of the campus. As outlined earlier, appreciation involves analysing and detailing the system or problem at hand. This process typically includes abstracting and breaking down events into a set of components or structures that are considered relevant to the situation.

BLD, a £7 million, state-of-the-art low-carbon facility, opened in 2018 and is equipped with smart building solutions and a specialised building management system (BMS) that controls heating, cooling and ventilation. It has become a notable venue for events, presentations, and a preferred space for students to collaborate on innovative projects.

While carbon dioxide (CO_2) monitoring through BMS is increasingly common, it is primarily used for energy savings and cost reduction rather than improving public health (Eykelbosh, 2021). Given that people spend nearly 90% of their time indoors (VELUX, 2022), research highlights the significant impact of indoor air quality on comfort, cognitive performance, health and well-being (Bluyssen, 2009; Sundell, 2004). Studies also suggest that improved air quality can enhance productivity by up to 10%.

There is growing concern over exposure to high concentrations of CO_2, with levels exceeding 5 000 ppm being harmful. Even lower concentrations, such as 1 000 ppm, can pose health risks after prolonged exposure. Jacobson *et al.* (2019) reviewed evidence of health risks linked to higher CO_2 levels, including inflammation, cognitive impairment, and more severe outcomes like kidney calcification and oxidative stress. The COVID-19 pandemic has further emphasised the urgency of addressing poor indoor air quality, as inadequate ventilation can increase the risk of transmission.

To fully *appreciate* the scope of the problem, a holistic understanding is essential. A two-hour workshop was held with three academics working at BLD to discuss and analyse the issue. This collaborative session aimed to identify the key structures influencing indoor air quality and to support the *retroduction* process. *Retroduction* involves postulating generative mechanisms and interactions between structures that, if present, would explain the observed events.

Systems thinking methods, such as 'rich pictures,' were employed to capture the complexity of the problem. Rich pictures facilitate team exploration of diverse perspectives, fostering creativity and enabling a visual representation of the problem to promote mutual understanding (Bell and Morse, 2013). Figure 8.13 illustrates the rich picture developed during the workshop.

While further probing and asking 'why' could uncover additional layers of explanation, it was important to define a clear system boundary. As DM2 is purpose-driven, the boundary and the selection of key structures shown in Figure 8.13 were shaped by the intervention's objective. The participants' input helped corroborate the understanding of the problem.

The hypothesised explanation of the problem includes

(a) *Structures*: The key components of the situation, such as BLD, its occupants, visitors, the events held there, indoor air quality, CO_2 levels, and the BMS (as shown in Figure 8.13).
(b) *Generative Mechanisms*: These are the causal tendencies of the structures that influence outcomes within the system. The arrows in Figure 8.13 depict these mechanisms and their interactions.
(c) *Interconnections*: These describe the systemic effects of the generative mechanisms when they interact – for instance, the question of which events could lead to overcrowding and poor air quality, despite the BMS's operation, reflects the need to explore the complex interrelations between the mechanisms involved. However, at this stage, providing detailed evidence-based answers regarding these interconnections remains a challenge.

Figure 8.13 Rich picture including key structures and mechanisms along with relevant datasets (author's own)

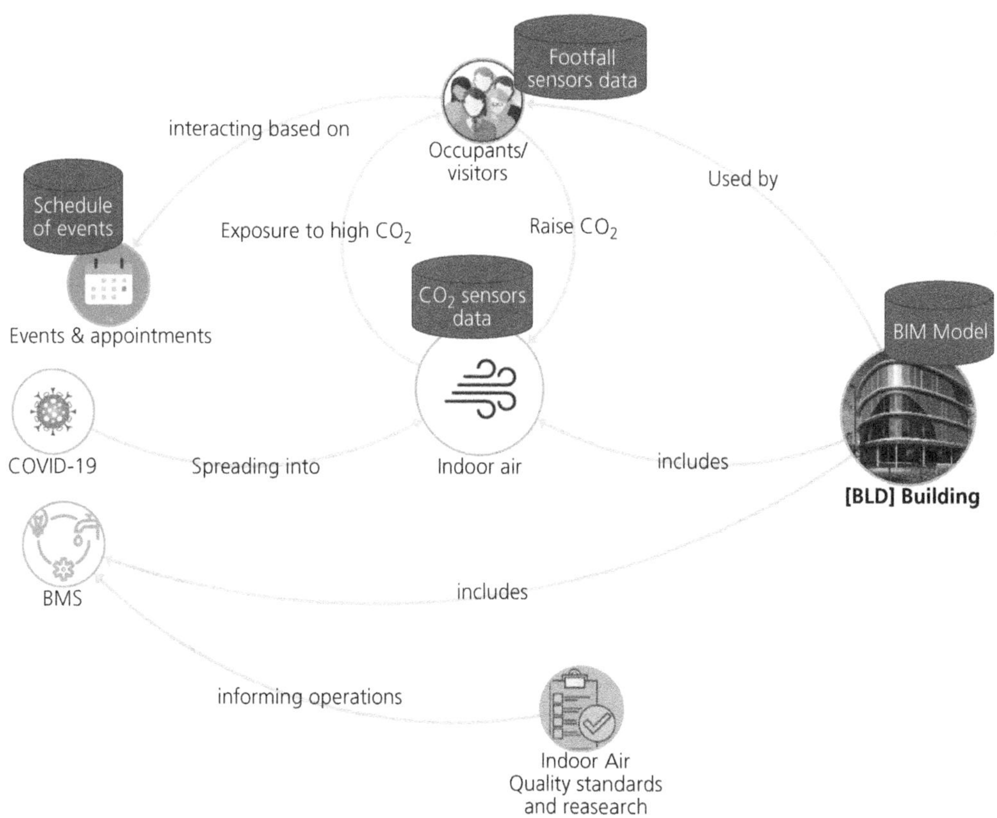

DIGITISATION [H-D]/MIRRORING [D]

Following the problem structuring and the creation of a rich picture outlining the causal mechanisms at play, the next phase involves the *digitisation* process. This entails developing a digital representation of the problem situation as perceived. The initial step is to articulate the GUC for the proposed DT. As derived from the preceding steps, the primary GUC for the potential DT is to 'improve indoor air quality at BLD'.

Next, it is crucial to identify the necessary digital datasets and allocate each dataset to the corresponding structures depicted in Figure 8.13. Each dataset serves as an appropriate digital representation of the mechanisms involved and their respective states.

The *digitisation* process further encompasses the design of the proposed DT and the modelling of the expected use cases it will deliver. To facilitate this, DTUCS Prong-B can be instrumental in specifying the uses of the DT necessary for addressing the identified problem. As outlined in Section 5.2.2, DTUCS Prong-B provides a comprehensive taxonomy of all potential DT uses. From this taxonomy, the specific uses required for this project have been identified and detailed in Table 8.5 below.

DTUCS Prong-A provides a framework for defining the various technical features and characteristics that the DT must possess to fulfil the uses outlined in Table 8.5. During the application of DTUCS Prong-A, it became apparent that there would be a natural overlap with the mirroring phase of DM2. This is due to the fact that the mirroring process (as discussed in Section 7.3.7) encompasses several aspects, such as the spatial and temporal specifications of the DT and the identification of its stakeholders, all of which DTUCS Prong-A can assist in clarifying. The distinct

Table 8.5 DT uses specified based on DTUCS Prong-B

GUC	Included use	Specialised use
Improve indoor air quality	*Mirror*	*Capture*: BIM model *Monitor*: CO_2 level at different building zones* *Quantify*: number of visitors per day*
	Analyse	*Compute*: mean value of CO_2 level per floor per day* *Mine*: patterns and correlations (i.e. demi-regs), if any, between datasets
	Communicate	*Visualise*: Spatial indoor air quality by navigating the BIM model and visualising the changes in CO_2 levels and number of visitors at different locations over time
		Document: Plot data using different charts and plotting methods
	Control	*Inform*: Gain insights to support decision-making to improve indoor air quality

*Constraints, such as existing installed and operating footfall and CO_2 sensors, have influenced the spatial and temporal resolutions specified for the DT.

features of the DT developed for this project are elaborated below and illustrated in Figure 8.14 based on the principles of DTUCS Prong-A.

(a) *Application area*: This DT is primarily focused on enhancing indoor air quality, thus categorising it within the 'environmental and carbon' application area.
(b) *Connectivity*: The DT is designed to operate at a 'systems' level of connectivity, integrating multiple datasets.
(c) *Layering*: The DT intersects three of the four urban layers. The first is the 'built environment,' encompassing the indoor environment of BLD. The second is the 'socio-economic' layer, which includes the traffic flow of BLD's occupants and visitors. The third is the 'natural environment,' which pertains to indoor air quality and CO_2 levels.
(d) *Spatial scale and resolution*: This DT operates on an 'individual' spatial scale, focusing on a single urban asset (i.e. the BLD building). In terms of spatial resolution, a floor-by-floor analysis is deemed sufficient to comprehend the state of the mechanisms on each level of BLD.
(e) *Temporality, temporal scale and resolution*: The DT must be 'dynamic' to accurately reflect fluctuations in air quality, human dynamics, and other pertinent contextual factors. Unlike a building management system (BMS) that receives sensor data in real time, this DT can effectively achieve its objectives by relying on historical data updated on a medium- to long-term basis, enabling proactive strategic decisions rather than merely operational responses.
(f) *Lifecycle stage*: The intention is to utilise this DT to inform the operational activities of BLD, thus situating it within the 'operation and maintenance' lifecycle stage.
(g) *DT actors*: The roles and responsibilities of the various actors involved in the DT process are mentioned in Figure 8.14.

DTUCS Prong-C is then used to model the expected DT use case scenario (UCS] (Figure 8.15).

Figure 8.14 DT features specified based on DTUCS Prong-A. (author's own)

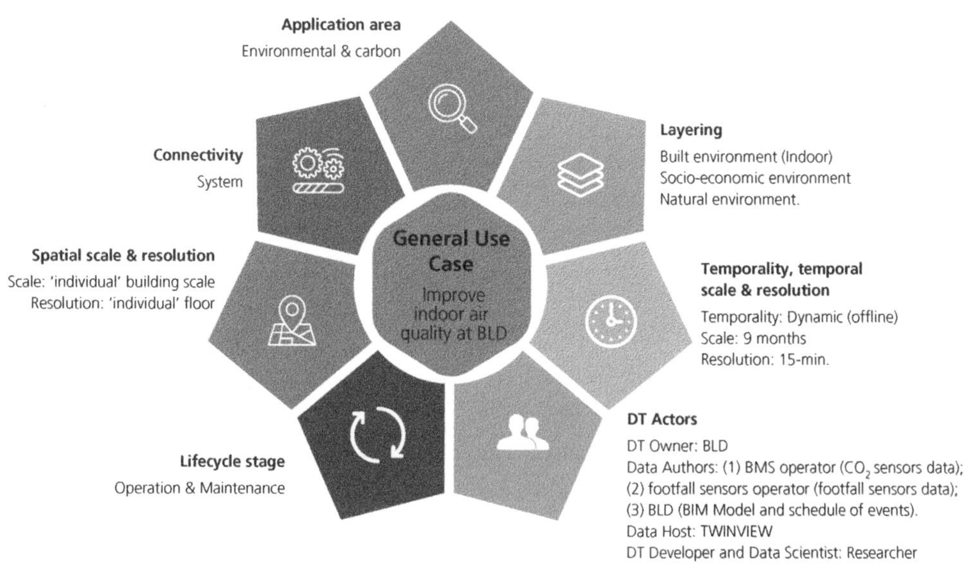

Figure 8.15 DT use case scenario (UCS] modelled using DTUCS Prong-C. (author's own)

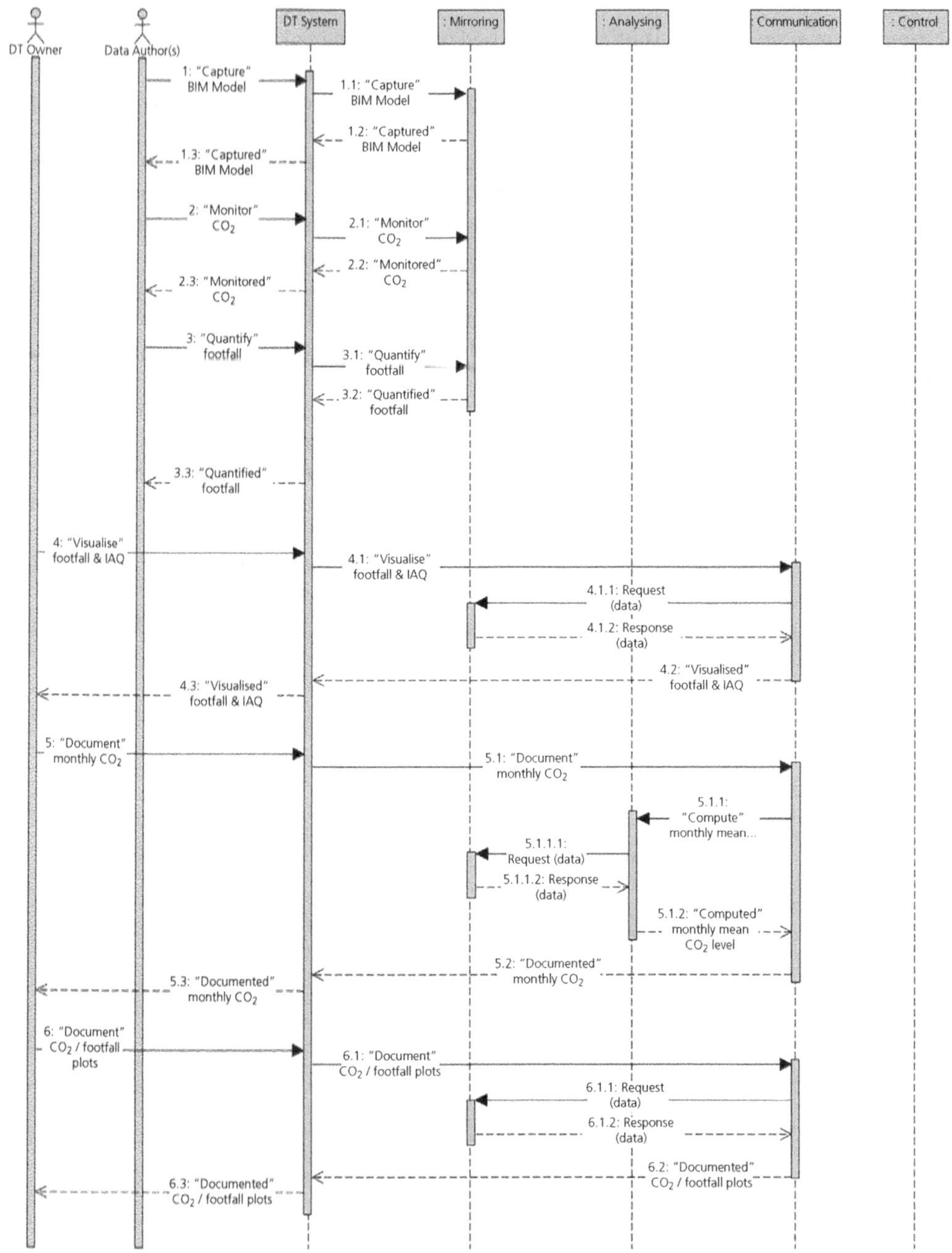

The collection of longitudinal data adds an extra dimension to how the problem situation is understood. It augments the ability to understand how events unfold over time and, thus, *corroborate* the posited explanation and the proposed set of mechanisms involved. The datasets required for the project (Figure 8.13) were collected. First, the static BIM model of BLD – including as-built architectural and mechanical, electrical and plumbing (MEP) models – is retrieved in 'Revit' project file format (.rvt.). Likewise, the CO_2 levels and the number of visitors at the 1st and 2nd floors (the ones hosting the vast majority of CIS users activities and interactions) were monitored by way of BMS (Figure 8.16) and quantified using footfall sensors data, respectively. The historical sensors data were collected for a total of nine months, from 1st of January 2021 to 30th of September 2021, at a 15-minutes temporal resolution. Moreover, schedule of key events held at BLD, including summer activities, were also collected in '.CSV' format.

COMMUNICATE [D-H]/ANALYSIS [D]/CORROBORATION [H]

A key phase in the *communication* process (Section 7.3.9) involves selecting the appropriate platform for integrating datasets, analysing them, and sharing the output from the DT. For this project, the 'Twinview' platform was chosen to host and communicate the data collected. The BIM model, CO_2 sensor data and footfall sensor data were all uploaded to Twinview, enabling the visualisation of CO_2 levels and footfall in different spatial locations over a set timeframe (Figure 8.17). The communication capabilities of the DT, facilitated by Twinview, made it easier to grasp the nature and patterns of the observed dynamic phenomena.

At this stage, *analysis* (Section 7.3.8) was vital for verifying the initial hypothesis and uncovering empirical evidence to support any proposed interventions. At the time of this project, Twinview offered limited data plotting techniques and analytical tools, such as correlation analysis or data mining, which could have provided deeper insights and helped identify underlying mechanisms (demi-regs). To overcome this, Excel was used for the initial evaluation of indoor air quality on

Figure 8.16 Building Management System (BMS] platform (author's own)

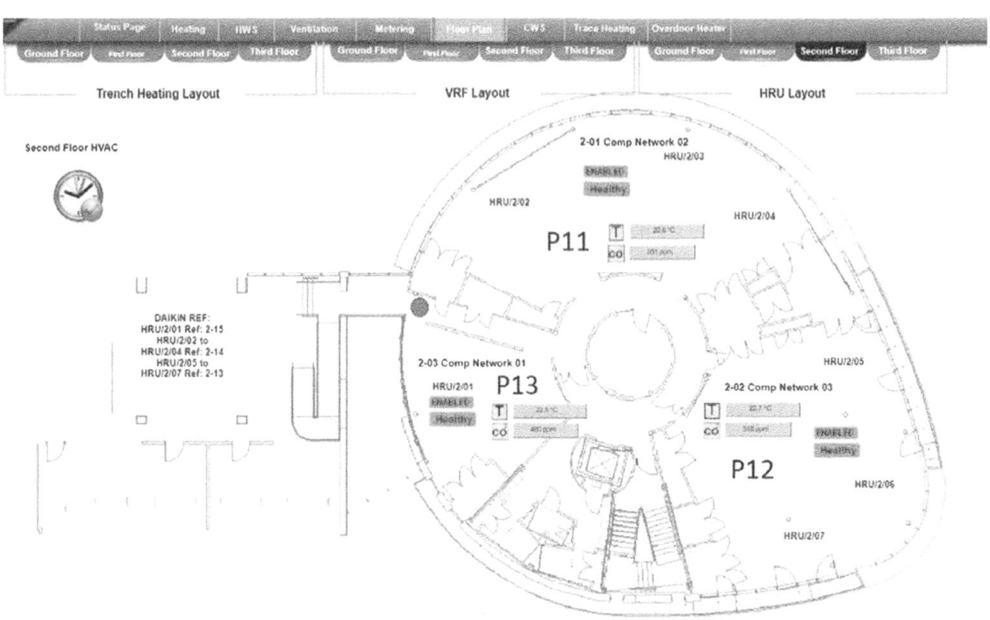

Figure 8.17 DT platform (TwinView) showing BIM model and sensors data (author's own)

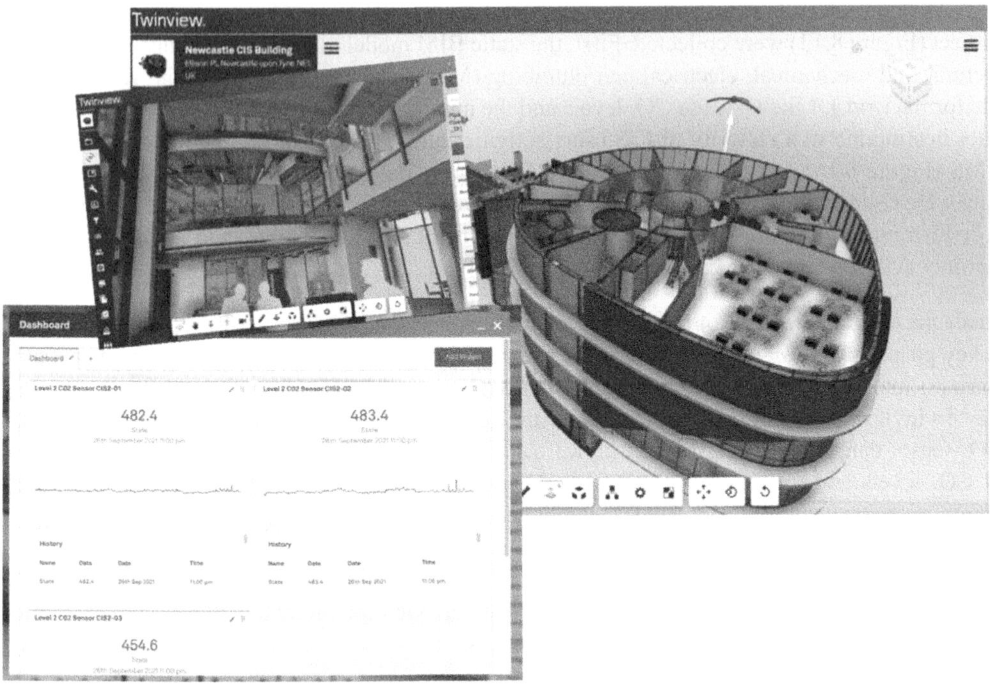

each floor. The arithmetic mean of CO_2 levels recorded at 15-minute intervals over the same day was calculated for both floors across the nine-month study period. These daily averages were then used to determine the monthly average peak CO_2 levels for each floor.

It became clear that the average monthly CO_2 levels on both floors ranged from 500 to 540 ppm, with the exception of September 2021, which saw a noticeable increase to 553 ppm on the first floor and 566 ppm on the second floor. This anomaly prompted further *analysis*. Although the nine-month averages remained below 800 ppm, well within acceptable indoor CO_2 levels (i.e. <1000 ppm) (Lowther *et al.*, 2021), relying solely on average values can be misleading. Averaging may obscure unusual or exceptional days with significantly high or unhealthy CO_2 peak levels. To avoid this, Python was used to create a swarm plot, *communicating* the peak CO_2 levels for each day on both floors (Figure 8.18). This approach allowed for a clearer understanding of the variations in air quality across the monitored period.

Upon reviewing the swarm plots, it became evident that, on most days, CO_2 levels remained below 550 ppm across both floors. However, there were occasional instances, particularly in September 2021, when CO_2 levels spiked above 700 ppm. From a CR perspective, these empirical observations, or 'demi-regs,' indicate the presence of active generative mechanisms within the deeper structures of reality. This observation prompted further analysis to confidently identify the primary generative mechanism responsible for these undesirable events.

The rich picture developed earlier (Figure 8.13) had already hypothesised the generative mechanisms at play. Based on existing scientific understanding, it was proposed that the building

Figure 8.18 Daily CO_2 peak levels at 1st and 2nd floors of BLD building (author's own)

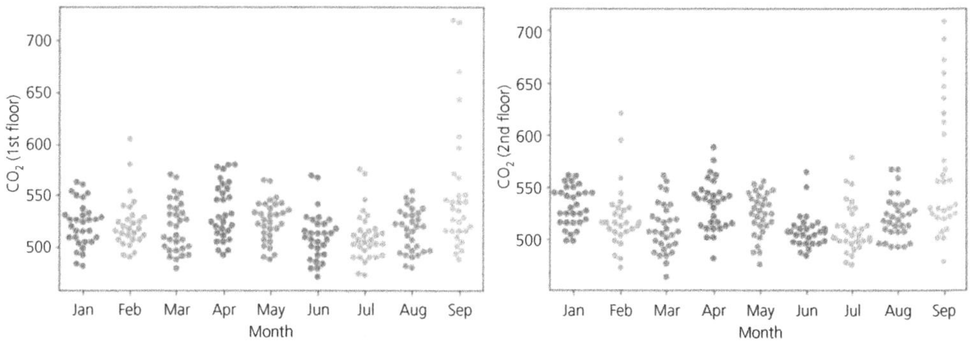

occupants possess a causal power capable of driving up CO_2 levels. To verify this assumption, further analysis was conducted to explore whether the significant rise in CO_2 levels observed in September 2021 could be attributed to an increase in footfall within BLD.

A boxplot chart was subsequently created using Python to illustrate the daily footfall on both floors over the nine-month study period (Figure 8.19). The chart reveals a similar pattern to that of the CO_2 levels, with compressed boxes indicating a consistent daily range of foot traffic, and a few outliers highlighting particularly busy days. This corroboration supports the hypothesis that the spikes in CO_2 were likely driven by higher occupancy levels during certain days.

Before drawing definitive conclusions, *corroborating* the mechanism relating the amount of people to CO_2 levels, it was essential to substantiate the mechanism linking footfall to CO_2 levels with robust empirical evidence, such as a strong correlation between the two variables. To achieve this, the specific dates on which footfall outliers were identified were cross-referenced with the corresponding CO_2 levels recorded on those same days.

Figure 8.19 Daily footfall traffic at the 1st and 2nd floors of BLD building (author's own)

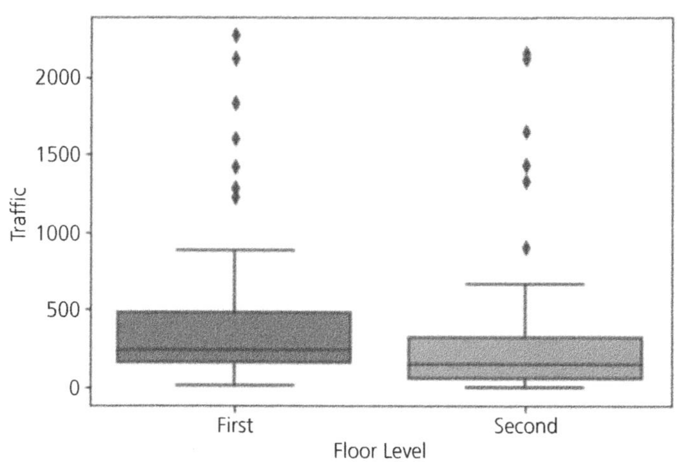

For the first floor, the seven days with the highest footfall outliers coincided with the days on which the top seven CO_2 daily peaks were recorded, although not necessarily in the same order (Figure 8.20). Similarly, five of the top six CO_2 peaks occurred on the same days as the highest foot traffic. This strong correlation uncovers a clear 'demi-reg' observed within the empirical domain, which directly *corroborates* the initial hypothesis: that building occupants possess the causal power to elevate CO_2 levels.

Further *analysis* was conducted to *corroborate* an additional hypothesis regarding the relationship between event scheduling and footfall, as posited in the rich picture (Figure 8.13). The event schedule data, gathered during the *mirroring* stage, was compared with the other datasets and *analysis* outcomes. This revealed that the week of 20th–24th September 2021, which saw both the footfall outliers and the highest CO_2 levels, was when the 'induction week' event occurred. This finding provides solid evidence that large-scale events, like induction week, have the causal power to influence foot traffic patterns. Consequently, incorporating the event schedule into the DT can help explain fluctuations in foot traffic and their impact on indoor air quality.

PLAN [H]

We can now leverage the findings from the analysis and the *corroborated* initial hypothesis to devise more proactive, systemic and sustainable solutions to the challenge of high indoor CO_2 levels, moving beyond reactive and short-term measures – for example, the current building management system (BMS) in operation at BLD only responds to elevated CO_2 levels by deploying mechanical ventilation, which has adverse effects on both the economy and the environment.

Figure 8.20 Matching traffic outliers with the corresponding CO_2 peak levels recorded on the same day (author's own)

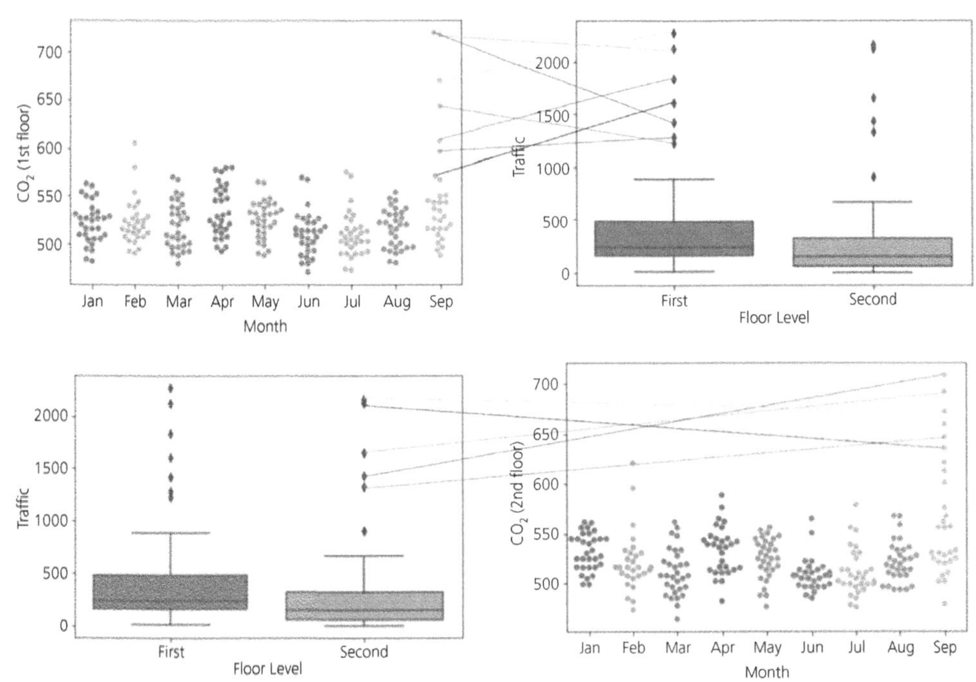

However, the DT developed as part of this project offers the opportunity to address the root causes of these undesirable events, such as high CO_2 levels, at an earlier stage.

One such approach could involve rethinking BLD's event scheduling, particularly for major events like induction sessions. By revising how these events are planned, hosted and organised, the occasional spikes in CO_2 levels can be mitigated, ultimately leading to a more efficient and sustainable way of managing indoor air quality.

VIRTUAL IMPLEMENTATION [D] AND REAL IMPLEMENTATION [H/D]

Ideally, the DT should have the capability to test alternative event schedules and locations within a virtual environment before the actual implementation of a specific plan. This would allow for a more thorough understanding of the potential implications and outcomes. However, at the time of this project, Twinview – the DT platform utilised – was still in development and lacked the functionality to support virtual implementation techniques, such as running simulations of proposed event schedules or locations. This limitation prevented a deeper exploration of their effects and consequences in the virtual domain.

8.4.3.3 Reflect

The DATUM-informed intervention successfully enhanced the understanding of indoor air quality (IAQ) dynamics within BLD by explicating the phenomenon in terms of its underlying causal mechanisms and generative factors. This led to the development of a DT, which enabled decision-makers to implement proactive, long-term and sustainable solutions to IAQ issues, such as revising event schedules and locations at BLD, rather than relying on conventional, reactive approaches that depend on mechanical ventilation – solutions that often have a negative environmental and economic impact.

Guided by the DM2 methodology, the project began with a pluralistic approach to appreciating the problem. Through triangulation, both secondary and primary data were gathered to support this initial phase. Secondary data was drawn from relevant literature, offering essential background knowledge on IAQ issues, while primary data was collected by way of a workshop, providing three key benefits. First, it incorporated the first-person perspectives of BLD users, allowing for a more critical examination of the current situation. Second, the involvement of multiple stakeholders helped minimise bias and recognised the limitations of individual worldviews. Lastly, the workshop provided crucial context-specific information, enhancing the understanding of BLD's existing operations.

Through the digitisation phase, DM2 helped establish an explicit trail of evidence connecting the datasets required for the DT with the underlying structures and mechanisms pertinent to the issue. This was instrumental in justifying the collection and accessibility of these datasets to all stakeholders involved.

DATUM, built on the principles of CR and pluralistic practices, facilitated the adoption of an 'exploratory research design' (outlined in Table 7.2). The initial qualitative analysis, conducted through appreciation and retroduction, was later corroborated by quantitative analysis using the DT. The flexible, non-linear nature of DM2 allowed for movement between different steps, such as appreciation and retroduction, as well as analysis and communication. This flexibility enabled a gradual increase in learning and understanding over time.

With sensitivity to ethical considerations, the footfall sensors dataset was preferred, as it provided valuable insights into building occupancy patterns without compromising privacy.

Table 8.6 DM2 steps and corresponding disciplines with potential value

DM2 step(s)	Discipline with potential value
Appreciation	Systems thinking
Mirroring / analysis / communication / virtual implementation	Computer science
All steps	Project management

However, DATUM, specifically the DM2 methodology, exhibited some inadequacies on two occasions. First, throughout the project, the team frequently had to draw on other disciplines, such as systems thinking, computer science, and project management, and utilise tools from these areas, including rich pictures, programming and project management techniques. While interdisciplinary approaches are inherent in real-world projects, it would be beneficial for DATUM to explicitly incorporate this trait. This could be achieved by adding a principle to DM2 that highlights the potential connection between its various steps and other disciplines, enriching each phase of the process (Table 8.6).

Second, the reflection on the intervention and DATUM itself was made possible primarily due to the action research methodology adopted in this study (Figure 8.12). In other words, DATUM lacks a built-in mechanism for self-reflection post-intervention. Without the reflexivity inherent in action research, the step of assessing the outcomes of the intervention – highlighting both the strengths and weaknesses of DATUM – could have been overlooked. Therefore, DATUM would benefit from incorporating an additional step of 're-appreciation,' allowing for post-intervention reflection and the capture of new knowledge gained through the process.

8.4.4 Cycle 2: DT of Glass House (GH)

8.4.4.1 Plan

As discussed in Section 6.4.2, the action research model employed in this study is focused on exposing the artefact under evaluation to a variety of challenges – much like testing a structure under different loading conditions – rather than repeatedly subjecting it to the same problem situation. Although the reflection on cycle (1) did not result in major changes to DATUM, this does not diminish the value of cycle (2), which exposes DATUM to a different set of challenges, offering an evaluation from a unique perspective. The problem addressed in cycle (1) involved a certain degree of structural complexity, featuring interconnected structures and mechanisms. Cycle (2), however, presents DATUM with a challenge characterised by not only a higher level of structural complexity but also pronounced people complexity. This type of complexity arises when individuals with varying worldviews and interests engage with the same problem situation (Jackson, 2019).

8.4.4.2 Act & Observe

EMERGENCE OF EVENTS [S]

At this initial step of DM2, the real-world structures give rise to a series of events that are fallibly observed and appreciated as detailed below.

APPRECIATION [H]/RETRODUCTION [H]/PLANNING [H]/CORROBORATION [H]

The 'Gardens' (GAR) is a UNESCO World Heritage site, and the GH, one of its most remarkable buildings, houses a priceless collection of plants. Many of these species are endangered, with some already extinct in the wild. Safeguarding this invaluable collection is not only a cultural necessity but also an ecological imperative. The last refurbishment of the GH took place in the 1980s, highlighting the urgent need for its renovation to both conserve the collection and maintain the structure. Additionally, glasshouses are typically energy-intensive buildings, characterised by high operational carbon emissions due to their predominantly single-glazed, metal-framed and often uninsulated design. GAR, acknowledging the climate crisis, has set ambitious climate science-based targets to rapidly decarbonise and aim to become climate positive by 2030. Their sustainability strategy requires a 46% reduction in carbon dioxide equivalent (CO_2e) emissions by 2030, compared to the 2019/20 baseline.

As previously noted, stakeholder engagement is critical for rich *appreciation* of the problem and conducting more accurate *retroduction*, particularly in situations involving multiple stakeholders and a high degree of complexity. The first step involved identifying and mapping out the various stakeholders before examining their perspectives on the situation (Figure 8.21). To ensure no group was overlooked or underrepresented – especially those with less influence – this task was carried out in collaboration with the project manager (PM) at GAR. This process facilitated a clearer understanding of each group's overarching objectives, as depicted in Figure 8.21.

Once the stakeholders had been identified, the next step was to explore their perspectives and opinions on how they could meet their interests and achieve their goals. Three commonly cited methods in the literature for this purpose are questionnaires, document coding and interviews (Hwang and Lin, 2012).

In line with the principle of triangulation (Table 5.3), the latter two methods – document coding and interviews – were employed to capture the multi-layered reality, which 'is composed of many types of structure' (Wynn and Williams, 2012: p. 803). On the document side, this included GAR's sustainability plan and the technical report produced by the consultant (CON). The report primarily contained CON's recommendations for a variety of interventions that could help the teams achieve their respective goals. Both documents were heavily focused on the technical and engineering dimensions of the project. However, the interviews were intended to focus more on the socio-technical aspects and capture the tacit knowledge of the various stakeholders. As one of CON's team members later noted during an interview, '*it has become clear that much of the knowledge around the optimal operation of the building is tacit, and difficult to assess or codify.*'

The interviews were designed following the pluralistic approach to triangulation and complementary research design (Table 5.3). The latter required interviewing as many of the identified

Figure 8.21 The involved stakeholders and their overarching goals (author's own)

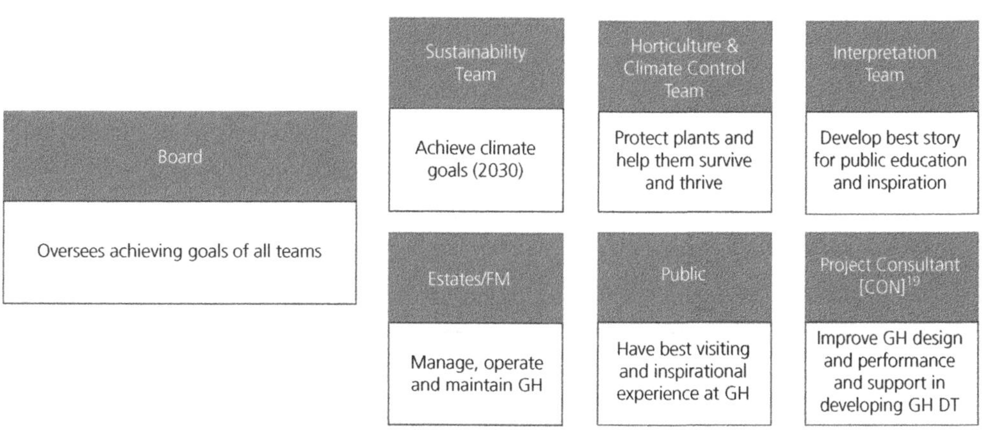

stakeholder groups (Figure 8.21) as possible, acknowledging that each group would have a limited perspective, focusing mainly on phenomena that aligned with their own interests and objectives. The principle of triangulation also informed the decision to invite multiple members from the same team, as each individual could bring a unique perspective. Conducting group interviews, rather than one-on-one sessions, helped mitigate bias. Table 8.7 includes the list of interview questions. The interviews were conducted with the groups and members detailed in Table 8.8 below[5].

Table 8.7 Interview questions

No.	Question
1	Please provide a brief overview of your role/experience in GAR?
2	What do you do in your role on a daily basis and how does it change throughout the year?
3	What are your priorities at GH?
4	What current systems/processes/resources/information you use to deliver your role?
5	What are some of the biggest challenges/risks you/your team are facing?
6	In future GH, what capabilities would you/your team like to have to better fulfil your role?
7	GAR is planning to achieve net-zero: How much (or in what way) do you think this impacts your team?
8	GAR is planning to achieve net-zero: How do you think your role can contribute to this agenda?
9	GAR is planning to achieve net-zero: In what ways would your way of working would need to change to accommodate this goal?
10	GAR is planning to achieve net-zero: What is the trade off when reducing carbon dioxide emissions – how would this impact on the plants and the way the plants and visitors behave?
11	What factors do you think impact the energy use at GH?
12	What information do you think would be useful to have about the GH environment or plants within to support your role further?
13	Do you think it would be useful for staff/visitors to have access to information that demonstrates the impact of certain actions on the GH achieving its sustainability goals – for example, there are currently 100 people visiting the GH today which creates x amount of heat which means we can reduce the heating by X and reduce our carbon footprint b Y?
14	Do you think it would be useful to have a dashboard or app that provided actionable advice like 'the weather for the next three days is going to be wet and 12 degrees, we noticed you left the back door wedged open we suggest you close it to reduce the amount of heat loss'?
15	What are your experiences / skills in terms of dealing with digital/computer devices/software?
16	Any restrictions regarding using tech inside GH (humidity/dirt/heat/security)?
17	Any other user groups you think we should be considering?
18	Any other thing you would like to add?
19	Are you ok with following up with you to clarify/validate the findings?

[5]Researcher could not reach representatives of the Estates (Facility Management) Team or public. Nonetheless, their needs were inferred from all other interviewees.

With multiple stakeholder groups involved, each group sought to propose various options for the interventions they deemed necessary to achieve their specific objectives. In the first cycle of the action research, the rich pictures method effectively illustrated assertions about the world and the interactions of its various structures. However, during this cycle, the focus shifted towards discussing proposed options and alternatives for interventions, indicating a significant transition to mode 2 of research. This was in contrast to the first cycle, which primarily concentrated on mode 1 to clarify and understand the problem situation. As a result, cognitive maps (CM) emerged as a more appropriate systems-thinking tool to facilitate the appreciation and retroduction steps. CMs can effectively illustrate both 'assertions about the world' and potential 'strategies' and 'options' for interventions (Figure 8.22).

While an in-depth exploration of CM, including its history and theoretical foundations, is beyond the scope of this discussion (for more, see Eden, 1988, 2004; Eden and Ackermann, 2004), a brief introduction is provided here to clarify its purpose, constituent modelling elements, and the connection between CM and CR, the underpinning philosophy of DATUM.

Cognitive mapping is a widely utilised method for 'exploring the mental structures of individuals and groups' (Ahmad and Xu, 2021: p. 4). It involves the task of mapping an individual's thoughts about a problem or issue to aid in 'problem structuring' and uncovering potential solutions (Eden, 2004: p. 673). As developed in existing literature, CMs are structured hierarchically, forming means/ends or 'connected options-outcomes chains' (Eden and Ackermann, 2004). As

Table 8.8 List of interviewees

No.	Stakeholder group	Interviewees
1	Board	Director of GAR and GAR's PM
2	Sustainability team	Sustainability director
3	Horticultural team	Two key senior members
4	CON	Architect, MEP, structural, and conservation senior engineers

Figure 8.22 CM model (adapted from Eden and Ackermann, 2004)

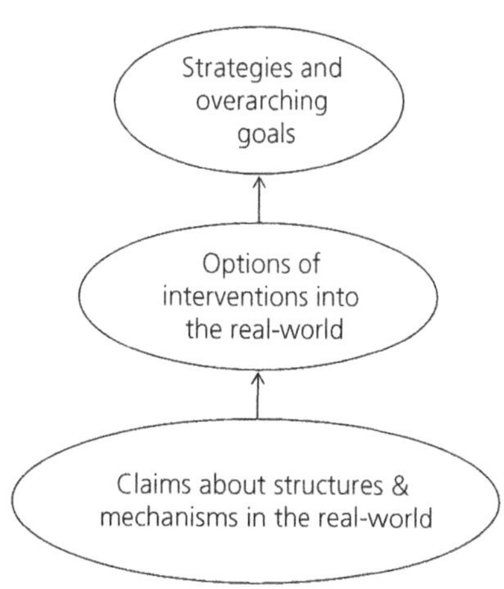

illustrated in Figure 8.22, created by Eden and Ackermann (2004), a CM requires the formulation of assertions about the world. These assertions are expected to have consequences or implications, as indicated by the arrows, suggesting possible actions that, if implemented, could lead to outcomes that either support or hinder the organisation's strategies. These strategies are considered interconnected in various ways with the overarching goals established by decision-makers. This particular CM model, depicted in Figure 8.22, is grounded in 'personal construct theory' (Kelly, 1991), which frames the researcher as a scientist 'constantly trying to make sense of the world in order to act within and upon that world' (Eden and Ackermann, 2004: p. 616).

'The theoretical basis for cognitive mapping, which enables the interpretation and analysis of these maps, is seldom explicitly stated' (Eden, 2004: p. 674). Therefore, it is important to clarify that the model applied in this book, illustrated in Figure 8.23, is grounded in CR. It encompasses the following elements.

(a) *Structures and mechanisms in reality*: This refers to the postulations made by various stakeholders during the interviews, akin to assertions about the world. Through the process of retroduction, the beliefs articulated by the interviewees provide explanatory accounts of reality, aligning with mode 1 of research. This involves hypothesising about the structures

Figure 8.23 CM model developed and implemented in this book (author's own)

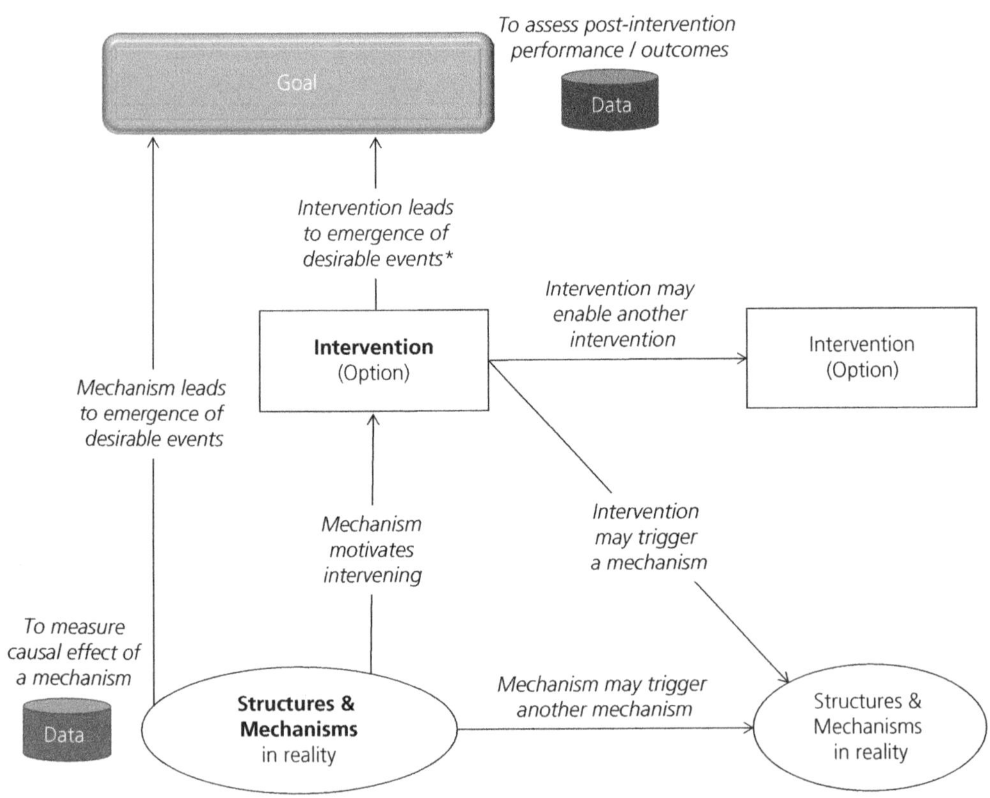

* For an event to emerge, there must be underlying active mechanism(s). Connecting 'Intervention' directly to 'Goal' is mere simplification for the mechanisms triggered by the intervention achieving the goal are either too obvious or too detailed for this level of analysis.

and generative mechanisms that exist and operate in the world, along with the causal powers they possess.

(b) *Interventions* (options): Shifting to mode 2 of research, this element supports the planning phase by representing the potential options for interventions in the real world, as suggested by different stakeholders. These options aim to influence the identified generative mechanisms to yield desirable outcomes.

(c) *Goals*: These are the overarching aims that have been predefined and articulated by the various stakeholder groups, which inform the range of proposed options or interventions.

(d) *Datasets*: Supporting the digitisation phase, this element identifies the suggested datasets that, when captured through a DT, can indicate the activity of a particular mechanism over time and measure its empirical effects, or assist in evaluating the intervention by assessing the outcomes in relation to the overarching goals.

In a CM, instances of the first three elements (i.e. structures and mechanisms, interventions, goals) are connected by unidirectional arrows, each representing the direction of causality. This means that an element at the tail of an arrow influences or causes a change in the element at the arrowhead. A negative sign over an arrow signifies a counter-causal effect.

It was now essential to gather the perspectives of each stakeholder group based on insights from the interviews and technical documents. The CM of the sustainability team was constructed first (see Figure 8.24). This map includes various technical options proposed to reduce heat loss through the glass of the GH or to lower overall energy demand and consumption in pursuit of climate goals. As depicted, it is asserted that the movement of people entering or exiting through the main doors of the GH serves as a mechanism contributing to heat loss.

Next, Figure 8.25 illustrates the CM for the horticultural team. This map highlights key mechanisms identified by the experts that significantly influence the condition of the plants. Additionally, it presents several suggested interventions believed to foster plant health. A central concept discussed is that of microclimates within the GH, which refers to the emergence of varying climatic conditions at different locations and heights. Consequently, the placement of a plant can affect its wellbeing. For instance, one interviewee noted, '*the hardy plants are the ones closest to the doors that are open all the time*,' creating favourable microclimate conditions for this type of plant.

Although the project team could not interview representatives from the Interpretation or Estates groups, insights were provided by the board of directors (the Board) and the GAR PM regarding the interventions both groups are willing to pursue. The CM for the interpretation team, which shares an overarching goal aligned with the interests of the public (see Figure 8.21), is depicted in Figure 8.26. Their primary objective is to relocate plants to craft the best 'story' for educational and inspirational purposes for visitors. In contrast, the estates team is primarily focused on maintaining and realigning the structure of the GH, which has shifted from its original position due to the long-term effects of wind loads (see Figure 8.27).

Regarding the Board of the GH, insights gleaned from the analysed data indicate that their primary interest lies in overseeing all groups and ensuring the overall efficiency of the GH system. The board's perspectives and interests are notably more holistic compared to those of other groups, highlighting that the problem situation is not only multi-perspectival but also hierarchical.

Concluding the *retroduction* step involves three fundamental elements of the comprehensive explanatory account: structures, mechanisms, and interconnections. The first two elements were

Figure 8.24 CM of the sustainability team (author's own)

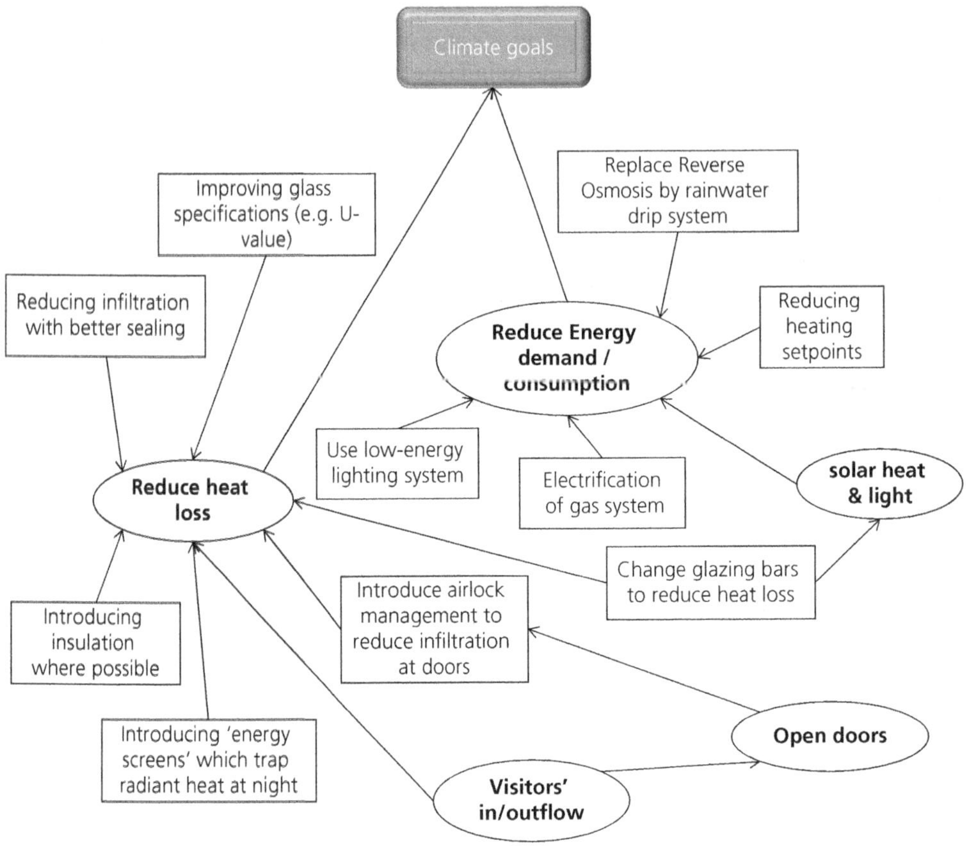

illustrated through the cognitive maps CMs presented earlier, which demonstrate the various structures involved and the causal effects they generate, as hypothesised by different stakeholder groups. However, to address the third element – interconnections – it is crucial to examine the potential interdependencies between the various mechanisms and the proposed interventions. This perspective is best attained from the board's vantage point. The Director of GAR succinctly encapsulates the board's goals and interests: *'it's about the structure, the plants, and the storytelling... and the energy supply.'*

From this holistic viewpoint, it appears that the board would derive greater benefit not from a local DT focused solely on their interests, but from integrating all local DTs. This approach would facilitate optimal decision-making at a global and systemic level (see Figure 8.28).

While local solutions tailored to a specific system may provide optimal, adaptive and straightforward approaches at that local level, global solutions within a system of systems can often be sub-optimal and maladaptive (Carhart *et al.*, 2020). Similarly, having separate specialised DTs for each team can result in inadequate solutions. Failing to acknowledge the interdependencies between various structures and creating isolated digital twins could ultimately lead to undesirable outcomes

Figure 8.25 CM of the horticultural team (author's own)

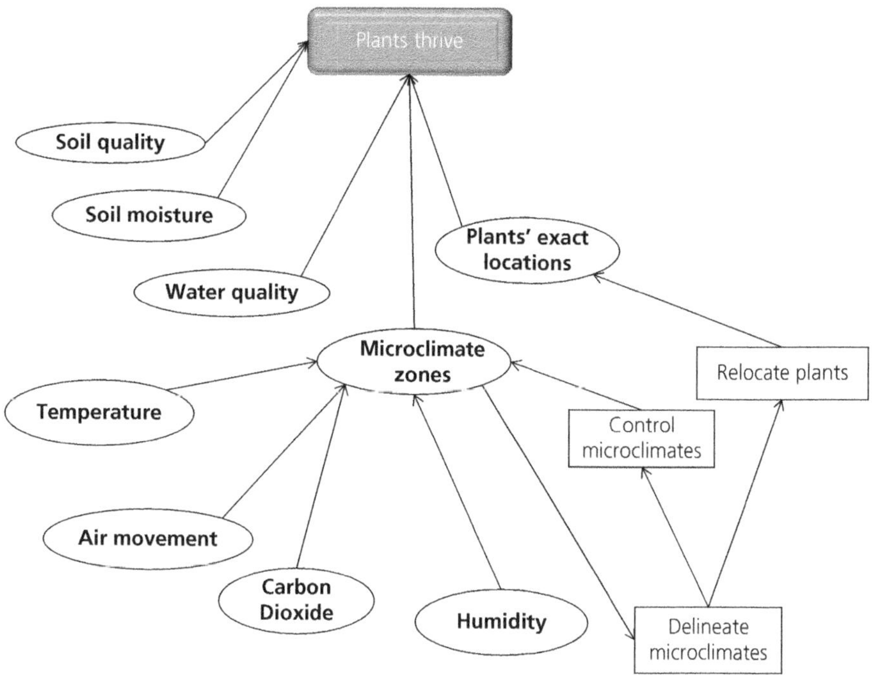

Figure 8.26 CM of the Interpretation team (author's own)

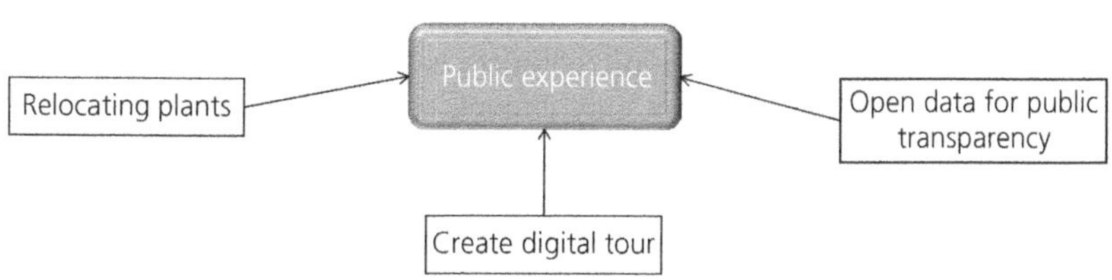

Figure 8.27 CM of the Estates team (author's own)

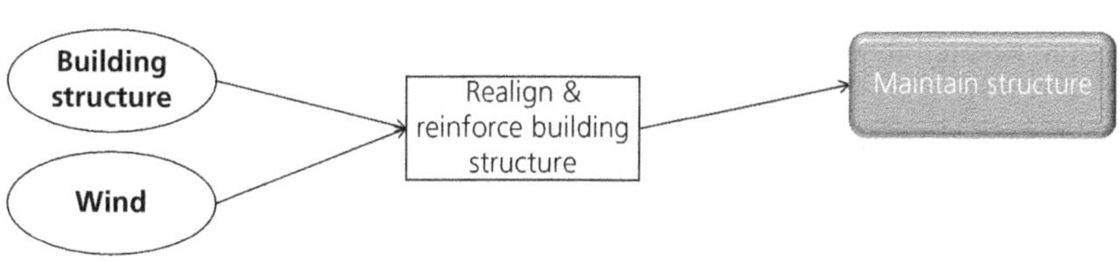

Figure 8.28 An ecosystem of connected DTs offering a holistic perspective (author's own)

at the GH system of systems level. The following examples, drawn from interviews, illustrate the limitations of siloed thinking and the concept of developing individual local DTs for each group.

The sustainability team is keen to implement various interventions to reduce energy consumption. However, they appear less mindful of the potential risks these interventions may pose to the plants – for instance, they propose lowering the heating setpoint, but the horticultural team is understandably reluctant to alter the climate conditions to which the plants have acclimatised. Another example is their intention to relocate plants based on microclimate conditions to optimise energy consumption. However, the interpretation team can only endorse specific configurations that convey the most compelling narrative to visitors.

Conversely, the horticultural team is less focused on the energy used or heat lost while creating the ideal climate for the plants. Their motivations for relocating a plant may vary – for example, they might do so to place it in a more suitable microclimate, to ensure it receives adequate sunlight, or simply for aesthetic reasons, particularly if the plant obstructs a desirable view. However, such interventions could adversely affect the storytelling and the message the interpretation team aims to convey.

As previously mentioned, certain plants can only thrive within specific microclimate zones inside the GH. Some plants may be at risk if the interpretation team fails to account for this when designing the layout to best tell their story to visitors. As GAR's PM noted

> *'one of the things that we want to be very clear with interpretation before they suggest where the horticultural team should plant their various plants and tell various stories is, as we said earlier, the Hardy plants are the ones closest to the doors that are open all the time. So we have to say, you know, in this area you need to think about what kind of plants that go in here, because they are going to be subject to a cold draft.'*

To adequately address these interdependencies and appreciate the entire problem situation holistically, the CMs created earlier (Figures 8.24 to 8.27) were merged into a composite group CM, as illustrated in Figure 8.29 below. Group maps, as described by Eden (2004: p. 674), are 'often developed by merging several cognitive maps derived from each member of a problem-solving team.' This composite map is constructed by identifying common nodes that refer to the same concept – be it a mechanism, intervention, or goal. The original CMs (Figures 8.24 to 8.27), representing each group's perspective, form 'clusters' or 'chunks' (Eden, 2004: p. 680) of the broader problem situation depicted in the group CM (Figure 8.29). This hierarchical clustering allows any node to appear in multiple clusters, demonstrating that the issue or problem comprises a system of interrelated sub-problems (Eden, 2004: p. 680).

To validate these outcomes, the group CM was circulated among all participants, allowing each to review and confirm the representation of their own perspectives as well as to verify the proposed mechanisms or interactions linking the different CMs. Sharing the group CM with all stakeholders fostered consensus and established a shared holistic understanding of the situation.

DIGITISATION [D]/MIRRORING [D]

Given the multi-perspectival nature of the project, the interviews provided a valuable opportunity to identify the key DT GUCs from the unique perspectives of each team. During the discussions, stakeholders were asked what kind of insights, useful information, or actionable advice they would seek to help achieve their objectives. The responses from each group revealed which DT use cases they valued most (Table 8.9).

Considering the multi-perspectival and hierarchal nature of the problem, its hierarchical structure, and the interdependencies between teams as illustrated in Figure 8.29, a connected DT is recommended, rather than isolated DTs for each team. The concept of a connected DT unlocks new value in two key ways.

(a) Holistic oversight: When all DTs are integrated, they provide access to the bigger picture, which is crucial for the board to oversee operations and the dynamics of all teams simultaneously.
(b) Context-awareness: Connecting local DTs allows each one to operate with a high level of context-awareness, ensuring that the actionable advice generated by the DT is aligned with the global, systemic goals.

For instance, as detailed in Table 8.8, most of the GUCs identified by each team either require input from other DTs to deliver context-aware advice or produce outputs that influence other teams. Following this, the proposed digital datasets were mapped to their relevant structures, as shown in Figure 8.30, laying the groundwork for developing a digital representation of the problem. The features of the connected DT were then defined using DATUM Prong-A (Figure 8.31), and the potential use cases were modelled through DATUM Prong-C (Figure 8.32).

8.4.4.3 Reflect

At the outset of the project, the principle of pluralism, central to the DM2 methodology, highlighted the importance of adopting a complementary approach, reinforced by triangulation. This complementary approach involved utilising diverse data sources: secondary data from technical reports prepared by the CON and primary data collected by way of interviews. These datasets allowed us to explore two key dimensions of the problem – explicit technical or engineering knowledge and implicit knowledge of daily operations within the greenhouse (GH). Interviews were conducted with multiple stakeholder groups to address the limitations of each group's worldview, as each

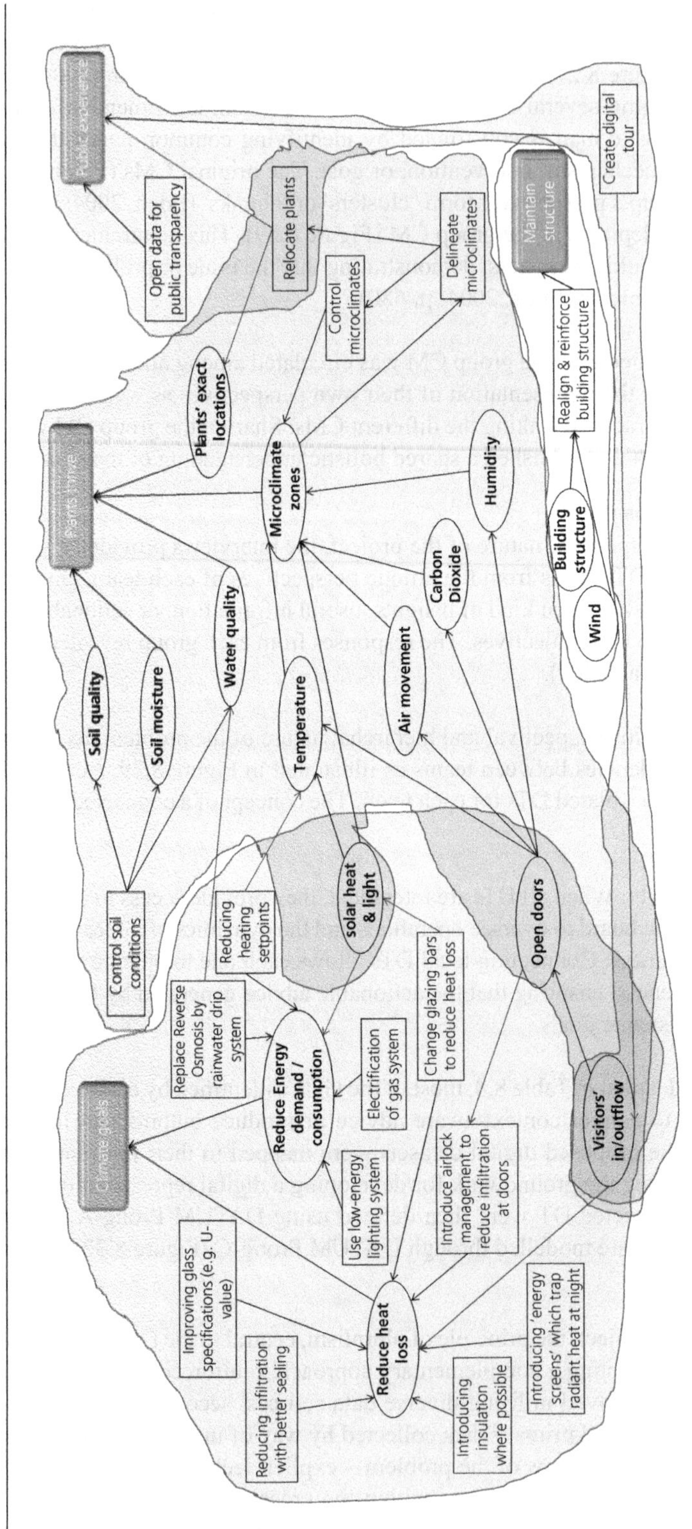

Figure 8.29 Group CM (author's own)

Table 8.9 Different DT GUCs as required by the various stakeholder groups (continued on next page)

DT GUC	Description	Evidence from collected data and interviews
View the bigger picture of current/ historic state	View number of visitors, actual carbon emissions (CO_2e); indoor climate conditions (temperature, humidity, CO_2, light, soil moisture, water quality); existing boundaries of microclimate zones inside GH; plants level of comfort; number of visitors today and discover correlations between any of them.	'it's about the structure, the plants and the storytelling ... and the energy supply.' 'I think it would be helpful if we did (monitor number of visitors].'
Evaluate impacts of interventions	Simulate any intervention and evaluate its impacts. Intervention may include changes to physical or thermal attributes of building to reduce heat loss or energy demand, or changes to number of daily visitors. Such changes can have impacts on emissions, indoor climate conditions, division of microclimate zones, plants level of comfort.	'During the future stage of the project, the environmental data will be used to calibrate our predictive energy and thermal models, to ensure that any design decisions are based on realistic data, and real-world performance and that environmental risks to the valuable plants are assessed and mitigated in detail.'
Auto-control climate conditions	Automatic control of climate conditions based on current plants level of comfort, within every microclimate zone, compared to pre-defined standard conditions.	'You know we'll look every day, every morning we'll come in and see what the temperatures are, if it's being very low overnight, there might have been an alarm. So then we'll contact the engineers to try and get things sorted out.'
Predict GH structure displacements	Predict GH structure displacements and movements due to winds, then realign and reinforce building structure accordingly.	'you can use inclinometers ... to work out how the building is actually moving in wind ... it could be quite a good idea ... in terms of predictive movements.'
Design plants layout	Propose a unique plants layout that tells an inspirational story by changing existing geo-locations. Consequently, evaluate impact of proposed layout changes on plants level of comfort with respect to climate conditions within existing microclimate zones. DT can further highlight opportunities or constraints related to reduce energy consumption in case the proposed layout is implemented.	'we have an interpretation team, so we're we're going to decamp the GH so that we can refurbish it when we, when we replant, it probably won't be put back in the in the manner that it's currently sorts out there may be a different focus to the Palm House.'

Table 8.9 Continued

DT GUC	Description	Evidence from collected data and interviews
Engage public	Engage public by providing open live stream of data relevant to current emissions (CO_2e), plants level of comfort, number of visitors.	'there's the whole area of transparency... Digital twin will allow us to be very transparent if we give access to other people, to our digital twin and our data ... it allows us to be very transparent of how much energy we're using, where it's going and what the buildings do etc.'
Virtual accessibility	Offer immersive virtual experience (e.g. virtual reality) to public who are unable to visit the glass house. The DT may provide information about the impact of the completed virtual tours on the indoor climate conditions and plants as less people visit the site and thus, less doors opening of doors.	'the mezzanine level of the GH, for instance, is not accessible for wheelchair users. Will never be accessible for wheelchair users, and the question will be how do we allow people who can't experience it to experience looking down on the canopy of the GH? It may be that digital tour can help.'

sees the problem from a unique angle. Triangulation was then applied to mitigate bias by including more than one representative from each group in the interviews.

The project presented a complex problem, involving various interconnected structures and mechanisms and characterised by its multi-perspective and hierarchical nature. Fortunately, the DATUM methodology proved adequate to manage this complexity.

First, the qualitative analysis conducted in line with the DM2 steps and principles revealed the multi-perspective and hierarchical nature of the situation. DATUM, with its methodological guidelines and principles, helped us avoid the pitfall of focusing solely on the sustainability team's interests and the achievement of climate goals. The technical report prepared by CON before our involvement focused heavily on technical solutions to achieve 'near net-zero' but paid less attention to how these solutions might affect the interests of other teams within GH. The report also offered little consideration for how teams like horticulture might achieve their goals. DATUM helped us embrace the problem's multi-perspective nature by proposing multiple DT GUCs that would satisfy the distinct interests of all teams.

Second, with CR at its core, DATUM views the world as stratified. This understanding – that events can arise at one level of reality due to generative mechanisms interacting at a deeper level – allowed us to take a holistic view of the problem and appreciate its systemic nature. We deliberately investigated how local systems related to different stakeholder groups might interact and create unexpected or undesirable outcomes at a global level. DATUM facilitated the capture of these operational complexities within GH, avoiding the reduction of the situation into siloed problems. This approach was recognised and validated by all stakeholders when confirming the project outcomes, including

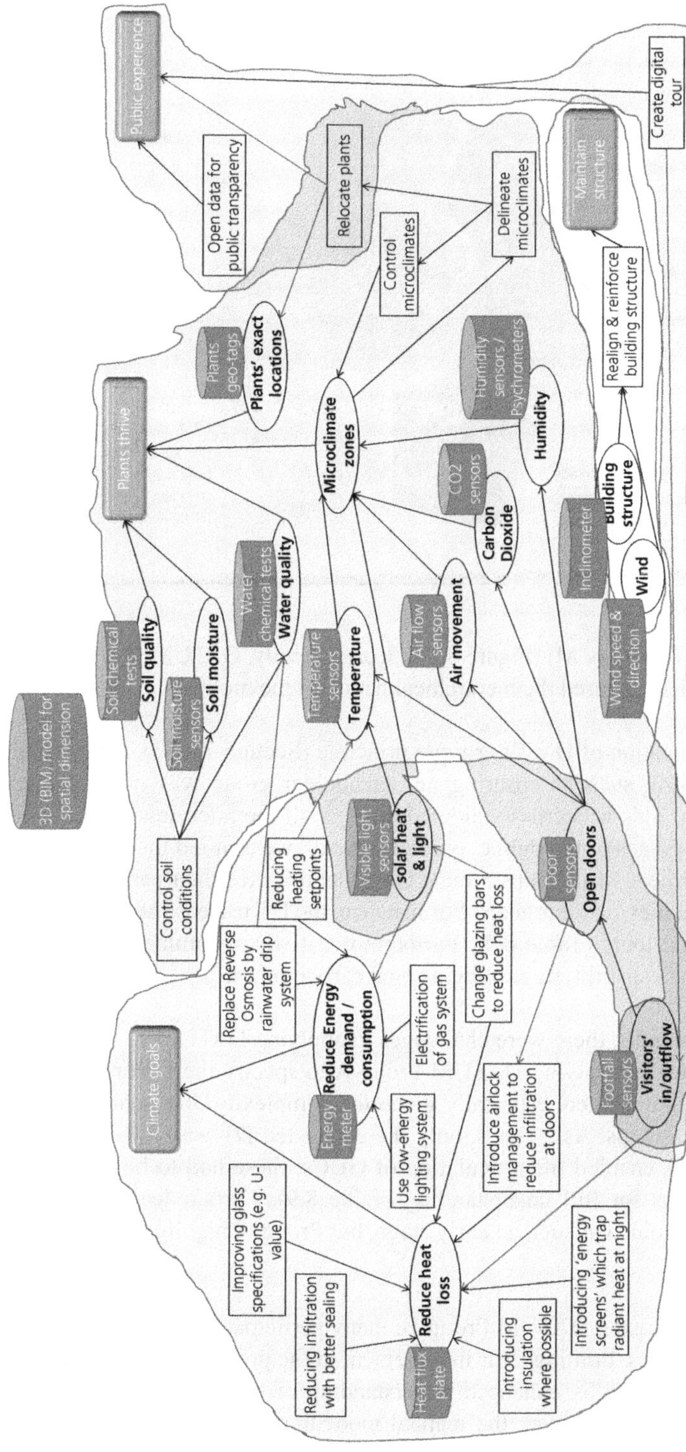

Figure 8.30 Group CM with relevant datasets associated with corresponding structures and mechanisms (author's own)

Figure 8.31 DT features of the connected DT specified using DTUCS Prong-A (author's own)

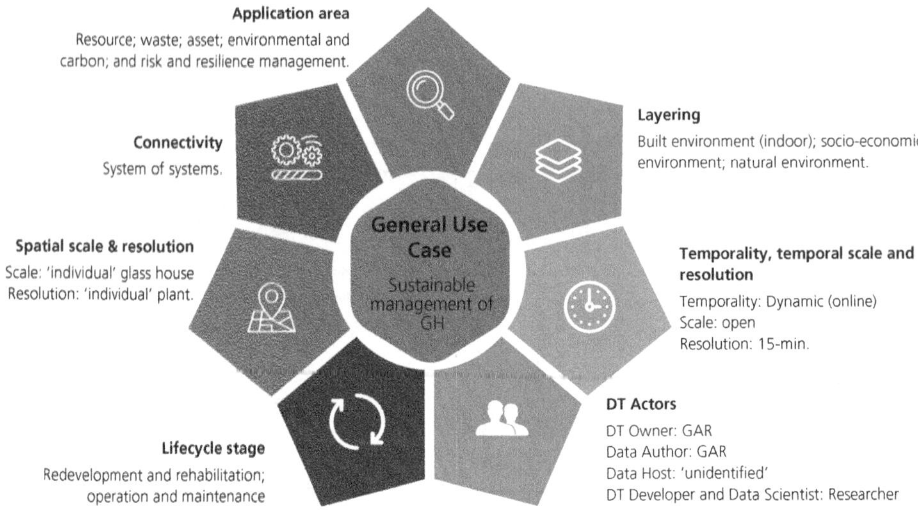

the group cognitive map (CM) (Figure 8.29). Consequently, DATUM also informed the design of a 'connected' DT that captured the hierarchical nature of the problem.

Through the application of DM2's critique principle (Section 7.3.13), we became aware of issues initially overlooked, such as ensuring no stakeholder group was underrepresented during the appreciation step. This was cross-validated through all the interviews. Even when access to certain groups, like estates or the public, proved difficult, we ensured their interests were represented throughout the project – from constructing CMs that reflected their perspectives to articulating DT GUCs aimed at achieving their goals. For instance, the DT use case addressing accessibility issues by creating virtual tours (Table 8.9). Furthermore, it was recommended that the connected DT should include the capability to display its own carbon footprint.

Despite these strengths, there were challenges in applying DATUM to this project. First, in terms of usability, manually applying DATUM Prong-A to specify the features of all GUCs was more challenging than anticipated, especially given the complexity of a connected DT involving multiple stakeholder groups. As a result, only the connected DT was fully developed (Figure 8.31). Although Prong-A enabled the articulation of GUCs, these had to be supplemented by detailed textual descriptions for full understanding (Table 8.9). Certain features, such as data hosting, could not be determined at such an early stage, but Prong-A highlighted the need to address these features later.

Furthermore, when using DATUM Prong-B, many participants were unfamiliar with the standard definitions of DT uses, limiting their involvement in the process. Finally, when employing Prong-C, the modelled DT UCSs enhanced understanding of how stakeholders were expected to use the DT (Figure 8.32). However, the manual modelling of UCSs was also time-consuming, and many participants, unfamiliar with UML, found it difficult to comprehend or actively contribute to the process.

8.5. Summary

After designing and developing DATUM in Chapters 4 to 7, its evaluation was necessary to ensure it could achieve its intended purpose. In this chapter, the philosophical and methodological elements of DATUM, as well as DATUM as a whole, were evaluated, as outlined in Figure 2.6, against the evaluation criteria set out in Figure 4.8.

The abstract research, conducted using focus groups, examined the unifiability of CR (Section 8.2). Specifically, this involved evaluating CR's ability to act as a pluralistic philosophical paradigm capable of accommodating and reconciling the various worldviews adopted by DT researchers and practitioners. The focus group discussions revealed that CR has a high degree of philosophical unifiability. It successfully underpins and justifies the different worldviews articulated by members of the discipline in a complementary way, except in the case of ethical DTs and the ethics of digital twinning. Most participants took a relativist stance, which contrasts with CR's objectivist position. However, the relativist perspective was challenged in Table 8.2, where an objective pluralistic approach was argued to be more robust.

Following the evaluation of the philosophical unifiability of DATUM, the generalisability of its methodological elements (i.e. DTUCS) was assessed through extensive research involving multiple DT case studies (Section 8.3). This research aimed to determine how adequate and generalisable DTUCS is in describing the various DT features and DT uses exhibited across diverse case studies within the paradigm of *DT for UM*, as well as in modelling different UCSs. DTUCS was rigorously tested by applying it to both real-world and hypothetical DT case studies, and it demonstrated a high degree of generalisability. It provided sufficient terms and tools to describe all the DT features and uses presented by the different case studies. Nevertheless, it was recommended to add a specialised sub-use to DTUCS Prong-B (i.e. pre-cast) to account for the possibility of using DTs to predict based on analogy.

While the abstract and extensive research supported the evaluation of the philosophical and methodological elements of DATUM separately, it was also necessary to evaluate DATUM as an integrated artefact. This was done to assess its usability and applicability in real-world problem situations. Intensive research, using a two-cycle bespoke action research model, was conducted to evaluate DATUM's usability and effectiveness when applied to two intrinsically different projects (Section 8.4). However, both projects involved developing DTs at the building scale. While buildings and the built environment are essential elements of urban contexts, it would be beneficial to explore, beyond this book, projects at a broader scale, such as neighbourhood or city-scale DT projects. Although this book focused on building-scale DT projects due to time and resource constraints, it is argued that the general process of implementing DATUM, and the value it brings, remains consistent and applicable across the entire field of *DT for UM*.

Upon completing both projects, it became evident that DATUM offers practical, valuable guidance to DT practitioners when addressing real-world problems. Most importantly, continuously drawing on DATUM's elements and DM2's steps and principles throughout the interventions helped justify decisions and improve the overall quality of the interventions. It ensured a pluralistic approach by engaging all stakeholders, utilising multiple data sources, and maintaining awareness of the ethical dimensions. In summary, it could be argued that both projects would not have been as successful without the guidance and insights provided by DATUM.

However, some persistent issues were identified that could hinder the application of DATUM or make it less usable. These include the time-consuming nature of manually implementing DTUCS and the lack of stakeholder familiarity with UML and DTUCS's frameworks and terminology.

Figure 8.32 Several DT use case scenarios (UCS] modelled using DTUCS Prong-C. (author's own) (continued on next page)

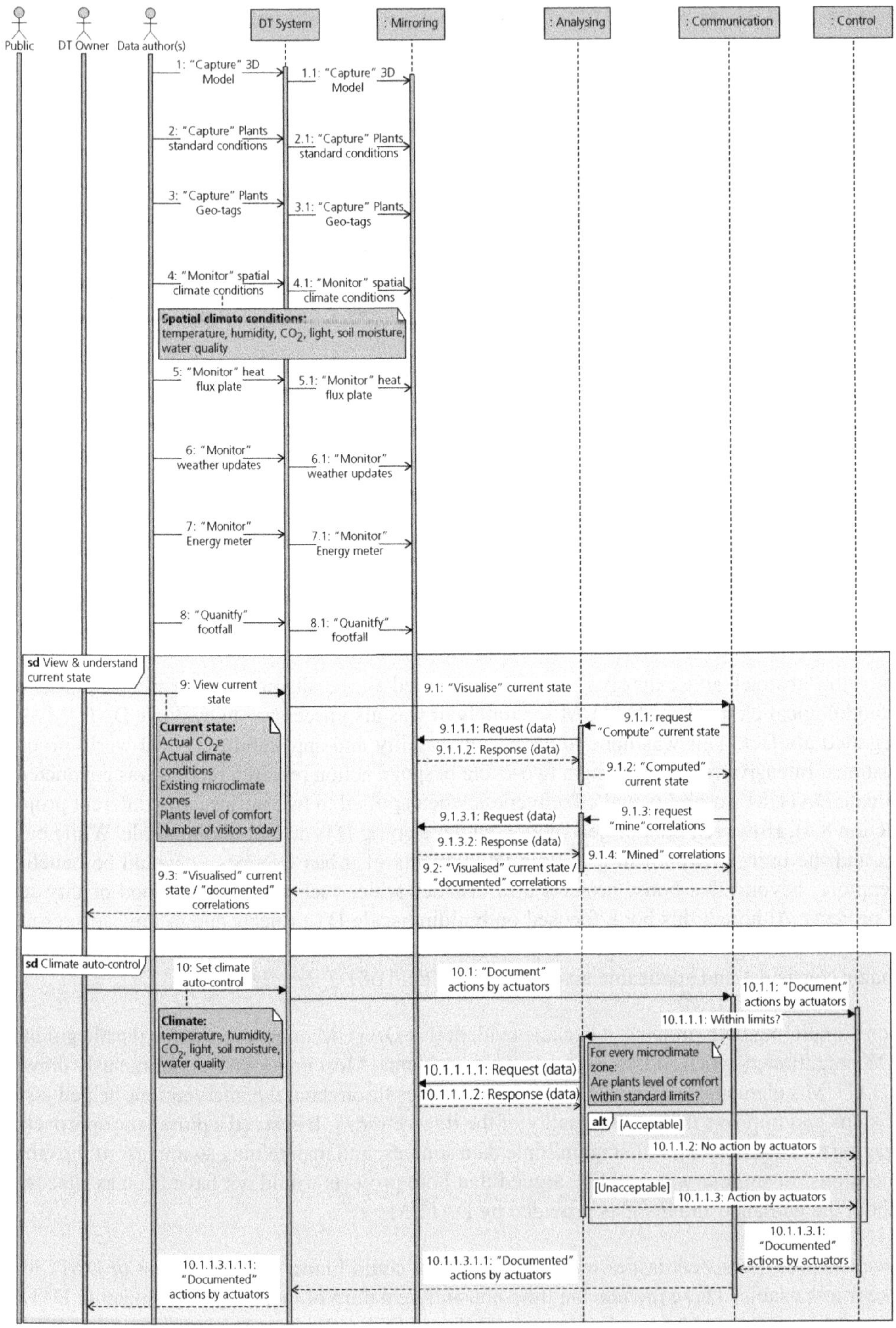

Figure 8.32. Continued

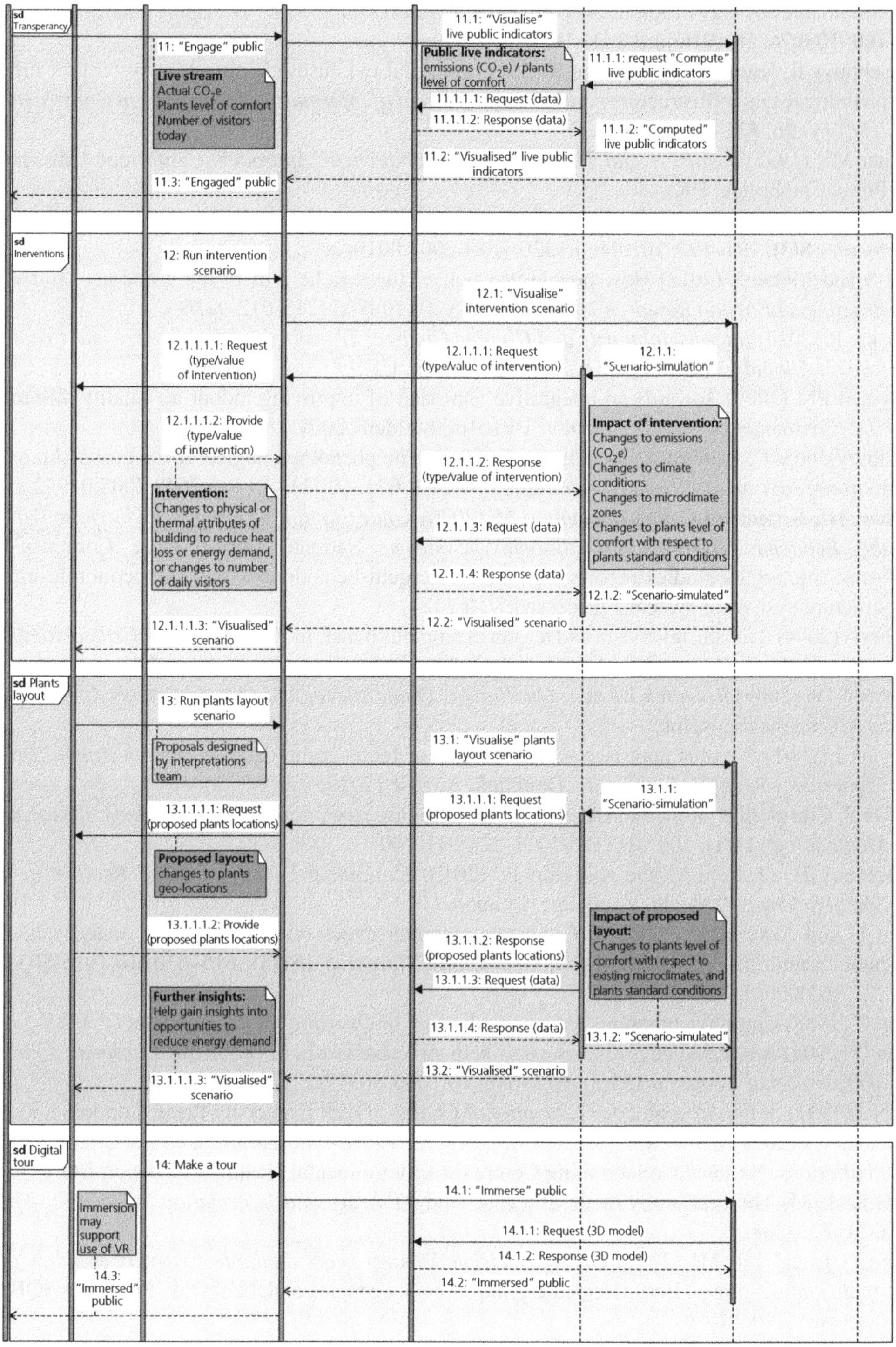

REFERENCES

Ahmad S and Xu B (2021) A cognitive mapping approach to analyse stakeholders' perspectives on sustainable aby way oftion fuels. *Transportation Research Part D: Transport and Environment* **100**: 103076. 10.1016/j.trd.2021.103076

Al-Sehrawy R, Kumar B and Watson R (2021) A digital twin uses classification system for urban planning & city infrastructure management. *Journal of Information Technology in Construction (ITCON)* **26**: 832–862.

Archer MS (1995) *Realist Social Theory: The Morphogenetic Approach*. Cambridge University Press, Cambridge, UK.

Barkway P (2001) Michael Crotty and nursing phenomenology: criticism or critique? *Nursing Inquiry* **8(3)**: 191–195. 10.1046/j.1320-7881.2001.00104.x

Bell S and Morse S (2013) How people use rich pictures to help them think and act. *Systemic Practice and Action Research* **26(4)**: 331–348. 10.1007/s11213-012-9236-x

Bhaskar R (2010) *Interdisciplinarity and Climate Change: Transforming Knowledge and Practice for Our Global Future*. Taylor & Francis, London, UK.

Bluyssen PM (2009) Towards an integrative approach of improving indoor air quality. *Building and Environment* **44(9)**: 1980–1989. 10.1016/j.buildenv.2009.01.012

Bradbury-Jones C, Sambrook S and Irvine F (2009) The phenomenological focus group: An oxymoron?. *Journal of Advanced Nursing* **65(3)**: 663–671. 10.1111/j.1365-2648.2008.04922.x

Carhart NJ, Rosenberg G and Pregnolato M (2020) *Understanding Emergent Behaviour Within the Economic Infrastructure System-Of-Systems*. National Infrastructure Commission. https://nic.org.uk/studies-reports/resilience/emergent-behaviour-within-the-economic-infrastructure-system-of-systems/ (accessed 7/10/2024).

Collier A (2004) Realism, relativism and reason in religious belief. In *Transcendence: Critical Realism and God* (Archer MS, Collier A, and Porpora DV (eds.)). Routledge, London, UK pp. 41–62.

Creswell JW (2009) *Research Design: Qualitative, Quantitative, And Mixed Methods Approaches*. SAGE Publications, Inc.

Crinson I (2001) A realist approach to the analysis of focus group data. In *The 5th Annual IACR Conference*, Roskilde University, Denmark, August 17–19.

Crowe S, Cresswell K, Robertson A *et al.* (2011) The case study approach. *BMC Medical Research Methodology* **11(1)**: 100. 10.1186/1471-2288-11-100

Danermark B, Ekström M and Karlsson JC (2019) *Explaining Society: Critical Realism in the Social Sciences*, 2nd edn. Routledge, London, UK.

Eden C and Ackermann F (2004) Cognitive mapping expert views for policy analysis in the public sector. *European Journal of Operational Research* **152(3)**: 615–630. 10.1016/S0377-2217(03)00061-4

Eden C (1988) Cognitive mapping. *European Journal of Operational Research* **36(1)**: 1–13.

Eden C (2004) Analyzing cognitive maps to help structure issues or problems. *European Journal of Operational Research* **159(3)**: 673–686. 10.1016/S0377-2217(03)00431-4

Elliot J (1991) *Action Research for Educational Change*. Open University Press, London, UK.

Eykelbosh A (2021) *Indoor CO_2 Sensors For COVID-19 Risk Mitigation: Current Guidance and Limitations*. National Collaborating Centre for Environmental Health, Vancouver, BC.

Fidel R (1984) The case study method: a case study. *Library and Information Science Research* **6(3)**: 273–288.

Hwang CL and Lin MJ (2012) *Group Decision Making under Multiple Criteria: Methods and Applications*. Springer Berlin Heidelberg. https://books.google.co.uk/books?id=F6r7CAAAQBAJ (accessed 7/10/2024).

Jackson MC (2019) *Critical Systems Thinking and the Management of Complexity*. Wiley, Hoboken, NJ. https://books.google.co.uk/books?id=1jmNDwAAQBAJ (accessed 7/10/2024).

Jacobson TA, Kler JS, Hernke MT *et al.* (2019) Direct human health risks of increased atmospheric carbon dioxide. *Nature Sustainability* **2(8)**: Article 8. 10.1038/s41893-019-0323-1

Karvonen A, Cvetkovic V, Herman P *et al.* (2021) The 'New Urban Science': towards the interdisciplinary and transdisciplinary pursuit of sustainable transformations. *Urban Transformations* **3(1)**: 9. 10.1186/s42854-021-00028-y

Kelly G (1991) *The Psychology of Personal Constructs: Clinical Diagnosis and Psychotherapy*. Routledge, London, UK. 10.4324/9780203405987

Kemmis S and McTaggart R (2000) Participatory action research. In *Handbook of Qualitative Research* (Denzin NK and Lincoln YS (eds)). SAGE Publications, Thousand Oaks, CA, USA, pp. 559–604.

Koshy V (2009) *Action Research for Improving Educational Practice: A Step-by-Step Guide*. SAGE Publications, Thousand Oaks, CA, USA. https://books.google.co.uk/books?id=53zLazswjQkC (accessed 7/10/2024).

Krueger RA (2014) *Focus Groups: A Practical Guide for Applied Research*. SAGE Publications, Thousand Oaks, CA, USA. https://books.google.co.uk/books?id=8wASBAAAQBAJ (accessed 7/10/2024).

Lowther SD, Dimitroulopoulou S, Foxall K *et al.* (2021) Low Level carbon dioxide indoors—a pollution indicator or a pollutant? A health-based perspective. *Environments* **8(11)**: Article 11. 10.3390/environments8110125

Mingers J (2006) *Realising Systems Thinking: Knowledge and Action in Management Science*. Springer US, New York, NY. https://books.google.co.uk/books?id=C-UcDQEACAAJ (accessed 7/10/2024).

Mingers J (2014) *Systems Thinking, Critical Realism and Philosophy: A Confluence of Ideas*. Routledge, Oxford, UK.

Næss P (2004) Prediction, regressions and critical realism. *Journal of Critical Realism* **3(1)**: 133–164. 10.1558/jcr.v3i1.133

National Infrastructure Commission (2017) *Data for the Public Good*. National Infrastructure Commission Report.

O'Brien R (2001) An overview of the methodological approach of action research. In *Theory and Practice of Action Research* (Richardson R (ed)). Universidade Federal da Paraíba, João Pessoa, Brazil. http://www.web.ca/~robrien/papers/arfinal.html (accessed 7/10/2024).

O'Leary Z (2004) *The Essential Guide to Doing Research*. SAGE Publications, Thousand Oaks, CA, USA.

Pawson R and Tilley N (1997) *Realistic Evaluation*. SAGE Publications. https://books.google.ws/books?id=qURsCgAAQBAJ (accessed 7/10/2024).

Porpora DV (2019) A reflection on critical realism and ethics. *Journal of Critical Realism* **18(3)**: 274–284. 10.1080/14767430.2019.1618064

Sayer RA (1992) *Method in Social Science: A Realist Approach*. Routledge, London, UK.

Sayer A (2000) *Realism and Social Science*. SAGE Publications, Thousand Oaks, CA, USA.

Spiegelberg H (1975) Phenomenology. In *Doing Phenomenology: Essays on and in Phenomenology* (Spiegelberg H (ed)). Springer Netherlands, Dordrecht, pp. 3–12. 10.1007/978-94-010-1670-4_1

Stake RE (1995) *The Art of Case Study Research*. SAGE Publications. https://books.google.co.uk/books?id=ApGdBx76b9kC (accessed 7/10/2024).

Sundell J (2004) On the history of indoor air quality and health. *Indoor Air* **14(s7)**: 51.

VELUX (2022) *Healthy Homes Barometer 2022. Sustainable Buildings for A Resilient Society*. VELUX A/S. https://velcdn.azureedge.net/-/media/com/healthy-homes-barometer/hhb-2022/velux-hhb-report-2022.pdf (accessed 7/10/2024).

Winter R and Munn-Giddings C (2013) *A Handbook for Action Research in Health and Social Care*. Taylor & Francis, Abingdon, UK. https://books.google.co.uk/books?id=HCLq5l3EGc0C (accessed 7/10/2024).

Wynn D and Williams CK (2012) Principles for Conducting Critical Realist Case Study Research in Information Systems. *MIS Quarterly* **36(3)**: 787–810. 10.2307/41703481

Yin RK (2009) *Case Study Research: Design and Methods*. SAGE Publications, Thousand Oaks, CA, USA. https://books.google.co.uk/books?id=FzawIAdilHkC (accessed 7/10/2024).

Ramy Elsehrawy and Bimal Kumar
ISBN 978-1-83662-235-2
https://doi.org/10.1108/978-1-83662-234-520251011
Copyright © 2025 by Ramy Elsehrawy and Bimal Kumar. Published under exclusive licence by Emerald Publishing Limited

Chapter 9
Reflections and future prospects

9.1. Introduction

In response to the growing urban challenges such as rapid urbanisation, escalating complexity, and climate change, the need for sustainable urban development (SUD) has become increasingly urgent. SUD offers a vision where people and nature can 'flourish together for generations' (A collaboration of leading figures in the built environment, 2021: p. 5). This book aims to explore recent efforts within the discipline of UM to harness advanced technologies, particularly the concept of the DT, to support research and practices in this field. The goal is to provide a structured framework for the emerging paradigm of *DT for UM*, facilitating its development, maturity, and effectiveness in addressing the wicked problems that characterise urban environments today.

The modern urban landscape has evolved into a system of systems, with an increasing level of complexity that has become both evident and problematic. Urban managers now recognise the importance of adopting a systemic perspective – one that views the urban environment holistically and accounts for the dynamic interactions between its socio-economic, physical, and environmental dimensions. Promoting systems thinking has thus become essential for any serious efforts aimed at achieving SUD.

The rapid advancement of ICT has introduced the novel concept of the DT, defined here as the connection of a physical system in the real world to its virtual counterpart in the digital realm via bidirectional communication, either with or without human intervention (Section 3.4.2). Although DTs initially emerged in industries such as manufacturing and aerospace, they have shown significant potential for application in UM. This potential, along with the increasing calls to leverage DT technology for tackling urban management challenges, inspired the work behind this book.

Recognising the importance of systems thinking in achieving SUD (Camagni, 1998), DTs are presented as enablers of this approach. As Mutch (2017: p. 499) notes, 'similar beliefs can result in significantly different practices depending on factors such as advances in technology'. DTs, with their ability to integrate data from multiple sources across organisational and sectoral silos, offer a unique means of representing and managing real-world complexity. This capability facilitates more advanced analysis, creating insights, knowledge, and deeper understanding of the intricate dynamics that characterise urban environments. These insights can then inform better decision-making.

In line with the aspirations of SUD, the field of UM is now seeing the rise of a new paradigm, referred to in this book as *DT for UM*. This paradigm is marked by an increasing number of DT-based pilot projects, proof-of-concept initiatives, and case studies that demonstrate the value of DT implementation in UM. When this project began, there was already considerable excitement within UM and the built environment sectors about the potential of DT, a momentum that has only grown stronger throughout the research process (Ferré-Bigorra *et al.*, 2022).

At the outset of this research, it was assumed that the main challenge hindering the wider adoption of DT in UM was largely technical. However, as the book progressed, it became clear that there were deeper, more fundamental gaps in the emerging paradigm of *DT for UM*. The focus gradually shifted from resolving technical issues to addressing the inconsistencies and lack of uniformity within the cultural framework of *DT for UM*. These challenges were evident across all levels, from abstract philosophical foundations to practical methodologies, tools and techniques.

This shift in perspective ultimately shaped the objectives of this book, which argues for the need to unify and systematise the *DT for UM* paradigm in order to support its growth and maturity, and to ensure its efficacy in addressing the wicked problems facing modern urban environments.

9.2. Contributions of this book

9.2.1 A novel research methodology

This work presents a novel methodological contribution by combining critical realism (CR) as a philosophical foundation with the DSR methodology. This approach proved to be both robust and practical in achieving the objectives set out in this book. By grounding DSR in a well-articulated philosophical framework like CR, the research gains greater rigour, soundness and transparency.

The adoption of a critical realist perspective not only provided the necessary depth for investigating the case studies of *DT for UM*, but also helped justify the selection of the various research methods and techniques used in evaluating DATUM. This, in turn, enhanced the repeatability of the research, ensuring that the findings can be scrutinised and built upon in future studies.

9.2.2 Literature gaps

The first objective of this work (O1) is addressed through an extensive review of literature relevant to *DT for UM*. The primary aim of this review was to identify the barriers that hinder the growth and maturity of *DT for UM*. The review process was conducted from a critical realist perspective, which assumes reality is stratified, consisting of structures that give entities their generative mechanisms, which interact to produce events.

The literature was analysed across three distinct levels (Figure 2.2). At the most abstract level, philosophy, lie the underlying philosophical worldviews that DT researchers and practitioners adopt – often unconsciously. These are shaped by ontological and epistemological assumptions, as well as beliefs about human nature. A practitioner's philosophical stance tends to influence their approach at the next level, methodology, guiding the form of practice or methodological approach adopted. This, in turn, shapes the methods and techniques used at the most concrete level of analysis, methods, where DT practices are observed.

This critical realist approach facilitated the identification of gaps in the literature across all three levels of analysis.

Philosophy: The philosophical worldviews underpinning and shaping DT practices are rarely declared or critically examined. However, when viewed through the lens of BMF, current DT practices show signs of theoretical inconsistency. Existing studies reveal that DT practitioners tend to mix incompatible philosophical paradigms within the same intervention.

Methodology: Most DT researchers and practitioners base their interventions on a general framework encompassing physical, digital and social systems. However, there is a lack of a detailed, clear and systematic methodology to guide these efforts, leading to a disconnect between abstract philosophical frameworks and concrete DT practices.

Methods: There are no standardised methods to bring coherence to the evolving field of *DT for UM*. Notably, the field lacks a common language for referring to various DT features or uses. Additionally, DT use cases are often documented only in narrative form, without a standardised way of modelling DT use case scenarios to facilitate knowledge exchange and dissemination.

9.2.3 Systematisation is the way forward

The second key contribution of this work is offering a recommendation on how the emerging paradigm of *DT for UM* can evolve and mature. The proposed path forward is through the systematisation and unification of the paradigm's CS. This recommendation is based on a three-layered analytical approach (Figure 4.1).

At the first layer, the analysis draws from a Kuhnian perspective (Section 4.4) to understand how UM research and practices have evolved over time. The second layer introduces Archer's M/M approach, which allows for the examination of the dynamic interplay between the discipline's body of theories and the actions and thoughts of its members. The third and final layer incorporates CR, where the M/M approach is informed by the concepts of independent reality and analytical dualism.

This layered analysis helps explain the current state of UM. With the increasing use of DT, UM is marked by a high degree of ideational diversity and heterogeneity but is also characterised by inconsistency and lack of uniformity. From a Kuhnian perspective, this indicates that UM is still in a *pre-paradigm* phase. Furthermore, the relationship between the various approaches to DT development and implementation is both contradictory and interdependent. The four 'ideal' approaches identified – tech-driven, disruptive, cognitive and humanistic – are each underpinned by different and incommensurable philosophical paradigms: functionalism, radical structuralism, interpretivism and radical humanism, respectively.

These approaches are contradictory because they are rooted in opposing worldviews, yet they are necessarily related because addressing the complex, multi-dimensional challenges of urban environments requires drawing on all of them simultaneously. If DT practitioners rely on only one or a subset of these approaches, their interventions are likely to fall short of the requisite variety (Ashby, 1961) needed to address real-world urban problems effectively. Archer (2005) describes this as a 'constraining contradiction,' which creates what this book refers to as the dilemma of pluralism (Section 3.8.5). The presence of these contradictions pushes members of the discipline towards a situational logic of correction (Archer, 2005), motivating the need for unification and reconciliation of the various worldviews and approaches.

In summary, the systematisation of the discipline's core structure is the most effective way to transition from the Kuhnian *pre-paradigm* phase to a fully established *normal paradigm*.

9.2.4 DATUM – a cultural system for *DT for UM*

To achieve the second (O2) and third (O3) objectives of this work, a CS for the paradigm of *DT for UM*, known as DATUM, was designed and developed. This represents a major contribution to the field, providing a framework that supports the systematisation and evolution of *DT for UM*.

DATUM serves as a conceptual structure, helping to organise and integrate the field, much like what Kuhn described as a 'conceptual box'. It offers both the necessary depth and breadth to meet the growing needs of the *DT for UM* paradigm. Its depth is reflected in its ability to connect various layers of abstraction, from philosophical foundations to practical DT methods. Its breadth lies in its generalisability and its intrinsic pluralism, accommodating the diverse range of DT research and practices that have emerged.

Moreover, DATUM has the potential to establish a distinct identity for this evolving field. With a solid and widely acknowledged CS, it brings value to both theoretical and practical aspects of UM. In essence, DATUM's contribution to the development of *DT for UM* is twofold: it enhances both practice and theory in meaningful ways as explained below.

9.2.4.1 Contribution to practice

Genuine pluralism – This work sheds light on the strengths and weaknesses of various DT practices, offering valuable guidance to practitioners in selecting the most appropriate methods for each stage of an intervention or to address specific aspects of a complex problem. By adopting a genuinely pluralistic approach, practitioners can apply a diverse range of DT methods and techniques, spanning from quantitative and tech-driven approaches to qualitative and humanistic ones. This allows them to tackle multi-dimensional real-world challenges more effectively.

Enhancing reflexivity – Linking theory to practice encourages practitioners to reflect on their methods and performance. This connection enables members of the discipline to explain, justify, advocate for, or critique their own research or interventions. By understanding why different approaches work in specific phases or contexts, researchers and practitioners can improve their own methods and share valuable insights with others, fostering continuous improvement.

Enriching *DT for UM* practices – Drawing on theoretical frameworks helps practitioners identify gaps and limitations in their work, enabling the development of new approaches – for example, the introduction of Mode 3 research and practice, which focuses on ethical and emancipatory DT practices, was proposed by drawing on CR's principle of ethics in life (Section 5.4.2.3). This allows DT practitioners to develop approaches that consider ethical implications alongside technological and functional concerns.

Clear and systematic methodology – The methodological component of DATUM, known as DM2, provides clear and structured guidelines for practitioners. From analysing the problem to applying DT interventions, DM2 supports a pluralistic and ethical framework. It creates a transparent evidence trail by linking the datasets necessary for a DT with the underlying structures and mechanisms of the problem being studied (e.g. Figures 6.9 and 6.26). This helps practitioners justify data collection efforts and strengthens the development of DT solutions.

Standardisation – The methodical element of DATUM, DTUCS, serves as a standard toolkit for designing, developing and documenting DT UCSs. It provides a consistent foundation for implementing DT projects and comparing actual results against initial goals. Additionally, DTUCS facilitates the dissemination of knowledge by offering a common language and standardised communication tools across the DT market (Figure 5.2). This reduces confusion and ambiguity, supporting the rapid growth and development of the entire field.

9.2.4.2 Contribution to theory

Initiating philosophical debate within *DT for UM* – So far, the paradigm *DT for UM* lacks any discussion about its philosophical foundations. This work disrupts that trend by sparking debate about the underlying worldviews that inform DT practices. These debates delve into theoretical assumptions regarding ontology and epistemology, exploring how DT practitioners perceive the roles and relationships between data, humans and the wider urban environment, including social, economic, natural and built aspects.

Pragmatisation of philosophical paradigms – By connecting various DT practices, such as the four ideal approaches (Section 3.8.4) or the pluralistic approach promoted by DATUM, to their corresponding philosophical paradigms, this work demonstrates the practical application of what might otherwise seem abstract ideas. This helps bring philosophical principles into the practical realm of DT, making them more accessible and actionable. Focus group discussions also played a key role in uncovering the philosophical assumptions held by practitioners, which are often implicit and unarticulated. This contributes to bridging the gap between philosophical theory and real-world DT practice, leading to theoretically informed practices and more pragmatic implementation of DTs.

Resolving the dilemma of pluralism DATUM not only promotes genuine pluralism by allowing practitioners to combine diverse approaches and methods, but also addresses the issue of paradigm incommensurability. Instead of basing pluralistic practice on contradictory paradigms, DATUM advocates for CR as an inherently pluralistic philosophical paradigm. This ensures that DT research and practice maintain theoretical consistency rather than falling into contradiction.

Proposing a new CR-informed methodology – This work contributes to the development of CR-informed methodologies by introducing DM2 (Section 5.4.3). While existing CR-based methodologies from the literature (Table 5.2) vary according to the type of research they guide, DM2 offers a comprehensive methodology tailored specifically for DT-driven interventions. It integrates all three modes of research (Section 5.4.2) in a complementary fashion, offering a holistic approach to DT practice informed by CR principles.

9.3. Recommendations

The fourth objective of this book (O4), set in Section 1.4, is fulfilled by evaluating DATUM. As a result, recommendations for research and practice that align with the overarching aim of this book, its general findings and aspirations were produced, including:

Philosophical level recommendations – To advance DT research and practice, practitioners should embody four key qualities: awareness, pluralism, consistency and transparency. Members of the discipline must be aware of how their beliefs and worldviews shape their work and its outcomes. Engaging in ongoing debates about the philosophical foundations of DT research and practice is essential. A pluralistic approach is also necessary, enabling the field to accommodate diverse practices and leverage a wide range of methods to address the multifaceted challenges of UM. However, this pluralism should not lead to theoretical contradictions. Practitioners should critically assess their philosophical positions, ensuring they are consistent, well-founded and conducive to producing sound, rigorous research. Additionally, they must transparently declare their philosophical stance, including assumptions related to ontology, epistemology, human nature and ethics.

Methodological level recommendations – A clear, systematic, and theoretically grounded methodology is essential for guiding DT-based interventions. This methodology must be anchored in a coherent philosophical stance, serving both to pragmatise abstract philosophical worldviews and to provide a heuristic framework for integrating various methods and techniques. By doing so, it ensures that the intervention is coherent, consistent and aligned with the intended outcomes.

Methodical level recommendations – Practitioners should aim to standardise how DT features and use cases are described, reducing ambiguity and fostering clarity. The articulation, documentation, and modelling of DT case studies or UCSs should be done using a common language that facilitates seamless communication and knowledge exchange across the field. This aligns with the vision of creating a systemic practice and supporting an interconnected ecosystem of DTs. Additionally, practitioners must be open to employing both quantitative and qualitative methods in the same intervention to ensure a more pluralistic and well-rounded approach.

9.4. Future work

There are several promising avenues through which future research could expand upon the findings of this work. Three areas, in particular, offer significant potential.

First, as highlighted by the CUT project team (Section 6.3.4), future research could explore the use of DTUCS to '*further develop a guidance to capture, document and communicate requirements of use cases in form of a questionnaire/checklist that will give us support to really comprehend a use case and its specifications at an early stage*'. This might take the form of a questionnaire or checklist that aids in understanding a use case and its specifications at an early stage. Additionally, the establishment of a standardised language and terminology across the field, as advocated by DTUCS, is essential. This consistency is also a prerequisite for creating machine-readable language, which could enable the automation of processes like classifying, detailing, publishing, searching for, and retrieving DT use cases and real-world projects. Future research could explore the development of machine-readable languages to automate these processes, leading to a codified knowledge management strategy (Al-Sehrawy *et al.*, 2022) that leverages IT to capture, disseminate and reuse the knowledge embedded in DT projects and case studies on a large scale.

Given that real-world UM challenges are inherently complex and require interdisciplinary interventions, Section 6.4.3.3 recommends, based on the first cycle of the action research, that DM2 incorporate a principle focusing on the interface between DATUM and other disciplines (Table 6.5). Future research could examine how CR can underpin and explain this principle, ensuring a robust connection between the methodological and philosophical elements of DATUM. Some critical realists have already addressed CR's ability to support interdisciplinary research, outlining a CR-informed general theory of interdisciplinarity that can enhance the integration of knowledge across fields (Bhaskar *et al.*, 2017; Danermark, 2019).

Lastly, the two action research projects conducted to evaluate DATUM (Section 6.4) exposed the framework to distinct and complex problem scenarios. However, there remain numerous opportunities to apply and test DATUM in other contexts. Future work could include the creation of DTs to address different application areas at varying scales, and the exploration of steps in DM2 that were underutilised in this study, such as *virtual implementation* (Section 5.4.3.11) and *real implementation* (Section 5.4.3.12).

REFERENCES

A collaboration of leading figures in the built environment (2021) *Our Vision for the Built Environment.* CDBB. https://www.cdbb.cam.ac.uk/news/industry-unites-behind-vision-built-environment

Al-Sehrawy R, Doukari O, Kumar B and Watson R. (2022) A knowledge management strategy for urban digital twins. *Proceedings of the 2022 European Conference on Computing in Construction* (Tagliabue LC, Hall DM, Chassiakos A, Nikolic D and Soman R (eds.)): 529–538, 10.35490/EC3.2022

Archer MS (2005) Structure, culture and agency. In *The Blackwell Companion to the Sociology of Culture* (Jacobs MD, Hanrahan NW (eds.)). Blackwell Publishing, Oxford, UK. pp. 17–34.

Ashby WR (1961) *An Introduction to Cybernetics*. Chapman & Hall Ltd, London, UK.

Bhaskar R, Danermark B and Price L (2017) *Interdisciplinarity and Wellbeing: A Critical Realist General Theory of Interdisciplinarity*. Routledge, Oxford, UK. https://doi.org/10.4324/9781315177298

Camagni R (1998) Sustainable urban development: Definition and reasons for a research programme. *International Journal of Environment and Pollution* **10(1)**: 6–27. 10.1504/IJEP.1998.002228

Danermark B (2019) Applied interdisciplinary research: A critical realist perspective. *Journal of Critical Realism* **18(4)**: 368–382. https://doi.org/10.1080/14767430.2019.1644983

Ferré-Bigorra J, Casals M and Gangolells M (2022) The adoption of urban digital twins. *Cities* **131**: 103905. 10.1016/j.cities.2022.103905

Mutch A (2017) Practices and morphogenesis. *Journal of Critical Realism* **16(5)**: 499–513. 10.1080/14767430.2017.1361289

Ramy Elsehrawy and Bimal Kumar
ISBN 978-1-83662-235-2
https://doi.org/10.1108/978-1-83662-234-520251012
Copyright © 2025 by Ramy Elsehrawy and Bimal Kumar. Published under exclusive licence by Emerald Publishing Limited

Index

Printed and bound by CPI Group (UK) Ltd, Croydon, CR0 4YY

11/06/2026

14897902-0001